History *of* Science
Selections *from* ISIS

GENERAL EDITOR, Robert P. Multhauf

Science
and
Technology
in
East Asia

Science
and
Technology
in
East Asia

Nathan Sivin

Editor

Science History Publications

New York · 1977

First published in the United States by
Science History Publications
a division of
Neale Watson Academic Publications, Inc.
156 Fifth Avenue, New York 10010

©Introduction, Nathan Sivin 1977
©Neale Watson Academic Publications, Inc. 1977

First Edition 1977
Designed and manufactured in the U.S.A.

Library of Congress Cataloging in Publication Data
Main entry under title:

Science and Technology in East Asia

    (History of science: Selections from Isis)
    Bibliography: p.
    1. Science—History—East (Far East)—Addresses,
essays, lectures. 2. Technology—History—East (Far
East)—Addresses, essays, lectures. I. Sivin, Nathan.
II. Isis.
Q127.E26S26   509'.5   76-26530
ISBN 0-88202-161-3 (paper)
ISBN 0-88202-162-1 (cloth)

Publisher's Note: We should like to express our
appreciation to ISIS for granting permission to
reprint the articles in this volume. If the
publishers have unwittingly infringed the copyright
in any illustration reproduced they will
gladly pay an appropriate fee on being satisfied
as to the owner's title.

Contents

For a period and a beginning:

The Fiftieth Anniversary of
The History of Science Society
and the Dedication of
The East Asian History of Science Library

Introduction

The history of science has more than once been characterized as a bridge propped between the Two Cultures. This is a suggestive metaphor, most of all in the image it presents of what will happen to that discipline as the two edges of the abyss continue to accelerate in opposite directions.

If the situation of the history of science is perilous, it is also strategic, so that there is scope for optimism. The separation of the technical and humanistic cultures—which take largely incompatible views of the means and ends of human activity, which are committed to mutually incomprehensible discourse, which have long been resigned to mutual illiteracy—is the work of human beings. There is hope that the trend can be reversed by human beings if enough of them value unity of mind and learn how to look for and build upon its badly disintegrated traces in modern culture.

It is easier to acknowledge this need for healing old breaks than to know where to start reasoning about it. The history of science—of science everywhere—can offer two kinds of help. One is to explain the gradual separation of science and the technology connected with it from other spheres of human concern. The other is to show what the mode of integration was in traditional societies—in Europe and in other civilizations before each had its Scientific and Industrial Revolution. Awareness of the multiplicity of such patterns in different times and places will not automatically generate answers for today, but can suggest which present possibilities may be worth exploring.

For this use questions of style and value in science become as important as those of content. For some decades (a period of accelerated discipline-building) the most influential students of early science have been preoccupied with content, and have tended to give little attention to how technical work was integrated in society. Historians of science have only recently begun again, as was customary in an earlier day when there were no professionals, to think of techniques, concepts, society, and the imaginations of people in society as one whole. Historians of technology, although fewer, have on the whole been less one-sided; but the relations between theoretical knowledge and technique remain cloudy in every period and every field of technological evolution.

When regarding cultures other than one's own it is much easier, of course, to see them whole, and to be aware of connections that over-familiarity has obscured in the universe of meaning one was taught as a small child. There is a great deal to be learned about the place of science in European civilization, for instance, from writings of non-European scholars such as Shigeru Nakayama and Seyyed Hossein Nasr who find the Western technical tradition exotic but who do not hesitate to tackle it whole.[1]

The study of Far Eastern science and technology is the antithesis of a narrow specialty. It encompasses every variety of technical thought and action by a majority of civilized humanity over the whole sweep of history. Most of those devoted to this study have mastered rigorous methodologies, but would not care to lavish forty years of study on seventeenth-century provincial French mathematics (the contributors to this volume are typical in this respect). The East Asian technical traditions are still open country—among the last remaining—in a built-up world of scholarship. Over it one may pursue problems in any direction they lead, so long as they remain worth the chase. This quarry, unlike the antiquarian delights that lure others on, is potentially interesting to almost every reflective person. Sharing it helps us to understand ourselves.

The scientific and technological traditions of China were so sophisticated and so richly documented that they allow study in depth of almost any issue of concern. Korea and Japan were socially quite distinct from China, although borrowing made their educated classes culturally similar—as similar, at least, as the Latin-writing elites of Eastern and Western Europe. Traditional Korean and Japanese science were, with considerable differences of strength and emphasis, variants of the Chinese model. In technology, although the influence was not so unidirectional and there was more scope for local uniqueness, the three states were to a fair extent also part of one extended system. Questions of how scientific style varies in different social circumstances can thus be examined by comparative study. The differing experiences of the three societies in their encounters with Western technology between the seventeenth and twentieth centuries promise to throw light on issues bandied about under the rubrics of modernization and technology transfer, discussion of which has so far been extraordinarily muddled by failure to examine fundamental assumptions in the light of history[2]. It is instructive, for instance, to relate the tone of early responses to science and engineering to the very different political, military, and economic threats that occasioned modernization in China, Japan, and Korea. The attempt to export current (or not too obsolete) technologies to the Third, Fourth, and Fifth Worlds is merely the latest instance of an exploitation of technical superiority that has continued at least since the invention of agriculture.

What do we mean by the science and technology of China and its neighbors? We mean a group of traditions very different from that of Europe.

Below is a list of the disciplines among which Chinese until recently divided their rational and abstract understanding of nature and man's relation to it (that is how I define science). These were disciplines in point of a methodology, a high standard of internal consistency, and a conscious cumulation of knowledge. Authors related fresh observations to theoretical foundations, and were expected to know their predecessors.

The division between quantitative and qualitative sciences, which is repeated in the arrangement of this volume, is my own; as I will show later, there was no tradition in philosophy or elsewhere of comprehensive discourse on the sciences considered as a single enterprise. This grouping does not violate Chinese conceptions as would, say, a division into physics and chemistry (or proto-physics and proto-chemistry).

I make no attempt to list fields of applied science or engineering. All of the sciences were applied, in ways indicated below, but not to economic production. On the other hand, manufacture and construction were the provinces of independent trades, uninfluenced by the scientific abstractions of the educated elite and only fortuitously linked to each other by occasional social circumstances. Reliable descriptions of the Chinese sciences are easily available, and need not be repeated here[3].

The Quantitative Sciences, concerned primarily with number and its application to physical reality

1. Mathematics, applied to astronomical prediction and to bureaucratic and commercial routines. It was almost wholly numerical or algebraic rather than geometrical in its approach, and was seldom wholly divorced from numerology.

2. Mathematical harmonics, concerned with the quantitative and numerological relations between the dimensions of musical instruments (those traditionally used in state ceremonial) and their scales and modes.

3. Mathematical astronomy, which applied instruments and mathematical techniques similar to those found in the Western tradition (although the mathematics was more Babylonian than Greek) to the same sky, with notable differences in style.

The Qualitative Sciences, applications of yin-yang, Five Phases (*wu hsing*), and other verbal concepts to different realms of human experience

1. Medicine and materia medica, in which knowledge of the human body and of therapeutic agents and hygienic practices was decisively molded by remarkably abstract theories and applied to the treatment of illness.

2. Alchemy, both the external alchemy of the workshop and the internal alchemy carried out within the body of the adept through a variety of ascetic and mystical practices. Both were esoteric, and the main goal of both was religious transcendence combined with physical immortality.

3. Astrology, which recorded and explained the significance of unpredictable celestial phenomena, just as mathematical astronomy predicted the rest. Astrology was applied to politics, for Chinese believed that an anomaly in the sky was a warning of inadequacy in the imperial virtue that would also result in social anomaly unless corrected. It was, in other words, inductive mumbo-jumbo quite on a par with the economic indicators and opinion polls which maintain free play for magical thinking in national politics today.

4. Geomancy, a topographic science of siting houses and tombs.

5. Physical studies, which applied the concepts of natural philosophy to particular instances of change and interaction. They were closer in style to Stoic physics than to the Aristotelian or other dominant components of the European tradition.

The essays on Chinese science that *Isis* has published over the last sixty years cover only part of this spectrum. There is still no satisfactory study in any language of geomancy as a science, although its anthropological and phenomenological aspects have recently drawn attention.[4] Physical studies were ignored by historians East and West until Needham and his associates began explaining their significance.[5] Although few and devoted to particular themes, the papers in this book reasonably sample the diversity of Chinese science. They provide a good impression of how the quantitative and qualitative sciences differ; there is a clear contrast between how much of the former seems familiar to us, and how little of the latter.

These sciences constitute the heritage that Japan and Korea drew on, moving in new directions to meet local opportunities and needs. Mikami's study of determinants in Japan pioneered in tracing divergence. In the last decade it has been supplemented by Nakayama's original and penetrating comparative study of Japanese astronomy and by Sang-woon Jeon's rich survey of Korean science[1,2] We still know very little about the reasons for differences in receptivity—for instance, the failure of alchemy to be established in Japan although practiced in Korea, and the perfection of movable-type printing in Korea although invented long before in China. Nor do we have a sound grasp of great variations in the later response to Western science: the immediate Chinese engagement with geometric celestial kinematics, the Japanese rage for anatomy and surgery while Chinese remained oblivious of European medicine. The issue of cultural interaction in science is so important that a special section of this book is given to it. The five essays in that section study Chinese contacts with many peoples, along a span of time from the late Middle Ages to the present day. Nakayama's essay on astrology and the contributions on the cannon also bear on relations between cultures.

The Chinese demarcation of the sciences is not directly reflected in most of this book. Until recently it has been almost entirely obscured by the tendency of those who write on East Asian technical history to impose modern Western rubrics on it. Isolated facts and theories that bear on the understanding of living phenomena are called "biology," perpetuating the mistaken notion that they formed a coherent discipline (as they did, more or less, in the ancient West). In the Chinese cultural sphere, to the contrary, they belonged to several unrelated disciplines and literary traditions. The understanding of living phenomena never emerged as a distinct issue in Chinese natural philosophy or in the operative sciences. "Proto-biochemistry" as a characterization of the breath-cycling exercises of "internal alchemy" (*nei tan*) is a splendidly audacious metaphor for the limited aspects that most directly concern the historian of science. Lu Gwei-djen and Joseph Needham introduce the term in a survey that does not ignore the non-physiological meditations (some of them ecstatic, and some concerned with the imaging of gods), the mystical beliefs and religious rituals with which physiological practices were integrated, as well as the goal of spiritual transcendence that sustained the alchemical quest and kept knowledge of the body an incidental by-product. One can be sure, however, that those who adapt the term "proto-biochemistry" to the popularization of Chinese science will not be so scrupulous[6]
Similarly, "mechanical engineering" suggests a coherent basis in principles and technics that was not found in ancient China. The weakness of the positivistic assumptions that prompt its frequent application to traditional China are even more obvious than in the case of biology, for there is considerable doubt today that the profession of mechanical engineering will continue to exist anywhere for another generation. It is surprising that the term persists in assumptions about other times and cultures. In the United States today, little besides basic scientific skills spans the gap between the power engineer, the textile engineer, the specialist in automatic controls, and the biomedical engineer. Specialists in the fields I have enumerated no longer claim the universal competence—or qualification for employment—of mechanical engineers a hundred years ago. We are perhaps readier than readers of an earlier time to keep in mind the difference between the instru-

ment-maker in the imperial workshops and the village wheelwright. What they shared was not a body of engineering principles and design skills, but art.

To insist on the historic differences between Chinese and European disciplines is a simple matter of not confusing what we understand about traditional China with what we know and assume about ourselves. I am not suggesting we lose sight of what is universal in modern science, or forget how much non-European peoples, while working along certain lines convergent with those of the Western tradition, have contributed to its formation.[7] That tradition was itself principally Islamic rather than European for centuries. My point is that comparative study is perhaps the most powerful tool available for examining our assumptions about which aspects of today's science are universal—constrained only by the objective reality of nature—and which reflect the values and perceptual habits of the culture in which they originated. Surely the current division of science and engineering into fields is exactly what it would have appeared to be to a traditional Chinese scientist: the outcome of European and Islamic tastes and particularities of institutional history. This scheme of knowledge has spread round the world not because it provides the only conceivable basis for organizing contemporary scientific work, but because the encounter between traditional and modern science in one society after another has been resolved by social change and political fiat, in view of which the comparative appropriateness of each system of science to the cultural environment is beside the point. Modern science, because of its worldwide ascendancy as part of a triumphant socioeconomic system based on instrumental values, has largely lost the exceptional capacity of earlier science to benefit from the fundamentally variant orientations of other cultures.

By identifying what is parochial in modern science we can come closer than philosophic speculation has yet led us to isolating the generative kernel that is truly universal, that lets us "discover how to discover"[8] exempt from limitations of time, place, language, or preconception. Many of the ideological positions current in public debate on modern science—those that claim it is universal in every essential respect, and those that deny it all universality—could hardly survive if the public to whom they appeal were familiar with how scientists worked and reasoned in other cultures.

This is an opportune place to discuss what I believe to be the chief issues in the study of Far Eastern science and technology (I will draw my examples from China):

1. What were the scientific and technological traditions?
2. What were the sources of new ideas, techniques, and problems?
3. How were the sciences related to other kinds of thought and activity?
4. What shaped and constrained movements in new directions? I have already remarked on one of the most important themes, the encounters of technological and scientific traditions of different cultures.

1. What were the scientific and technological traditions? Where were they located in society? What made them traditions? These issues illustrate the inseparability of society and the individual intelligence in forming science.

No problem of comprehension seems more obviously "internalist," concerned predominantly with relations of ideas, than that posed by the traditional Chinese unconcern for integrating all the sciences. Those who did science drew on a common fund of philosophic concepts, but, as I noted earlier, these were not generally seen as connected across the divisions between sciences. Often, in addition to its everyday acceptances, a word would have certain technical meanings in medicine, others in geomancy, and still others in alchemy. Philosophers, far from synthesizing or imposing order on all these meanings, tended to create special senses of their own.

I suggest that the contrast with the Western view of *scientia* is also a contrast of institutions.[9] The European natural sciences were related to each other and kept subordinate to philosophy in educational institutions, from the Academy and Lyceum to the proliferating universities of the nineteenth century. In China astronomy and astrology were done primarily in the imperial court. Classical medicine was passed down through master-disciple relations diffused over the upper reaches of society (priests and folk healers treated the majority of Chinese). The major alchemists were marginal figures, whether dropouts or the retainers of wealthy patrons. The geomancers who left written records were the upper crust of a popular art, which they transmuted into forms more or less compatible with world-views and esthetic orientations (especially views of ideal relations between man and his environment) current among the educated minority.

In what forum could these practitioners meet? There was none. Without it, occasional attempts by individuals to think out a personal synthesis of the sciences had little transforming influence.

Current philosophy could not provide a forum. The sciences were founded on myths that represented new insight as a rediscovery of truths and significances fully revealed in a golden age (before history quite began) and long since lost or uncomprehended. These were social myths. They legitimized schools, master-disciple relations, and traditions in which practitioners situated themselves. It is irrelevant that nearly two centuries of critical study has affirmed that the founding texts were not really passed down from that golden age, but were mostly products of the great flowering of thought in the last four centuries B.C. and the first two of our era. Their scriptural authority ensured that every scientist had to come to terms with the ideas the founding texts perpetuated. What bearing current philosophy might have was a matter of much less importance.

Considering the great intellectual influence of neo-Confucianism over the past five hundred years or so, I continue to be puzzled by how little of its special philosophic perspectives can be detected in the sources of science. One can find exceptions, as in mathematics and astronomy from about 1700 on, when they were largely revived by members of recognizable neo-Confucian groups. To consider a more representative case, attempts to prove the neo-Confucian character of late medicine have depended on vague comparisons and have failed to concentrate on specific convictions or patterns of them that originated in neo-Confucianism; they have fallen flat. More critical study will no doubt demonstrate some neo-Confucian influence on medicine, but my own reading of the sources indicates that few mutations of late traditional medicine can be explained this way—or, for that matter, by any other lopsidedly internalist or externalist hypothesis.

In sum, because the philosophic assumptions of the sciences were so solidly rooted in traditions, to which authors constantly referred their work, new philosophic syntheses—which in any case tended to center on the problem of moral self-cultivation—could not provide the integration that *scientia*, eventually embodied in faculties of philosophy, imposed in Europe.

Technological traditions are just as problematic. They were crafts, passed down privately from artisan to apprentice. The historical Chinese reluctance to invest wealth in what we would call development meant that projects of great magnitude or duration could be financed only in the court, at the whim of the monarch, until recent times. It is well known, for instance, that the tastes of two Korean kings in the fifteenth century were decisive in culminating the long development of cast metal printing types.[10]

Were the creators of endless mechanical marvels in the courts of two thousand years the high officials who were usually credited with them or illiterate artisans whose names we will never know? That is an open question. We do know the names of a few technicians of great talent and historic importance whose backgrounds denied them official standing or recognition, but who were certainly literate to some extent before literacy had spread far beyond the great families.[11] They were not, characteristically, lettered enough to leave us books. Most of what we know of early engineering comes from the reports (formal or informal) of officials about work done for them or about processes or artifacts they had observed. Such writings were used not to train technicians but to help civil service generalists prepare to supervise artisans and peasants as bureaucratic duties required.

2. What were the sources of new ideas, techniques, and problems? This does not appear at first sight to be a major issue with respect to technology, since the manufacturing traditions were so largely self-contained. Nevertheless the influences of peoples on each other were not at all negligible, and greatly affected techniques.

Elite T'ang culture (at its height in the first half of the eighth century) was as cosmopolitan in its style of life as the United States today—more so, if we consider what pains being cosmopolitan cost before the era of gasoline-powered transportation. Foreign objects, ways of doing things, diversions, for that matter people, flowed in and out of every Chinese border. Not only the tastes of patrons but the esthetic responses of craftsmen to objects from abroad prompted new manufactures and new approaches to old arts.

In the sciences the foreign influence in China was substantial, although not at all evenly distributed; in Japan and Korea it was overwhelming, although as time passed novelties were more thoroughly adapted to local needs and habits of thought.

There is a still more fundamental question to be asked about the springs of innovation.

We know practically nothing about ordinary thought and practice, which can hardly be identical with what we are told of the high tradition by the two or three greatest figures of each science in each age, or with what is recorded about imperially sponsored technical work. What we know about the sciences comes mostly from books written to expound what were considered classical writings (that is, the canons of a conscious tradition), to embellish them, to further their aims, or even to become classics. The documents of each science reflect constant interaction with traditions of a more popular kind, and generally suggest that each high science sat at the end of a spectrum. Based on the conceptions of the literate minority—which a thousand years ago was still practically confined to the great families, and by the nineteenth century was on the order of one Chinese in ten—each classical tradition represented the most sophisticated forms out of the variety created at every level of society to come to terms with experience of nature.

This generalization about the social ranges of science is least true of the exact sciences, to which the uneducated could contribute little or nothing. They did not need special types of practice. Still, as merchant culture developed over the last millennium, it evolved a characteristic mathematics, centered to a considerable degree about the use of the abacus, which we have the documents to study.[12]

There are other social distinctions to explore, even within the governing elite. For instance, although mathematical astronomy was meant to be a monopoly of the imperial court (and in a few periods could not legally be practiced elsewhere), the contributions of outsiders were responsible for a great deal, perhaps the bulk, of crucial innovation. In Tokugawa Japan, astronomy and astrology were true monopolies of designated families; but it is generally agreed that the success of the Shibukawa clan as astronomers to the military dictator (including their leadership in the response to Western cosmology) depended at more than one juncture on the adoption of talented young men.[13]

With the qualitative sciences the links between the thought and practice of the various levels of Chinese society are clearer and stronger. Classical medicine interacted with, and was constantly invigorated by, many kinds of healing that historians of Chinese medicine have explored only superficially—not only local herbal remedies, but an endless variety of magical, religious, and ritual cures that do not accord in obvious ways with classical theory but grow directly out of vitalistic and animistic popular views of the world.[14] The alchemist, the geomancer, and the investigator in one or another kind of physical study over the centuries were equally drawing on what ordinary people believed and did. At the same time that they drew on them—never without adaptation—the high traditions enriched the popular arts and, I believe, kept alive popular ideas and practices that might have died out or changed beyond recognition without this indirect access to writing.

The situation was more complicated in Korea and Japan, where, although folk traditions flourished, medical amateurs and pedants educated on foreign ideas in a foreign written language usually ignored popular symbolic therapies prevalent in their own societies, and seldom attempted synthesis.

The work needed to understand the flow of concepts and methods across these various social spectra is hardly under way. The disciplines of social and cultural anthropology are as essential to further it as are those of history and philology.

3. How were the sciences related to other kinds of thought and activity? I have already suggested a contrast between the more or less shared metaphysics of modern science and the divergent disciplines of the early Far East. Another contrast is hardly less striking.

The ground of modern science and technology has been progressively isolated from other human activities.[15] Science has no direct concern with the creation and reexamination of values that characterize the humanities. Nothing links modern physical science to the social sciences but computational techniques and tortured analogies invoked to preserve social science and political science from being considered mere social studies, politics, and so on.

In the traditional Far East, the autonomy of the sciences would have been repugnant if anyone had even taken the possibility seriously. Even the most skeptical thinkers did not envision a mechanical universe (their boldest stroke was to consider it a spontaneous system free of anthropomorphic feeling or motivation). The assumptions about the cosmos shared by scientists were the same as those of Chinese painting, and for that matter were congruent with the foundations of ethical conduct. Some thinkers in the Confucian tradition believed that the good life was attainable only in human society, but when they thought about nature their ideas were shaped by yin-yang, the Five Phases, and similar broadly diffused concepts, just as those of Europeans have been shaped so long by notions of causality that played negligible roles in the thought of other cultures.

Nor did those who did science, any more than Confucian moralists or orthodox Taoists, believe that their perceptions could lead eventually to complete knowledge of the natural order. One could learn a great deal about that order, for the sake of wisdom or practical benefit, up to a point. Its texture was too fine (*wei*), too obscure (*hsuan*), to be fully penetrated by induction, and there was no pure deduction from rational first principles. What has often been mistaken for the latter (although recent scholarship has been more critical in this respect) is a process rich in philosophic subtlety but centered in intuition, introspection, concentration, meditation, and other highly disciplined forms of insight. These were always, even in Confucian self-cultivation, spiritual in content or reference (although considered out of context they are often atheistic). Most Chinese schemas of the sacred embraced notions of fate and destiny, and indeed a multiplicity of divinatory techniques made it possible to circumvent blocked insight (just as in popular religion they were used to determine what were the current policies of the celestial bureaucracy). Physical investigation, insight, and divination were complementary. The sciences took a due place among them, but had no basis on which to claim hegemony.[16]

Some may consider it unreasonable to ask students of Chinese science to apply craft and subtlety to religious issues, and even to what by modern standards is gross superstition. But the alternative is a very one-sided picture of what was important to early scientists, and of how their lives and thought were integrated.

4. In what ways did the styles that emerged in the sciences and productive arts shape and constrain movements in new directions? This is a more delicate question than it appears to be; it is the closest I can come at the moment to one

that can help us concretely to understand (to the extent that investigation from the Chinese side can be productive) why the Scientific Revolution happened elsewhere. I say delicate because in early science established working and thinking habits may have constrained conventional people, but did not prevent bold reorientations from appearing on the margins of one field or another, brought off by people whose positions were often as not ambiguous or marginal in society.

It is a simple fallacy to assume that the anti-scientific attitudes of "Confucian" humanists could have ruled out the passage to a mathematical physics of modern type. It is not at the center that revolution is generated, either in politics or in science. The fallacy becomes obvious when we note the enormous similarities between such negative attitudes of average intellectuals in China and those of the Schoolmen of Galileo's and Descartes' time. The Schoolmen could delay the dissemination of the new sciences, but could not block them or create alternatives as intellectually compelling. The Scientific Revolution held the initiative. It proceeded past them by constructing a new intellectual community (and careers for its members) outside the establishment of school and church, assuming independent authority to formulate the laws of nature, and displacing the prerogatives of ancient institutions in order gradually to form what has now become a technical establishment.

Reducing elite and alternative institutions, traditions, and views of reality to lists of factors which seem favorable and unfavorable to the evolution of science, as Robert K. Merton pointed out long ago, is not a way of answering questions about the conditions of scientific revolutions, but rather of atomizing such questions.[17] Working out an understanding of the Asian side of the Scientific Revolution problem calls for seeking out the potential Galileos and Descartes of the Far East and studying them closely against their own backgrounds in the light of their own values and associations. They will be found, I believe, as often among people committed to Confucian ideologies as to Taoist. I think of China's greatest scientific and technological polymath, Shen Kua (1031–1095), whose personal philosophy was influenced primarily by the *Mencius*, and the brilliant astronomer Wang Hsi-shan (1628–1682), who played a major part in the critical reception of European exact sciences, and who "accepted the orthodox Confucian tradition of the Chu Hsi line as his personal mission."[18]

To sum up, the study of Chinese science need not continue as a last refuge for questions like the following, whose answers are dictated by questionable assumptions and which therefore have been long since rejected in older fields: Was the social standing of physicians in traditional China high or low? (As though physicians formed a single profession!) Was the failure of the Scientific Revolution to take place first in China due primarily to socioeconomic or intellectual causes? (As though the two might be independent!) Should Chinese alchemy be accorded temporal priority because the Hellenistic art was not true alchemy? (This question seems a parody of the formulations that used to rationalize high-minded ignorance of non-European traditions.)

An open-minded history of science is bound to be cross-cultural and comparative. Constant travel of ideas and techniques since very early times means that the single closed tradition, the only one that can be studied as a self-suffi-cient whole, is the technical tradition of all mankind. The idea of a linear Western tradition or a linear Chinese tradition is an abstraction for the purpose of defining the coherence of local styles and structures. There is a great deal to be learned by juxtaposing cultural patterns, primarily if the comparisons are of wholes rather than of superficial characteristics torn out of context. Cross-cultural under-standing must grow, in other words, out of comprehending the sciences as parts of each culture's much wider system for experiencing nature and making sense of it, and as the creation of groups of people with common values and interests.

   Science and Technology in East Asia is one of four books published to cele-brate the fiftieth anniversary of the History of Science Society. The essays in this volume are those published in *Isis*, over a period of sixty years, which in my judg-ment retain their value for the study of Far Eastern science and technology. They represent roughly half of all those which have appeared in *Isis*, a high average for a field which is not a discipline. They are reliable as well as cogent, and have not been superseded (although I regret not having been able to persuade either of the scholars who have contributed so many bits and pieces of the firearms story to provide a few pages of synthesis). The high general standard must be credited to editorial insistence on careful refereeing; too many journals, by relaxing their customary standards in order to indulge new fields or eminent amateurs taking a flyer, maintain an altogether unnecessary rate of obsolescence.

   Some pioneering studies in the pages of *Isis* could not be included in this col-lection because present understanding, having proceeded from them, has moved so far. This is especially true of the writings of Tenny L. Davis and his Chinese collaborators, who more than anyone else in the late 1930s and early 1940s began making the esoteric literature of Chinese alchemy (external and internal without distinction) accessible to Western readers. A couple of choices, notably a study of Korean astronomy by Rufus and Chao, have been ruled out by technical prob-lems of reproduction.[19]

   Most of the papers in this book are concerned with identifying some aspect of the special style of Far Eastern science and technology or its effect upon the en-counter between East and West. The earlier articles on the whole were written too close to the sources to address general significances, and are valuable mainly for their accurate characterizations. A higher level of generalization emerges only gradually—characteristically striven for with some success in Willy Hartner's remarks on mathematical harmonics (1938) and, I suppose, first rising out of ex-haustive study in Nakayama's essay on Chinese astrology (1966). Despite the fre-quent attention to East-West interaction, it is not until 1972, with Salaff's study of a contemporary scientist, that the ambiguities of that interaction receive ex-plicit attention.[20]

   Isis's sustained interest in the Far East—as part of a general interest in science and technology as world phenomena—deserves notice. No other Euro-pean or American journal approaches *Isis* in this respect. In view of the parochialism of most historians of science, only an uncommon vision of the cultural universality of science could have shaped the journal that way. In *Isis* the informing ideal was that of George Sarton, its first editor, just as *Tōhō gakuhō*, a similarly distinguished Kyoto journal of sinology, bears the ecumenical imprint of Kiyosi Yabuuti and several like-minded colleagues.

It has recently been asserted that Sarton's view of technical history not only lacked intellectual depth but was too broad to serve as the basis of an emerging profession.[21] I agree, noting that with few exceptions Sarton's critics have lacked his intellectual breadth. The history of science eventually established a presence in the academic world by building strength in the intellectual history of European science, with little rigorous study of cultural or social issues (aside from the formation of scientific organizations). It is remarkable that Sarton's successors as editors of *Isis*, as they adapted it to the needs of a growing profession, did not discard his ideal despite its lack of currency among their associates. Until recently research and writing on the Chinese traditions have largely been left to occasional sinologists and sinophile scientists with a bent in that direction, but who have been unprepared to relate the subject to issues being raised by historians of science exploring more familiar and less problematic traditions. Hartner and Needham were the first major European exceptions. The study of Far Eastern science and technology has begun to involve sensible numbers of people only in the last decade, mainly as Needham's work has gradually succeeded in reaching over the heads of the professional technical historians and sinologists (who praise his work but seldom let it deflect their own) to engage the imagination of what used to be called the common reader, old and young. There are now over fifty European and North American scholars working in the primary sources,[22] and more graduate students than can be accommodated who are willing to master two fields in order to prepare for research and teaching on Chinese science. In the absence of a comparable stimulus, the traditional Japanese and Korean sciences remain practically unstudied outside Asia.

The generation that narrowed its conception of the history of science in order to build a profession has attained that end. The next generation, which can take professional status for granted—and choose, if need be, the freedom of the amateur—appears ready to act upon a much more general curiosity and concern for issues (as historians of technology, whose very different problems I lack space to discuss, have been doing all along). Although there will always be historians of science who consider only the European tradition the real thing, and more and more of them pointedly lack interest in any history before the nineteenth century, many of the most lively minds are becoming engaged in the effort to build on a solid base a history of science that encompasses every period and all peoples.

Notes

[1]Shigeru Nakayama, *A History of Japanese Astronomy. Chinese Background and Western Impact* (Harvard-Yenching Institute Monograph Series, 18; Cambridge, Mass., 1969); Seyyed Hossein Nasr, *The Encounter of Man and Nature* (London, 1968).

[2]I have discussed this issue in the prefaces to Sang-woon Jeon, *Science and Technology in Korea, Traditional Instruments and Techniques* (MIT East Asian Science Series, 4; Cambridge, Mass., 1974), and Masayoshi Sugimoto and David L. Swain, *Science and*

Culture in Japan, Vol. I (forthcoming as Vol. 6 in the MIT East Asian Science Series).

[3]I have described them at greater length in the preface to Shigeru Nakayama and Nathan Sivin (eds.), *Chinese Science. Explorations of an Ancient Tradition* (MIT East Asian Science Series, 2; Cambridge, Mass., 1973). This volume also includes a selected and annotated bibliography of works in Western languages useful for the study of Chinese science.

[4]Andrew L. March, "An Appreciation of Chinese Geomancy," *Journal of Asian Studies*, 1968, *27:* 253−267, a phenomenological study, and Stephan D.R. Feuchtwang, *An Anthropological Analysis of Chinese Geomancy* (Vientiane, Laos, 1974). Steven J. Bennett's article "Patterns of the Sky and Earth: A Chinese Science of Applied Cosmology" (forthcoming) will be the first introduction to the subject based on close study of a group of early texts and adequate understanding of geomantic concepts.

[5]There is much discussion of resonance and systems for categorization throughout *Science and Civilisation in China* (7 vols. projected; Cambridge, England, 1954−); see also Ho Ping-yü and Needham, "Theories of Categories in Early Mediaeval Chinese Alchemy," *Journal of the Warburg and Courtauld Institutes*, 1959, *22:* 173−210.

[6]Lu Gwei-djen, "The Inner Elixir (Nei Tan); Chinese Physiological Alchemy," pp. 68−84 in Mikulas Teich and Robert Young (eds.), *Changing Perspectives in the History of Science. Essays in Honour of Joseph Needham* (London, 1973); Lu and Needham, *Science and Civilisation in China*, Vol. 5, Part 5 (forthcoming). Even with established translations, scrupulousness about their limitations does not preclude very frequent misapprehension by those too busy to read carefully. When Needham justified his use of the misleading "five elements" to translate *wu hsing* on account of long usage, he added a clear warning that it is not an accurate translation (*ibid.*, 3, p. 244); nevertheless popular books on Chinese science and medicine continue to speak of the *wu hsing* as material constituents of the human body and of other things. They do not, of course, discuss the fact that medical documents do not claim the body is composed of wood, metal, and so forth. As an attempt to end this confusion, "Five Phases," which is much closer to the Chinese, is gaining currency among students of Chinese science. See Sivin, "Chinese Alchemy and the Manipulation of Time," in this volume.

[7]Needham, "Science and China's Influence on the World," pp. 234−308 in Raymond Dawson (ed.), *The Legacy of China* (Oxford, 1964), reprinted in Needham, *The Grand Titration. Science and Society in East and West* (Toronto, 1969), pp. 55−122.

[8]The phrase is A.C. Graham's, in "China, Europe, and the Origins of Modern Science: Needham's *The Grand Titration*," pp. 45−69 in Nakayama and Sivin.

[9]My point of departure for this line of thought is a series of discussions with Nakayama. See his "The Empirical Tradition. Science and Technology in China," pp. 141−150 in Arnold Toynbee (ed.), *East Asia: Half the World* (London, 1973), and Nakayama's much more extended argument in *Rekishi toshite no gakumon* (Academia as history; Tokyo, 1974).

[10]Jeon, *Science and Technology in Korea*, pp. 167−184.

[11]For instance, Ma Chün (fl. 260), on whom see Needham, *Science and Civilisation in China*, Vol. 4, Part 2, pp. 39−42 and 158; and Wei P'u (fl. ca. 1070), whose association with the polymath Shen Kua is discussed by Sivin *s.v.* "Shen Kua" in *Dictionary of Scientific Biography*.

[12]This topic has not had more than cursory treatment in Western languages. See Li Yen, *Shih-san, shih-ssu shih-chi chung-kuo min-chien shu-hsueh* (Chinese popular mathematics in the thirteenth and fourteenth centuries; Peking, 1957); Kodama Akito, *Jūroku seki matsu Minkan no shuzan sho* (Ming imprints of abacus handbooks from the end of the sixteenth century; Tokyo, 1970).

[13]Details are given in a general study of Japanese science by Nakayama to be published in 1977 by Scribners.

[14]See the forthcoming study by Sivin provisionally entitled *Ailment and Cure in Traditional China*. For a report of an early stage in this investigation, see William C. Cooper and N. Sivin, "Man as a Medicine: Pharmacological and Ritual Aspects of Traditional Therapy Using Drugs Derived from the Human Body," pp. 203–272 in Nakayama and Sivin, *Chinese Science* (see note 3).

[15]Most readers will be aware that the prevalent understanding of the relations between the two cultures, as well as the relations between the parts of modern science (e.g., the correspondence between the ideas underlying physics and biology) are very much in flux.

[16]See the discussion in Sivin, "Shen Kua."

[17]This point has been well made by Robert K. Merton in *Science, Technology and Society in Seventeenth-Century England* (New York, 1970), p. x.

[18]Sivin, "Shen Kua," and "Wang Hsi-shan," *s. v.* in *Dictionary of Scientific Biography*. The quotation is from Wang's epitaph.

[19]L.C. Wu and Tenney L. Davis, "An Ancient Chinese Treatise on Alchemy Entitled *Ts'an t'ung ch'i*," *Isis*, 1933, *18:* 210–289; Davis, "The Dualistic Cosmogony of Huai-nan-tzu and its Relation to the Background of Chinese and European Alchemy," *ibid.*, 1936, *25:* 327–340; "Pictorial Representations of Alchemical Theory," *ibid.*, 1938, *28:* 73–86; Carl W. Rufus and Celia Chao, "A Korean Star Map," *ibid.*, 1944, *35:* 316–326. For other alchemical publications of Davis and his collaborators, see Henry M. Leicester and Herbert S. Klickstein, "Tenney Lombard Davis and the History of Chemistry," *Chymia*, 1950, *3:* 6–16. Another article on alchemy that retains some value if read critically is Roy C. Spooner and C.H. Wang, "The Divine Nine Turn Tan Sha Method, a Chinese Alchemical Recipe," *Isis*, 1948, *38:* 235–242.

[20]The complexity of the seventeenth-century Chinese response to European learning was taken up in 1965 in Sivin, "On 'China's Opposition to Western Science during Late Ming and Early Ch'ing,' " *Isis*, 1965, *56:* 201–205, but that note has been superseded by "Copernicus in China," *Studia Copernicana*, 1973, *6:* 63–122.

[21]Arnold Thackray and Robert K. Merton, "On Discipline Building: The Paradoxes of George Sarton," *Isis*, 1972, *63:* 473–495.

[22]A directory is provided in the first issue of the journal *Chinese Science* (1975).

Quantitative
Sciences

On the Japanese theory of determinants.

I. — The determinant theory of Seki Kôwa and subsequent commentaries and corrections.

It is very remarkable that the theory of determinants had developed in Japan independent of, and even preceding, the progress in the West. The subject was considered some years ago by Prof. T. Hayashi [1], who was the first historian of mathematics to take notice of it [2]. In my *Development of Mathematics in China and Japan*, Leipzig, 1913, p. 191-199, I also mentioned something of the matter. But it will not be without interest to try a further account in the following lines.

The oldest of the documents concerning the subject before us is doubtless the *Fukudai* of Seki Kôwa, which bears the date of 1683. In this manuscript and in all other treatises which we have examined the Japanese theory of determinants was invariably applied to the elimination of an arbitrary quantity between two equations containing it. This is a feature that is most interesting, and that must not be overlooked.

It will be sufficient to treat only of the case where the two equations are of the same degree, and let them be given in the form

$$a_1 + a_2 x + a_3 x^2 + .. + a_n x^{n-1} = 0 \qquad \text{(a)},$$
$$b_1 + b_2 x + b_3 x^2 + ... + b_n x^{n-1} = 0 \qquad \text{(b)}.$$

---

[1] *The Proc. of the Tokyo Math. Phys. Soc.*, 2nd Series, vol. 5, p. 254-271.

[2] For this fact we must humbly acknowledge his great merit, and so I have the honour of dedicating this short article to him. It is said, according to Mr. R. Higashi, that Mr. Endô Masanosuke had early noticed the subject and told him about it in 1892.

Here SEKI considers the cases where n is 3 or 4; whereas his successors took up cases of higher degrees, and in the anonymous *Taisei Sankyô*, a rule is given for the general case. But the different authors differ little from one another in their principles of treatments. To make our statement, therefore, once for all, let us multiply (a) by b_n and (b) by a_n and subtract the one from the other, obtaining

$$\begin{vmatrix} a_1 b_n \\ - a_n b_1 \end{vmatrix} \begin{vmatrix} + a_2 b_n \\ - a_n b_2 \end{vmatrix} x + \dots \begin{vmatrix} + a_{n-1} b_n \\ - a_n b_{n-1} \end{vmatrix} x^{n-2} = 0 \quad (1).$$

Again multiply (a) by b_{n-1} and (b) by a_{n-1}, and subtracting each from the other, we have

$$\begin{vmatrix} a_1 b_{n-1} \\ - a_{n-1} b_1 \end{vmatrix} \begin{vmatrix} + a_2 b_{n-1} \\ - a_{n-1} b_2 \end{vmatrix} x + \dots \begin{vmatrix} + a_{n-2} b_{n-1} \\ - a_{n-1} b_{n-2} \end{vmatrix} x^{n-3} + \begin{vmatrix} a_n b_{n-1} \\ - a_{n-1} b_n \end{vmatrix} x^{n-1} = 0,$$

which, when x times (1) has been added to, gives

$$\begin{vmatrix} a_1 b_{n-1} \\ - a_{n-1} b_1 \\ + a_1 b_n \\ - a_n b_1 \end{vmatrix} \begin{vmatrix} + a_2 b_{n-1} \\ - a_{n-1} b_2 \\ + a_{n-3} b_n \\ - a_n b_{n-3} \end{vmatrix} x + \dots + \begin{vmatrix} a_{n-2} b_{n-1} \\ - a_{n-1} b_{n-2} \end{vmatrix} x^{n-3} \begin{vmatrix} + a_{n-2} b_n \\ - a_n b_{n-2} \end{vmatrix} x^{n-2} = 0 \quad (2).$$

Again multiply (a) by b_{n-2} and (b) by a_{n-2}, subtract each from the other, and add thereto $(2) \times x$, and we get an equation of the $(n-2)$th degree as before. This process may be continued further, until we arrive in the end at a system of $n-1$ equations each of degree $n-2$. When arranged in detached coefficients, it is at once represented in the form of a determinant, which is in fact the resultant of eliminating x from the two original equations.

The elegance of this way of elimination no one can deny; it is perhaps more elegant than the process followed by SYLVESTER in England two centuries later than SEKI. The Japanese mathematicians of the 17th and 18th centuries were certainly led to such a skilful device by their study of the old Chinese way of solving linear simultaneous equations [1].

In Japan, as in China, an algebraical equation was usually recorded in a vertical column, the absolute term being placed in the uppermost row, and the powers of the unknown coming successsively below it. In this way the powers were distinguished by means of their respective

---

[1] See for this Chinese method the *Development*, etc., p. 18-21.

positions in the column so that it was not needed to invent any special symbols for them. Further the terms were all placed on one side of the sign of equality, which they were consequently accustomed to always omit. It therefore sufficed to write simply the coefficients in superposition in a vertical column, when an equation was to be represented. Thus the n equations of the form

$$a_1 + a_2 x + a_3 x^2 + \ldots + a_n x^{n-1} = 0,$$

etc., could be arranged in this way :

k_1	....	c_1	b_1	a_1
k_2	....	c_2	b_2	a_2
k_3	....	c_3	b_3	a_3
k_n	....	c_n	b_n	a_n

Here it will be seen that the equations are to be counted from right to left.

When represented in this form it is not necessary to take any other notation for the determinant resulting from the required elimination. The development of such a determinant was the task which the Japanese endeavoured to accomplish in various ways.

In the first place mention should be made of the devices SEKI makes in his *Fukudai* of 1683. SEKI already knew that a factor common to all the elements in a vertical column or in a horizontal row of a determinant might be removed. He distinguished the cases where the common factors are numerical and literal ([1]).

([1]) HAYASHI's article, p. 260.

SEKI first gives the results of expanding the determinants of the second, third and fourth orders, and then describes the mechanical methods called *kôshiki* and *shajô*, with which the expressions and the signs of the terms in the development may be determined, methods which were certainly established in a tentative way. He states that the development may be obtained by a process called *chikushiki kôjô*, but he does not explain it. Something will be considered of this matter later on. Meantime we shall proceed to account for the two methods of *kôshiki* and *shajô*.

An idea of the latter of these methods may be obtained from the figures 1 to 4, where the dotted lines have originally been drawn in red ink (¹). The dotted and real lines, or the red and black lines in

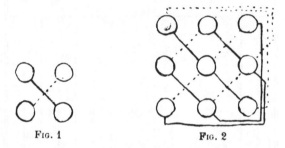

FIG. 1   FIG. 2

the original manuscript, are used to indicate the signs which the products of the elements connected by these lines, will take in the development, the dotted lines corresponding to the positive sign and the real lines to the negative sign, if all the elements be positive. But in reality, these rules will not strictly hold, for the determinant may contain elements which are different in their signs. Consequently SEKI employs, instead of the terms positive and negative, other terms *sei* (creative) and *koku* (destructive), the former meaning what will produce a positive and the latter a negative product, when all the elements are positive. Thus the dotted lines represent creative products, while the real lines represent destructive products.

SEKI considers the diagrams of the *shajô* or literally « oblique multiplications » as presenting two cases. If the number of equations be odd, the products formed along and paralled to the diagonal from the right upper corner will be creative, while the oppposite are destruc-

(¹) HAYASHI's article, p. 269-271.

tive. In the case of an even number of equations, the creative and destructive products will both appear parallel to the two diagonals (¹).

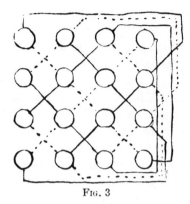

FIG. 3

The figures shown above have sometimes been represented in different forms, though expressing the same thing (²).

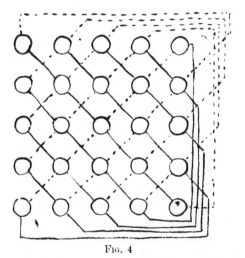

FIG. 4

(¹) HAYASHI's article, p. 271. It will be seen that the above description is not correct, but we shall refer to this question later on.

(²) So for instance see SEKI's *Yendan kai Fukudai no Hô*, TOITA YASUSUKE's *Sei-koku Impô Den* of 1759, KWANNO GENKEN's *Hoi kai Fukudai Sei Koku Hen* of 1798, KWANNO's description of the *shajô* or oblique multiplications is very noteworthy, because he was the first mathematician, so far as we know, who noticed the fault committed in SEKI's method. This point will be described later on.

As will be seen from the figures 1 to 4, the *shajô* or oblique multiplications will give 2*n* terms of the expansion in the case of a determinant of the *n*th order. But the total number of terms in such an expansion must be evidently much greater. There was still to be invented a way in which such a total number could be exhausted. The *kôshiki* was the process that answered this purpose. The word *kôshiki* means literally « interchanging of equations », the process actually consisting in the interchanges of the columns of the determinant. It is hardly necessary to add that of such interchanges a certain number only are to be retained.

In the determinants of the second and third orders the oblique multiplications will exhaust all the requisite terms in their expansions, and so there is no need for the *kôshiki* process. It is only required for the determinants of the fourth and higher orders. Here SEKI's process lies in deducing the case of the fourth order from that of the third order, the case of the fifth order from that of the fourth order, and so on (¹). SEKI gives the orders in which the equations are to be taken in the *kôshiki* of the third, fourth, and fifth orders as in the figures 5-7 (²).

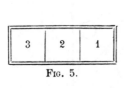

4	3	2	1
2	4	3	1
3	2	4	1

FIG. 5.

FIG. 6.

When the orders, in which the equations forming a determinant are to be taken, are once determined, the *shajô* or oblique multiplications may be effected for each of these orders, and thus we shall get all the requisite terms of the expansion. But we know nothing as to the principle by which SEKI has been led to select these orders.

_____

(¹) In respect to this passage SEKI's description is exceedingly obscure, but we have no doubt as to the correctness of our diciphering in the sense mentioned in the text.

(²) HAYASHI's article, p. 267 and 269. In these figures each horizontal row represents an order of the *kôshiki*, where the numbers are to be counted from right to left. Though there are added in these figures the ideograms *jun* (regular order) and *gyaku* (reverse order), we are not yet enabled to decipher them correctly.

SEKI's description of his method of expanding determinants terminates with the case of the fifth order, but he expressly indicates that the same may be extended to the case of any order.

5	4	3	2	1
4	5	2	3	1
3	2	5	4	1
2	3	4	5	1
3	5	4	2	1
5	3	2	4	1
4	2	3	5	1
2	4	5	3	1
4	3	5	2	1
3	4	2	5	1
5	2	4	3	1
2	5	3	4	1

FIG. 7.

SEKI does not state in his *Fukudai* of 1683 how many terms will be in the expansion of a determinant; but this is stated in a manuscript entitled *Kai Fukudai Kôshiki Shajô no Genkai*, which is usually given as revised by MATSUNAGA RYÔHITSU in 1715, but sometimes as written by SEKI himself ([1]). Here the number of terms are stated to be 2 for the 2nd order, and for the successive higher orders as follows :

$$3 \times (\text{number for 2nd. order}) = 3 \times 2 = 6,$$
$$4 \times (\text{number for 3rd. order}) = 4 \times 6 = 24,$$
$$5 \times (\text{number for 4th. order}) = 5 \times 24 = 120,$$

. . . .

etc., which shows that $n!$ is the required number for the nth order ([2]). From this number follows how to determine the number of the

---

([1]) Prof. HAYASHI gives this manuscript as anonymous and without date. See his article, p. 255.

([2]) HAYASHI's article, p. 267.

kôshiki or different arrangements arising from the permutations of the columns. Thus the number of products obtained by a set of oblique multiplications being $2n$, we have only to take

$$\frac{n\,!}{2n} = \frac{(n-1)\,!}{2} = 3 \cdot 4 \cdot 5 \ldots (n-1)$$

different arrangements of the columns. For the 3rd, 4th and 5th orders these numbers are 1, 3 and 12, respectively, which show that Seki has taken sufficient numbers of the permutations in his *kôshiki* tables in his *Fukudai*. Further the law of their formation is indicated as follows.

First take the *kôshiki* for the 3rd order, which is 321, where of course the figures are to be read from right to left according to the Japanese custom, add 1 to each of its members, and adjoin a new 1 to the right side of the system, thus obtaining 4321. This is one of the column interchanges required for the 4th order.

But now there being required three different arrangements, we must form two others. For this purpose a kind of mechanical derivation is attempted, starting from the one already obtained, according to the scheme shown in figure 8, where the horizontal rows represent the derived arrangements ([1]).

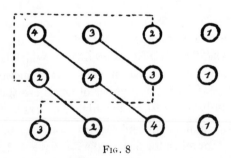

Fig. 8

Take one of the arrangements for the 4th order and apply the same way of derivation as just tried, we shall obtain four ways of arrangements, and thus we get in all 12 different column interchanges representing the *kôshiki* of the 5th order. The results are as shown in figure 9.

([1]) Hayashi's article, p. 268.

The different arrangements for the 6th order may also be derived similarly from the results for the 5th order, and so on.

5	4	3	2	1
2	5	4	3	1
3	2	5	4	1
4	3	2	5	1
3	5	4	2	1
2	3	5	4	1
4	2	3	5	1
5	4	2	3	1
4	3	5	2	1
2	4	3	5	1
5	2	4	3	1
3	5	2	4	1

Fig. 9.

Here we cannot proceed without noticing the disagreement found between the results in figure 9 and those given by Seki in his *Fukudai* (see fig. 7) [1]. Arima Raidó [2] and Kwanno Genken [3] used the same method as we have described. Moreover the latter mathematician noticed Seki's way as unintelligible because of its incorrectness [4], and stoutly maintained the correctness of that used by Matsunaga Ryóhitsu and Yamaji Shujû. This latter refers doubtless

---

[1] Hayashi's article, p. 268. He has simply mentioned that the old Japanese mathematicians used these different sets of arrangements.

[2] In his manuscript *Kaihó Yóshi* of 1762. Arima was a feudal lord of Kurume in Kyûshû.

[3] In the manuscript of 1798, already referred to. Hayashi mentions this manuscript.

[4] This point has been misinterpreted by Prof. Hayashi. See his article, p. 268.

to the arrangements in figure 9 (^{1}). The question of correctness or incorrectness of the two systems of arrangements may be discussed in full in accordance to the contents of a manuscript composed by an old Japanese mathematician. By this I mean TOITA YASUSUKE'S *Seikoku Impô-den* of 1759.

In the case of the 3rd order, let the arrangement 321 be cyclically interchanged, each time transfering the right most figure to the left most position. We thus obtain the three arrangements 321, 132 and 213. Suppose the determinant arranged according to these three schemes, and let us take the products that are formed along the two diagonals of these arrangements. The six terms thus obtained are the same as those arising from the *shajô* or oblique multiplications. Thus the latter may be considered as formed in the above way.

This same explanation holds also for the case of the 4th order. Here it is remarkable however that the products along the two diagonals are of the same sign, and that when the rightmost column is transfered to the leftmost, the signs of the diagonals are altered to the opposite. But in this way a system of oblique multiplications arises by taking the diagonal products of the four arrangements resulting from the cyclical interchanges of the columns. This appears at once when we compare the results with those of a *shajô* diagram.

TOITA does not proceed in his explanations beyond the 5th order.

From what has been considered we infer that the *shajô* or oblique multiplications will not be affected, save for the signs, if we interchange the columns cyclically. It follows therefore that the different arrangements of the columns arising from cyclic interchanges cannot be considered as independent of one another for the purpose of the oblique multiplications.

An arrangement is also not independent of another with the columns in the reversed order.

Starting from these considerations TOITA proceeds how to determine an independent set of different arrangements of the columns of a determinant (^{2}). In the case of the 4th order there are in all 4!

(^{1}) The manuscript that treats of this mode of analysis is signed by the name of YAMAJI at its end, following the signature of MATSUNAGA as revising it. Prof. HAYASHI has misrepresented MATSUNAGA for TAKAHASHI SHIJI, because KWANNO mentions him by his literary name TÔKÔ, a literary name shared also by TAKAHASHI. See HAYASHI's article p. 255.

(^{2}) This is a subject that has not been explained in SEKI's *Fukudai* and in the manuscript given as revised by MATSUNAGA. In these works the results only are given.

= 24 different ways of arranging the columns. These 24 arrangements he forms in six groups, which arise by cyclically interchanging the columns from the orders 4321, 3421, 4231, 2431, 3241, 2344. Of these groups each consisting of four permutations, the first and the sixth will be found to be equivalent, the one being in the reverse order of the other. Thus one of them, the sixth group say, may at once be dispensed with. Similarly we may omit the second and third groups. In each of the remaining three groups only a single arrangement is to be retained, the others not being independent of the retained ones. In each group we may of course select any one of the four cyclically interchanged arrangements, but for convenience take we shall take the one from which we have started. We thus obtain the three orders 4321, 2431 and 3241, which form the independent *kôshiki* or column-interchanges for the 4th order.

In the case of the 5th order TOITA applies the same mode of reasoning and establishes the twelve independent interchanges as shown in figure 10. This is different from the result in figure 9. But it will

5 4 3 2 1	3 5 4 2 1	4 3 5 2 1
4 5 2 3 1	5 3 2 4 1	3 4 2 5 1
2 3 5 4 1	2 4 3 5 1	2 5 4 3 1
3 2 4 5 1	4 2 5 3 1	5 2 3 4 1

Fig. 10.

at once appear that by making suitable selections among the equivalent groups in our procedure we are easily led to this latter. It is needless to say that there may arise different systems which are legitimate. We are however unable to arrive by a similar treatment at the result which SEKI mentions in his *Fukudai*, and TOITA shows its fallacy by indicating that there are only six of independent arrangements in it, for instance the first and the fourth arrangements in figure 7 being not independent of one another ([1]).

In the *kôshiki* or column-interchanges in figure 10 TOITA shows that

---

([1]) Though I have been able to examine some ten copies of SEKI's *Fukudai* directly or indirectly, yet in all of them the same fault has been discovered, thus showing it not to have been commited in the course of transcriptions.

the law of their formation may be indicated by the diagram or figure 11.

Toita was not able to discover any fault in Seki's scheme of the *shajô* or oblique multiplications. This discovery was reserved for another mathematician, Kwanno Genken. Kwanno's manuscript entitled *Hoi Kai Fukudai Sei-koku Hen* is of two versions, both bearing the date of 1798. The two are somewhat different, but nothing differs in the main features in their contents. It is in this work that Kwanno condemns the fault in Seki's *shajô* process, which has been followed for more than a century after him by those distinguished mathematicians

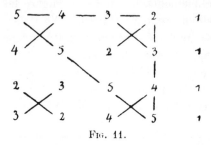

Fig. 11.

among whom we may count Matsunaga, Yamaji, Arima and Toita. It was not sufficient for Kwanno to distinguish between the two cases of odd and even orders as done by Seki, but he considered the matter in four separate cases with the order numbers of the form

$$2 + 4n, \quad 3 + 4n, \quad 4 + 4n, \quad 5 + 4n.$$

In the *shajô* or oblique multiplications of the order $2 + 4n$, the diagonal drawn from the right upper corner is of the creative sign, while the other diagonal is of the destructive sign ([1]), and the products taken parallel to these diagonals are alternately of the creative and destructive signs. An example is given in figure 12.

For the case of the order $3 + 4n$ the products along and parallel to the diagonal drawn from the right upper corner, are all creative, the others being all destructive ([2]). In different *kôshiki* the signs will be altered.

([1]) As to the use of the *sei* (creative) and the *koku* (destructive) signs we have referred to them in connection with Seki's work.

([2]) Kwanno's description of the signs for the orders $2+4n$ and $3+4n$ seems to be reversed, taking the creative to be destructive and vice versa. But he has drawn his figures correctly.

For the case of the order $4 + 4n$, the two diagonals are both creative, other products being successively alternated as to their signs.

For the case of the order $5 + 4n$ the products are all creative; in different sets of oblique multiplications different signs occur.

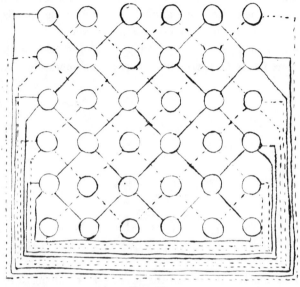

FIG. 12.

Though KWANNO's description goes no further than this, the legitimacy of his results is out of question. But we have to prove it by dint of some results considered by old Japanese mathematicians. For this purpose the contents of TOITA's *Sei-koku Impô-den* will best answer.

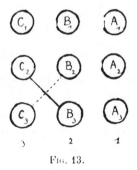

FIG. 13.

To explain the method of expansion of a determinant of the third order TOITA prepares three separate rods on which he records the three columns and arranges them as in figure 13. Here he tries the operation of expansion of the second order for the second and third rows of the second and third columns, and multiplies the results by the first row of the first column. He then transfers the first rod to the left of the third rod, and carries out the same operation for this new arrangement. Evidently this may be repeated for a third time. The results of these three operations being taken together we get the terms of the expansion of the given determinant.

This way of analysis is also applicable to the case of the fourth order. Here figure 14 represents a part of the operation. And again

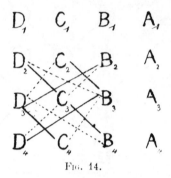

Fig. 14.

it must be remarked that when the first rod is transfered to the left side, the partial products of the minor of third order, now to be taken into account, are to be reversed as to their signs. TOITA has known the relation of signs arising from cyclically interchanging the columns. When expressed in modern notation, his results are equivalent to the formula

$$\mid A_1\ B_2\ C_3\ \dots\ K_n\ \mid\ = A_1\ \mid\ B_2\ C_3\ \dots\ K_n\ \mid$$
$$+ (-1)^{n-1}\ B_1\ \mid\ C_2\ D_3\ \dots\ K_{n-1}\ A_n\ \mid$$
$$+ \dots$$

TOITA imagines this way of analysis to be the same as SEKI's method of *chikushiki kôjô*, which remains until now unknown to us.

The above analysis at once affords us a reliable means of determining which is true of the results of oblique multiplications as given by SEKI and by KWANNO. For instance, take the determinant $\mid E_5\ D_4\ C_3\ B_2\ A_1\ \mid$, where of course we have to read from right to left, then the products formed parallel to the diagonal drawn from the left upper corner are all to be *koku* or destructive according to SEKI's scheme (see fig. 4), but all *sei* or creative according to KWANNO's scheme. Our question will therefore be answered if we take one of the said products, $A_1 E_2 D_3 C_4 B_5$ say, and examine its sign by means of TOITA's alleged formula. This at once establishes the correctness of KWANNO's diagram setting that of SEKI at fault. Here is no need to fully account for the consideration of KWANNO's results. It is only to be regretted that TOITA has not proceeded so far in his analysis as to correct the diagrams of oblique multiplications. This is a passage that has hitherto escaped the notice of historians.

We have already mentioned SEKI's obscure employment of the terms indicating regular and reverse orders in connection with his description of the *kôshiki* or column-interchanges of a determinant. This is explained by TOITA to mean the fact that the minors are to be taken with unaltered or altered signs when the operation is carried out which he imagines to be SEKI's method of *chikushiki kôjô*, and which we have already partly described. But this explanation does not seem appropriate, for the terms in question are stated by SEKI, in connection with the *kôshiki* and not with the *chikushiki kôjô*. If I am right in my conjecture they must necessarily concern the derivation of the *kôshiki* tables from those of the preceding orders, the signs of the resulting arrangements being to be altered or unaltered according to circumstances. We are however left uncertain, because we cannot verify this on account of the incorrectness of SEKI's results.

Something more may be said as to SEKI's unknown method of *chikushiki kôjô*, for in the anonymous manuscript *Taisei Sankyô* there is given a method called *kojo-ho*, which we cannot pass over without comparing with TOITA's discussion. The manuscript before us consists of twenty books, of which the 17th and the 19th concern the determinant-theory. Though the work is sometimes referred to as by SEKI himself, it is probably more correct to ascribe it to his pupil TAKEBE KENKÔ ([1]), who died in 1739 at the age of 75. The *kôjô* process treated of in this manuscript does not widely deviate in its main features from the considerations of TOITA. But in the former the columns are not cyclically interchanged as done by TOITA. The multiplication of the elements in the first row into the expansions of what are practically their complementary minors, which TOITA tries, is here effected with the elements in the first column and their complementary

---

([1]) ARIMA RAIDÔ, among others, was of this opinion in his manuscript *Kaihō Yôski* of 1762.

minors, the determination for the sign in each step being clearly indicated. The two processes of the *Taisei Sankyô* and of TOITA might be considered as the same, if the Japanese had known the theory according to which the rows and the columns of a determinant might be treated as its columns and its rows, respectively; but this does not appear to be so, so far as our limited knowledge now goes, to have been an established fact in those times; the Japanese mathematicians have always considered the columns as separate equations from which the determinant is formed. It is quite unknown which one SEKI has employed of the two processes just laid down, whether he has not meant an entirely different process by the *chikushiki kôjô*, or whether he actually possessed the knowledge corresponding to such a technical term.

II. — *The Sampô Hakki of 1690.*

If it be very remarkable that SEKI has considered a kind of determinant theory in one of his secret-held writings, it will be no less so to discover allied considerations in some printed works that appeared contemporarily with this noted master. Of these works the most noteworthy is none other than the *Sampô Hakki* of 1690, which was printed at Osaka bearing the name of SHIMADA SHÔSEI's pupil IZEKI CHISHIN [1]. It is rather astonishing that this important document has hitherto escaped the notice of the students of mathematical history [2]. Mr. T. ENDÔ [3] mentions it merely as a work devoted to able considerations of the celestial element method [4], an account which is far from being satisfactory. Its contents in reality consist exclusively of the determinant theory and its application for the solution of some algebraic-geometric problems such as considered by the abacus algebra

---

(1) Though the *Wakan Sampô* of 1693, published by MIYAGI SEIKÔ, has treated something of the same subject, its contents are not of much value.

(2) Here the author of the present article must acknowledge his deep indebtedness to Mr N. OKAMOTO who has permitted him to consult this valuable material in his library.

(3) ENDÔ, *History of Japanese Mathematics* (in Japanese), Tokyo, 1896. Book II, p. 31.

(4) This is the Chinese algebra which was manipulated by using the calculating pieces; it is therefore to be called a sort of abacus algebra. Descriptions thereof are given in Y. MIKAMI, *Development of Mathematics in China and Japan*.

of the Chinese.   Now we must compare the treatise before us with SEKI's theory.

SEKI's *Fukudai* being dated as revised in 1683, we must admit that the *Sampô Hakki* was printed seven years after this date.   But this does not prove the indebtedness of the latter to the former.   The connection between the two must be carefully considered.

Of SHIMADA SHÔSEI's career we know little, but according to FURU-KAWA UJIICHI's *Sangaku Dai Keizu* (Lineage of Japanese Mathematicians), he was a pupil of GOTÔ KAKUBEI, one of SEKI's numerous disciples.   If so, he may have had some chance of being initiated into the determinant theory of SEKI.   In this case, however, how could he, or his pupil IZEKI, dare print the treatise before us against his master's will, without his permission, indeed in his life-time ?   Besides, SHIMADA's

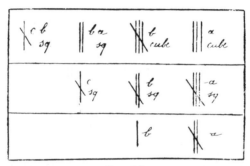

FIG. 16.

connection with GOTÔ is very doubtful, so also that of GOTÔ with SEKI. GOTÔ KAKUBEI is sometimes identified with SAWAGUCHI KAZUYUKI, the author of the *Kokon Sampôki* of 1670, while at times this identification is denied.   About this enigmatical personage practically nothing is known.   If he was really SEKI's disciple, it does not necessarily follow that he was instructed by his master in the subject referred to.   The information now in my possession has not helped me in the slightest to make any further step towards the solution of our question.

If we consider, however, the circumstances upon which the determinant theories of the two authorities have been started, namely the fact that the determinants are formed equally by the consideration of eliminating the arbitrary between two equations containing it (¹), we

---

(¹) About this matter the *Sampô Hakki* is more detailed in its descriptions than in SEKI's work.

cannot help guessing at the same origin for the two. But there are some differences in these works, which no one can overlook. The algebraic notation adopted by SEKI and his followers was an extension of the old Chinese way; while in the *Sampô Hakki* are envolved different principles. Figure 16, for instance, is the expression with which SEKI records in his *Fukudai* the equation

$$(3a^3 - 3b^3 + 2a^2b - b^2c) + (-3a^2 - 2b^2 - c^2) \, x$$
$$+ (-2a + b) \, x^2 = 0,$$

while in the *Sampô Hakki* the equations.

$$(af - be) + (ag - ce) \, x + (ah - de) \, x^2 = 0,$$

and

$$(a^4 - 2a^2b^2 + b^4) + (2a^2 - 2b^2) \, x + x^2 = 0$$

are represented as in figures 17 and 18. Both agree in expressing

a	b		a	a	b	
f	c		3rd.	sq.	3rd.	
pos.	neg.		1	b	1	
			pos.	sq.	pos.	
				2		
a	c			neg.		
g	c					
pos.	neg.		a	b		
			sq.	sq.		
			2	2		
			pos.	neg.		
a	d					
h	e		1			
pos.	neg.		pos.			

        Fig. 17.   Fig. 18.

equations by writing the coefficients only, besides the vertical rods, in superposition, the absolute term coming topmost and the highest term in the lowest position, but the sums and products are quite differently denoted. The system of notation used in the *Sampô Hakki* was followed in subsequent ages by the Takuma Shool ([1]). This disagreement in the notations used must be evidence of the *Sampô Hakki* being independent of SEKI. In any case the true author of the analysis followed in it had been endowed with the meritorious talent of devising or adopting this sort of notations, which was not the one prevailing in the mathematical circle of his days.

As to the considerations of the determinants, the *Sampô Hakki* does not follow the same method as SEKI. To expand a determinant of the *n*th order, to make our statement for the general case, the minors are formed which are complementary to the elements in the first column, there being accompanied illustrations as shown in figure 19. In

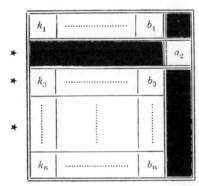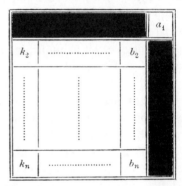

FIG. 19.

forming the products of these minors multiplied by the elements respectively complementary to them, the rule of signs is indicated to be effected according to the scheme

$$a_1 \mid b_2\, c_3 \ldots k_n \mid -a_2 \mid b_1\, d_3 \ldots k_n \mid +a_3 \mid b_1\, c_2\, e_4 \ldots k_n \mid -\ldots,$$

where we have employed the modern method of notation. In the first place this process is applied to the case of the 2nd order. Then the case of the third order is tried by applying the result of the preceding case, and so on.

This mode of expansion is exactly the same as explained in the *Taisei Sankyô*, save for the illustration by diagrams. If the process in the *Taisei Sankyô* could be considered as originative from SEKI, then the two works *Fukudai* and *Sampô Hakki* would appear more intimately related to each other than otherwise thought. In the latter work the term *kôjô* or *chikushiki kôjô* is not made use of. The technical terms found therein mainly differ from those of SEKI employed in his *Fukudai*.

It is most interesting that the above method of expansion is demonstrated in the *Sampô Hakki* though only for the case of the third order. It is accompanied by the statement that it may be generalised for higher orders. Thus let us first form the expansion for the nth order according to the relation to be proved. Then let us assume it to be true for the $(n-1)$th order and apply it to verify the expansion just obtained. Such is the essence of the demonstration attempted in the *Sampô Hakki*. Thus for this purpose, of the n equations of the $(n-1)$th degree, let us multiply one of each successive two, 1 and 2, 2 and 3, 3 and 4, ... alternately by the coefficients of the highest terms of the other and subtract each other, resulting in $n-1$ equations of the $(n-2)$th degree. In this way we have reduced the problem to the case of the $(n-1)$th order, which has been assumed as possible to be expanded. Compare the result of this expansion with that obtained by dint of the assumed relation, when the two will be found entirely identical, except for a superfluous factor. This mode of proof may first be applied to the case of the third order upon the basis of the actual expansion for the 2nd order, then to the case of the 4th order, the next to the case of the 5th order, and so on, successively for the next higher orders.

In this demonstration we may safely recognise the germ of idea that will lead to the establishment of the mathematical induction. The *renjutsu* or *chikusaku-jutsu*, that thrived in subsequent ages among the Japanese mathematicians, has certainly developed out of this source.

In the *Sampô Hakki* are calculated the terms of the expansions up to the 5th order by the above mentioned process.

Such a demonstration as we have just described has never been found hitherto among the writings of SEKI and his followers, and yet by its existence we are fortunately led to conjecture accurately how the Japanese mathematicians have been brought to effect the elimination of the powers of the arbitrary quantity from the set of n equations of $(n-1)$th degree. The *Sampô Hakki* plainly shows it as effected according to the same spirit as in solving the linear simultaneous equations after the old Chinese manner, the same process by which the set has already been arrived at from the original two equations. SEKI and his followers have always simply striven to form the expansion of a determinant ever dispensing with the formation of it as granted so to say. Though the elimination is effected in this way by FURUKAWA KUNSHÔ, it is of only too late a date, his manuscript *Kôshiki Shajô Yendan Shinkai* being written in 1791.

In the *Sampô Hakki* we do not encounter with those mechanical process, which SEKI has given by the name of *kôshiki* and *shajô*. The author states, however, that there are still other processes for expanding the determinants, which he does not choose to give on account of their too complicatedness. We are not certain whether these are meant to be the same as SEKI's ways.

When we reflect on what we have described we feel we are pressed to believe in the origin of the Japanese theory of determinants as assuredly derived from the method of solution of the linear simultaneous equations, which the Japanese have learned from their Chinese masters. This view will remain the same whether the theory has come from SEKI, from the author of the *Sampô Hakki*, or from the common source from which these two authorities may have drawn their information, at least in their rudiments. The application of the term *yuijô* to the expansion of a determinant or to the elimination from the equations forming it seems to confirm this view very strongly. In FURUKAWA's work above referred to, the process of elimination is called by this name. In AIDA AMMEI's manuscript the same word is used to indicate the expansion, whereby diagrams are drawn to illustrate the process, meaning the *kôshiki* and *shajô* applied simultaneously. The word *yuijô* appears to have been employed from very early times in China, mostly applied to some process occurring in connection with the solution of simultaneous equations, originally meaning multiplying two equations alternately by their highest coefficients and then subtracting each from the other. We are not yet sure of the extended use of the word to be found in China or not. In AIDA's sense the word has been prevalent in Japan from before his days, for BAN SEIYEI calls the considerations of determinants by the name of *yendan yuijô-hô*. It is to be noted that his treatise *Sampô Gakkai* was a printed work of 1782. He has also applied the name of *tengen yendan jutsu* to the same subject. The use of these names must be very significant for the further exploration of the historical development of the subject, for they expressly refer to the peculiar algebraical

forms studied in China.   In INAGAKI HÔKYÔ's *Sankei* [1] of 1791 it is
stated that there have been a variety of processes called by the name
of *yuijô*.

Moreover a passage in TAKEBE KENKÔ's *Hatsubi Sampô Yendan Genkai*, a printed work of 1685, seems to indicate more clearly the alleged
relation of the Japanese theory of determinants to the Chinese mode
of solving simultaneous equations.   Concerning the algebraical analysis of problems, mostly geometric in nature, he says :

« Again we may construct two equations (containing an arbitrary
quantity) and proceed as in solving the simultaneous equations, or we
may employ the process of *yuijô*.... »

Thus far it is very evident, or very probable, that the Japanese
theory of the determinants has sprung or originated from the considerations of equations after the manner encountered in the simultaneous equations, but curious enough the determinant has never been
applied to the solution of simultaneous equations, or at least we have
not as yet had occasion to discover any document concerning the subject.

III.— *Theories described by Kurushima and Kwanno.*

The manuscript entitled *Kyûshi Ikô*, wherein are collected some of
the results obtained by KURUSHIMA GITA (deceased in 1757), gives a
passage treating of a new feature of the determinant theory.   In the
case of the third order KURUSHIMA's expansion is the same as given in
the *Sampô Hakki* of 1690.   In the case of the 4th order, $| A_1 B_2 C_3 D_4 |$, which we have expressed in modern notation, he gives the result
in a form equivalent to

$$| A_1 B_2 | \times | C_3 D_4 | + | A_1 B_3 | \times | C_2 D_4 | + | A_1 B_4 | \times | C_2 D_3 |$$
$$+ | A_2 B_3 | \times | C_1 D_4 | + | A_2 B_4 | \times | C_1 D_3 | + | A_4 B_3 | \times | C_1 D_2 |.$$

The expansion of $| A_1 B_2 C_3 D_4 E_5 |$ is given to be

$$(12) (345) — (13) (245) + (14) (235) — (15) (234)$$
$$+ (23) (145) — (24) (135) + (25) (134) + (34) (125)$$
$$— (35) (124) + (45) (123),$$

---

[1] This work has been erroneously given as the *Sansen* in my *Development*,
p. 185.   The author's name given there as INAGAKI SAKUHO is not correct,
though this is in accordance to ENDÔ's *History*.   His true name should be
called INAGAKI HÔKYÔ.

where we have represented by $(r\,s)$ and $(i\,j\,k)$ the determinants $|\ A_r\ B_s\ |$ and $|\ C_i\ D_j\ E_k\ |$, respectively. And the expansion of $|\ A_1 B_2 C_3$ $D_4\ E_5\ F_6\ |$ is given to be the sum of the products of the 20 minors each consisting of 3 rows out of the three columns A, B, C, multiplied respectively by their complementary minors or those that are formed by taking the remaining three rows from the remaining three columns. The signs of the minors are to be taken such that, if $|\ A_1\ B_2\ C_3\ |$ be creative, then $|\ D_4\ E_5\ F_6\ |$ should be taken with the destructive sign, and $|\ A_4\ B_5\ C_6\ |$ and $|\ D_1\ E_2\ F_3\ |$ both with the creative sign. Here the descriptions appear to have been incorrectly carried out reversing the sign of the whole ([1]).

Thus we believe we have rendered clear what KURUSHIMA gave in an exceedingly obscure way, merely writing down the terms of expansions without any further reference. The same way of expansion was studied some years afterwards by KWANNO GENKEN somewhat more systematically, his results being laid down in a manuscript work entitled *Kôshiki Shajô Shôhô*, dated 1798. Here KWANNO describes in a general way the method of expanding determinants as tried by KURUSHIMA, whereby he makes reference to a new sort of the *kôshiki* and *shajô* processes. He himself recognises his indebtedness to KURUSHIMA though for the special cases only, but he expressly states that he has not been able to examine KURUSHIMA's work, which he takes as lost. In any case KWANNO's work is of great value, as therein are carried out the descriptions in a generalised and systematic form.

First of all KWANNO distinguishes the determinants of even and odd orders. In a determinant of the even order $(2n)$ KWANNO separates it into four minors each of the nth order as in figure 20, and considers the result as a determinant of the 2nd order, whose elements consist of the developments of these minors, whereby a rule is to be observed to the effect that if n be even, the sign for one of the diagonals should be changed from the destructive to the creative. This is the process which KWANNO calls the *shajô* or oblique multiplications in the extended sense for the determinant of the even order.

A determinant of order $2n + 1$ will be divided into four parts as in figure 21, such that the part at the right upper corner consists of a square of n columns and the opposite part of a square of $n + 1$ columns.

([1]) This must be compared to our description of KWANNO's method which just follows.

The *shajô* in this case will consist of the multiplication of the expansions of these squares considered as independent determinants.

F_6	E_6	D_6
F_5	E_5	D_5
F_4	E_4	D_4

C_6	B_6	A_6
C_5	B_5	A_5
C_4	B_4	A_4

F_3	E_3	D_3
F_2	E_2	D_2
F_1	E_1	D_1

C_3	B_3	A_3
C_2	B_2	A_2
C_1	B_1	A_1

Fig. 20.

The number of partial products resulting from this way of oblique multiplications, though not expressly stated by Kwanno, will be easily seen to be $n! \, (n+1)!$ in the case of order $2n+1$, and $2 \times n! \, n!$ in the

E_5	D_5	C_5
E_4	D_4	C_4

B_5	A_5
B_4	A_4

E_3	D_3	C_3
E_2	D_2	C_2
E_1	D_1	C_1

B_3	A_3
B_2	A_2
B_1	A_1

Fig. 21.

case of order $2n$. But the total numbers of terms are $(2n+1)!$ and $(2n)!$, respectively, so that these *shajô* or oblique multiplications must be repeated for

$$\frac{(2n+1)!}{n! \, (n+1)!} \quad \text{and} \quad \frac{(2n)!}{2 \times n! \, n!}$$

sets of times, respectively, in order to obtain the total number of terms in the respective expansions. This purpose will be answered by repeating the process after variously interchanging the orders of the columns. Thus we have only to determine the different ways of the *kōshiki* or column-interchanges, whose numbers should be equal to the numbers given above. But if we replace n by $n + 1$ in the case of an even order, we shall have

$$\frac{[2(n+1)]\,!}{2(n+1)\,!\,(n+1)\,!} = \frac{2(n+1)\,(2n+1)\,!}{2(n+1)\,n\,!\,(n+1)\,!} = \frac{(2n+1)\,!}{n\,!\,(n+1)\,!}\,,$$

which is equal to the number for the $(2n+1)$th order. Kwanno has therefore first to form the *kōshiki* or column-interchanges for the odd order, and therefrom to deduce those for the even order.

For the case of the $(2n+1)$th order Kwanno proceeds as follows :

Of the $2n+1$ columns we may take n and arrange them in their natural order in

$$_{2n+1}C_n = \frac{(2n+1)\,!}{n\,!\,(n+1)\,!}$$

different ways. But this is the number of the required column interchanges. Hence for any one of the above arrangements the remaining $n+1$ columns are sufficient to be arranged in a single way. Let these be arranged, therefore, in the natural order following the columns already arranged, with the convention that..., $2n$, $2n+1$ should be continued by 1, 2. 3,... In the arrangement thus formed Kwanno calls the n and $n+1$ columns as the first and second divisions.

Thus far the necessary column arrangements have been all obtained, but it remains to determine their signs. For this purpose Kwanno classifies the column interchanges into some groups, such as shown in figure 22, where each horizontal row represents a column-interchange. Here the separate interchanges are arranged in order of the numbers coming at the rightmost positions, and are classified according to the changes in these numbers below the $(n-1)$th, inclusive, as will be seen from the figure. In each group Kwanno takes the first column-interchange to be *ki* or additive, the second to be *shō* or subtractive, and the successive columns being taken alternately additive and subtractive. By the terms additive and subtractive here mean that the results of the *shajō* or oblique multiplications arising from these column-interchanges are to be added and subtracted, respectively.

3

This process described by KWANNO is evidently correct in the main, but his rule for determining the additive and subtractive signs appears somewhat incorrect, probably it being necessary to distinguish between the cases n is odd or even. This KWANNO neglected to do, thus he who

SECOND DIVISION				FIRST DIVISION				
7	6	5	4	3	2	1	additive	I group
3	7	6	5	4	2	1	subtractive	
4	3	7	6	5	2	1	additive	
5	4	3	7	6	2	1	subtractive	
6	5	4	3	7	2	1	additive	
2	7	6	5	4	3	1	additive	II group
4	2	7	6	5	3	1	subtractive	
5	4	2	7	6	3	1	additive	
6	5	4	2	7	3	1	subtractive	
3	2	7	6	5	4	1	additive	III group
:	:	:	:	:	:	:	:	:
5	4	2	1	7	6	3	additive	XII
3	2	1	7	6	5	4	additive	XIII
6	3	2	1	7	5	4	subtractive	
5	3	2	1	7	6	4	additive	XIV
4	3	2	1	7	6	5	additive	XV

FIG. 22.

had pointed out the fault in the old method of oblique multiplications was not free from commiting a similar fault in his new method of forming the column interchanges.

The column interchanges for the order $2n+2$, where as has been already mentioned the number of such arrangements is the same as for the order $2n + 1$, may be derived from those of the latter case. If namely we take the column-interchanges for the $(2n + 1)$th order, add unity to every figure in each arrangement, and write a new figure 1 to the right of it, we have at once the required result. Here too, though equally erroneous, the same way is indicated for the determination of signs as in the former case.

KWANNO's methods so systematically laid down and described for the general case are not essentially unrelated to the results given by KURUSHIMA before him. But despite all the resemblance of the principles underlying the methods used by these two scholars, there is an important difference which we cannot overlook. As we have described, KWANNO attempted the expansion of determinants by considering the interchanges of columns, but in KURUSHIMA's results the interchanges are done with the rows and not with the columns. The two are just in the same relation as TOITA's manuscript of 1759 is to the *Sampô Hakki* of 1690. They have unknowingly established, when taken together, the relation that the columns and rows of a determinant may be replaced by one another without effecting its value. In a word, if we are allowed to make our own interpretation, KURU-SHIMA has extended the method of expanding a determinant by using the products of the elements in a column and their complementary minors as set forth in the *Sampô Hakki*, and has started the operation of forming the products of complementary minors taking them from certain predetermined columns. Then KWANNO was able to do the same upon the complementary minors taken on some given rows. Though KURUSHIMA and KWANNO would not have imagined it in any way, yet their processes have been necessarily an extension from their predecessors' works, and indeed an extension or generalization of great importance.

If the determinants theory, thus far advanced in Japan during the 17th and 18th centuries, had continued to be cared for some time longer, it would no doubt have undergone a wonderful progress, but such was not the case. From the beginning of the 19th century onward, it was destined to be gradually neglected, and curious enough, in the closing days of the old Japanese school of mathematics it was forgotten altogether ([1]). This is a fact that must be explained. If we

---

([1]) The *shôchô yendan* method, which is intimately related to the determinant theory, was taught and followed in the OGAWA School until the last days of the

do not err in our judgment, we may infer thus : In the earlier days of the development of Japanese mathematics the equations of wondrously high degrees had often been preferred as being elegant, but by and by simplicity had come to be sought for, causing equations to be desirably obtained in as low a degree as the case might admit.   When, however, a problem is solved by applying the determinant theory, the result cannot satisfactorily fulfil this requirement, sometimes unnecessarily raising the order of the resulting equation on account of some superfluous factors, which might be easily avoided when analysed in the ordinary algebraical way.

Consequently such a process could not escape being considered as very tedious or awkward, nor the elegance of the theory ever retain attention of the Japanese mind.   Therefore it must have disappeared through neglect and decay.

YOSHIO MIKAMI.

Tokyo, July 19, 1913.

---

old Japanese school of mathematics. For this information I am indebted to Mr. Y. SANTÔ. Hitherto nothing has been known of the OGAWA School or of its founder OGAWA Kôkei, who flourished in the 18th century.

Some notes on Chinese musical art

He who knows music is close to the rites.
To know music and rites, means to possess virtue.

SSǓ-MA CH'IEN

The scarce literature on the history of Chinese music, its theoretical foundations and the development of a proper technique of musical composition, has been recently enriched by the publication of a thorough scientific study of the subject : JOHN HAZEDEL LEVIS' *Foundations of Chinese Musical Art* (1). The young author, who is well known as a lecture-recitalist in China as well as in the United States, was born and brought up in China, and lived there for the greater part of his life — more than a quarter of a century; his approach to the spirit of Chinese music is naturally essentially different from that of his European predecessors. Among the latter, half a century ago, J. A. VAN AALST wrote a treatise (2) which is remarkable from the purely scientific point of view, but he does not make it a secret that he is by no means favorably impressed by the beauty of Far-Eastern music; similarly, GEORGE SOULIÉ (3) in his work strictly denies the existence of any art of composition in China. No doubt that both VAN AALST and SOULIÉ have made valuable contributions to our knowledge of scientific facts, but it must always be considered a dangerous, not to say preposterous, attitude if an author discusses his artistic subject as a mere monstrosity without making an attempt of penetrating beyond the surface, in other words, if only his scientific interest participates in the matter, while his heart remains unaffected.

---

(1) HENRI VETCH, Publisher, The French Bookstore, Peiping, China, 1936, XIII and 233 pp., illustrated. Selling price in China: Ch. $ 15.00, in North America : U.S. $ 6.00, in England : 21/—.

(2) *Chinese Music*, Published by order of the Inspector General of Customs. Imperial Maritime Customs, II. — Special Series : No. 6 (1888). Re-issued and sold by the French Bookstore, Peiping, under the authorithy of The Inspector General of Customs, 1933.

(3) La musique en Chine, *Bulletin de l'Association franco-chinoise*, Paris, 1910-11.

A musically educated European who has never listened to Oriental music, will undoubtedly not enjoy the first Chinese recital he attends; he will find it disgusting, a strange and tiresome cacophony, and, physically and mentally worn out, he possibly may need several hours of Western classical music, to find rest and consolation, and to forget about his bitter experience. Our friend will hardly be able to refrain from comparing the monotonous and rhythmically poor Chinese melodies with the abundant richness of European melodic movement and rhythmical structure, or comparing consistently offending progressions in parallel fourths, fifths, or octaves (all untempered, of course), produced on two simultaneously played strings of the *ch'in* (2109) or *sê* (9599), with the beautiful polyphonic pattern of Western compositions based on our tempered scale and worked out according to the rules of the theories of harmony and counterpoint.

However, such a comparison is of no use at all. Let us frankly admit that the European music is more highly developed than the Chinese, but let us also, at the same time, carefully avoid the question of superiority or inferiority, as this necessarily leads away from those points which are of true interest. Undoubtedly, Chinese music serves just the same essential purpose as European, that of talking immediately to the hearts of those whom it is destined for. And we may be sure that there is no measurable difference in the effect produced by Chinese music played before a Chinese audience, and that which European music exercises on Western audiences. But do not let us try to convince one or the other of our Chinese friends of the eternal beauty of Western music by inviting him to listen to a WAGNER opera or BRUCKNER symphony. In case it is his first trial, he will certainly suffer just as much as we did in our first Chinese musical recital. Of course, I do not want to pretend that there is no bridge between Oriental and Occidental music, no way of mutual understanding, but it may be taken for granted that real understanding and intuitive perception of the hidden beauty of foreign music is not attained in one day.

The present article is partly based on the very fertile theories developed in Mr. LEVIS' book. Many of his discoveries, at which he arrived through a careful study of rare documents of Chinese musical notations, are of such an importance and general interest that this concise analytical summary, which appears instead of a critical review, needs no justification. Some explanatory sections are added; critical remarks will mostly be found in the footnotes.

The salient point in the treatment of his subject is the author's theory of poetry and music having been an inseparable unity from the earliest times in history. This is certainly true also of countries other than China, but due to the peculiar "musical" character of the Chinese language itself (the tones or "neumes", in LEVIS' terminology), the musical composition is in principle preconceived by the poet; all that is left to the composer, is to carry out in a musically satisfactory way the rules inherent in the poet's words.

In consequence of this important statement, which will be illustrated later by several examples, the author divides his work into three main parts :

I. The origin of the musical art, which springs from the musical elements in the Chinese language;

II. The development of the art : How the art grew from ancient times through the understanding and use of the elements of the language;

III. Examples of highest development : The actual use of the musical elements of the language in the composition of music for poetry (analyses of composition and forms).

A concluding fourth part is devoted to the question of the value of the art in the modern world : The ancient art as a living form of musical creation. This part, though very stimulating, is of a more speculative nature, dealing with the possibilities of amalgamating Far Eastern and Western music, and superimposing the foreign element of harmony on the ancient system of musical forms. It is Zukunftsmusik in the concrete sense of the word rather than historical deduction, and this is therefore not the place for discussing it in extenso. However, it may be mentioned here that Mr. LEVIS himself has applied the ancient Chinese forms and principles of musical structure to his own compositions (which are unfortunately unknown to me). We are looking forward with the greatest interest to their publication, as this appears to be the first serious attempt made in this direction. Indeed, other attempts of the kind have been made, as for instance, more than a century ago, by CARL MARIA V. WEBER in his *Turandot Overture* (1809) where he uses a Chinese melody limited to the pentatonic scale (an interesting, though somewhat superficial experiment). GUSTAV MAHLER's unique *Lied von der Erde* (1911, posthumously published 1912) composed over half a dozen poems from the T'ang era — undying flowers of lyric poetry, though badly mutilated by the translation — lies on another line. A European audience would readily accept the artist's strange tonal effects as "Chinese" (just as the orang-outang's cries, in POE's *Murders in the Rue Morgue*, are identified with all possible languages unknown to the various witnesses) but they would hardly sound particularly familiar to a Chinese audience and, thus, immediately touch their hearts.

1. Musical tradition in China comprises thousands of years. Music was counted among the six Confucian arts, and the earliest manifestations of Chinese poetry, the odes of the *Shih ching*, are clearly recognizable as *songs*; they were inseparably joined with music. Also the beginnings of theoretical considerations about music must be dated back to a very high antiquity. The earliest theory of tonal proportions, which in principle was never changed during the whole of Chinese history, shows a striking similarity with that of the Pythagoreans. It is based exclusively on the consonance of the fundamental tone (say *C*) with its perfect fifth (*G*); all other tonal proportions are considered dissonances (4). In order to determine such proportions, the Chinese did not use the monochord, as did the Greeks, but twelve tubes of different lenght called *lü* (7520) which means "rules", "principles", or in this connection, "pitch-pipes". The longest which emitted the fundamental note (its absolute pitch is, of course, unknown and also irrelevant (cf. p. 78, note 8); we shall identify it with the *C* of our scale), is reported to have measured 9 inches. A tube reduced to two-thirds of its length, i.e. 6 inches, will then emit the

(4) On this, most of the ancient peoples seem to have agreed. A remarkable exception from the rule is formed by the Persian-Arabic *mathal* "messel" theory, as described in AL-FĀRĀBĪ's treatise on the theory of music (*Kitāb al-mūsīqī*) which recognizes the major and minor thirds, and even the major and minor sixths, as consonances. As a metaphysical reason for the Chinese theory of the fifth being the sole consonance, we read that two sounds which are in the proportion of 3 to 2 harmonize as perfectly as heaven and earth, because 3 is the attribute of heaven, and 2, that of earth.

perfect fifth (*G*) of the fundamental note. The next step yields a tube of four inches emitting *d*; as this lies above the octave of the fundamental, the length of the pitch-pipe is taken double (8 inches), and thus *D* is obtained (5).

It appears that, to begin with, this procedure was repeated only four times and that its sixfold application which yields a heptatonic gamut belongs to a somewhat later period (6). At least it can be stated that the pentatonic (anhemitonic) scale : *C. D. E.. G. A..(C)*, has played, throughout the ages, a particularly important part. In fact, to this day, ancient temple melodies as well as popular songs, still give a marked preference to the pentatonic gamut (6*a*).

Of course, the tones produced in this way are not tempered. To the Western ear, some of them sound too flat, some too sharp, and the octave in particular exceeds the ratio of 2 to 1 by that of $\dfrac{531441}{524288}$ (i.e. $\dfrac{9^6}{8^6} : \dfrac{2}{1}$), the well-known "ditonic comma" of the Pythagoreans. It must, however, be stated that the Chinese musicians themselves were aware of the imperfectness of their untempered scale. After a number of unsuccessfull endeavours made during more than sixteen centuries (6*b*), the prince CHU

---

(5) The length of 9 inches for *huang-chung* was, of course, chosen for practical, mathematical, reasons only. The method of determining the relative pitch of the *lüs* according to the ratio 2 : 3, or 4 : 3, is found for the first time in the "Spring and Autumn" of Lü Pu-wei (*Lü-shih Ch'un-ch'iu* 7520, 9978, 2854, 2302) dating from the third century B.C.; this book, however, reports that *huang-chung* was a thick bamboo pipe measuring 3 inches 9 lines (3.9"). The first mathematical calculation of the relative lengths of the twelve *lüs* appears, a century later, in the *Shih-chi* (9893, 922), the work of the great historian Ssŭ-ma Ch'ien 10250, 7576, 1711).

(6) The heptatonic scale : *C. D. E. F♯ G. A. B* (*C*), obtained according to this method, is reported to have been introduced during the early Chou era. It is mentioned for the first time in the work *Kuo Yü* (6609, 13626) of the fourth century B.C., and has been in use since as a basis of musical composition. This does not stand in contradiction with the fact that all the twelve *lüs* (our "chromatic" gamut) were known at the latest about the same time, or even in a much earlier period. Their invention is attributed to HUANG TI. In literature, their names and arrangement occur equally for the first time in *Kuo Yü*. But the chromatic scale never served as a basis of composition; it was used exclusively for the purpose of transposing music-pieces from one key into another.

(6*a*) This is true of Southern Chinese popular music at least. In the North, the heptatonic scale predominates over the pentatonic.

(6*b*) Cf. WANG KUANG KI, Über chinesische Musik, *Sinica*, Vol. II (Frankfurt a.M., 1927), pp. 136-38.

Tsai-yü (2544; 11485; radical 32 + 13659) of the ruling Ming family, in the year 1596, set forth a theory of tempered intonation, by which he anticipated ANDREAS WERCKMEISTER's famous discovery of 1691 (7) by almost a century. Unfortunately, CHU TSAI-YÜ's fundamental innovation seems to have fallen into oblivion pretty soon, and it certainly never came into general use.

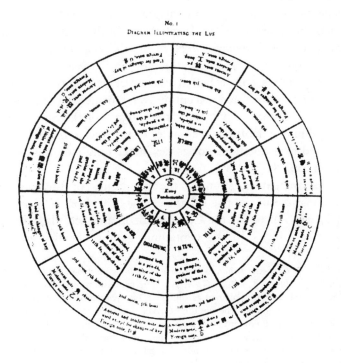

Diagram Nº 1.

But at least as far as octaves are concerned, Chinese string instruments [as the lute, *ku-ch'in* (6188, 2109)], nowadays use perfectly tuned intervals. Too sharp octaves, or other intervals offending our ear, will only be perceived in certain keys, when it is not possible to make recourse to the open strings; in corresponding cases, however, our Western guitars, lutes, or mandolines,

(7) *Musikalische Temperatur, oder deutlicher und wahrer mathematischer Unterricht, wie man durch Anweisung des Monochordi ein Klavier, sonderlich die Orgelwerke, Positive, Regale, Spinetten u. dgl. wohltemperiert stimmen könne.*

suffer from just the same weakness; for the sake of justice this must be admitted.

The twelve *lüs* forming the twelve half-steps of the chromatic scale, were, of course, divided into *yin* and *yang lüs*, those of even numbers being *yin*, the odd numbers, *yang* (7a). Furthermore, they were attributed to the twelve moons of the year, the twelve hours of the day, and so on. Their names, attributes, and Western equivalents, are illustrated by·diagram no. 1 (8).

It has been asserted by Western authorities that the names of the twelve *lüs*: *huang-chung*, *ta-lü*, *t'ai-ts'u*, etc., "are solely used in practice to indicate the key-note of a composition". This, however, is only a half truth, as LEVIS points out, because at least one remarkable exception was formed by the musical notations of CHIANG K'UEI (8a) (1233, 6507) of the Sung dynasty, where the first syllables of the names of the twelve *lüs* were consistently used as a notation, not as key-notes only (see the examples given on pp. 85 and 87).

2. The ancient notation of the five tones of the pentatonic gamut consisted of the five syllables *kung* (*C*), *shang* (*D*), *chiao* (*E*), *chih* (*G*), *yü* (*A*). (For the Chinese ideograms see diagram no. 1, in the outer rim). In the heptatonic gamut, the two additional

---

(7a) Thus, if *C* be the fundamental, the *yang lüs* would be : *C*, *D*, *E*, *F*$\sharp$, *G*$\sharp$, *A* , and the *yin lüs* : *C*$\sharp$, *D*$\sharp$, *F*, *G*, *A*, *B*. The generation of the twelve lüs from the fundamental is then explained as follows : The *yang* tone *C* "marries a wife", i.e. generates its dominant, the *yin* tone *G*; this, in turn, "gives birth to a son", i.e. generates its subdominant, the *yang* tone *D*, and so on. In the special case of the pentatonic gamut, we thus arrive at the "grandson" *E*, but this unfortunate, for purely euphonious reasons, is not allowed to marry its *B*. This drawback is remedied in the heptatonic scale only.

(8) From VAN AALST, *op. cit.*, p. 9. For practical reasons, the fundamental note is again called "*C*". Père AMIOT assumes *F* to be the equivalent of the Chinese fundamental (*huang-chung*), although his reasons for this choice do not seem to be very convincing. In the modern flute, *hsiao* (4321), the ground note approximates to *D*, on the horizontal, *ti-tzŭ* (10939, 12317), to *A*. As results from the numerous measurements carried out by Professor SCHÜNEMANN of the Staatliche Hochschule für Musik in Berlin, the fundamental tone of all kinds of musical instruments of a great many different peoples, from Southeastern Asia to South America, corresponds to our *F*$\sharp$. From this fact, H. TREFZGER, Die Musik in China, *Sinica*, Vol. XI (1936), p. 187, concludes that *F*$\sharp$ also, already in the remote times of the beginnings of Chinese Music, represented the "diapason normal" or absolute pitch of *huang-chung*. Of course, such an hypothesis can neither be confirmed nor refuted.

(8a) LEVIS, *op. cit.*, consistently writes "KUEI" instead of "K'UEI".

tones, *F♯* and *B*, are named *pien* (9210) *chih*, "modified *G*" (8*b*), and *pien kung*, "modified *C*", respectively, the term *pien* indicating the half-tone step. These two "modified notes" were mostly used as passing tones only, but there also exist compositions in which the same importance is conceded to each of the seven tones of the scale. This heptatonic scale, like our major diatonic scale, consists of five full- and two half-tone steps; but, while in the Western gamut the first half-tone stands between the third and fourth notes, it is found in the Chinese between the fourth and fifth degrees, as in the Lydian church-mode (9) :

During the Mongolian invasion, a new scale was brought to China by the conquerors which was in principle identical with our major scale, or also the Ionian mode, the first half-tone step taking place between the third and fourth degrees :

For this new scale, another notation came into general use which, however, contrary to the opinion of earlier authorities (as VAN AALST), was already known and used for the ancient heptatonic scale during the Sung era, considerable time before the Mongolian period. It consisted of nine simple characters (see diagram no. 1, outer rim) : *ho* (*C*), *ssŭ* (*D*), *i* (*E*), *shang* (*F*), *ch'ih* (*G*), *kung* (*A*), *fan* (*B*), [*liu* (*c*), *wu* (*d*)]. The two last denote the octaves of *ho* and *ssŭ* respectively; they were introduced for the sake of convenience because most of the musical compositions used to be restricted to the interval of a ninth. In exceptional cases, when higher tones are required, these are indicated by prefixing the radical no. 9 (*iên*, "man") to the character which,

(8*b*) LEVIS, *op. cit.*, p. 69, obviously confuses pien [4] (9210), "to change", "modify", with pien [1] (9192), "side", and therefore translates "side" *G*, "side" *C*.

(9) LEVIS consistently speaks of "Ancient Greek modes", when he means *church-modes*. It is sufficiently well known that the Ancient Greek and the church-modes have nothing but the names in common.

when standing isolated, would denote a tone of the fundamental octave (10).

In order to reconcile the new Mongolian scale with the ancient Chinese, the Mongolian rulers decreed that a kind of hybrid gamut mixed of the two be used, in which *F* as well as *F♯* should appear; for denoting *F♯*, a new sign, *kou* (6135), was introduced:

ho ssŭ i shang kou ch'ih kung fan liu wu (11)

This artificial and unnatural scale, however, apparently never became popular among the Chinese musicians. During the centuries subsequent to the Mongolian era, the Mongolian scale came into general use, and to this day it forms the basis of practically all popular Northern Chinese melodies.

The following diagram no. 2 (12) gives a synopsis of the main musical notations and their Western equivalents (13).

CHINESE WITH APPROXIMATE WESTERN EQUIVALENTS

1. Chromatic Scale	C C♯ D D♯ E F F♯ G G♯ A A♯ B
2. *Li* notation	黄大太夾姑仲㽔 林夷南無 應
3. Ancient 5-tone notation	宮 商 角 徵 羽
4. Ancient 7-tone notation	宮 商 角 變徵徵 羽 變宮
5. Modern 7-tone	合 四 一上 尺 工 凡
6. Modern 12-tone	合亽四乙乙上勾 尺工工㕧 凡
7. Sung notation	ㄥ ▽ ▽ 一 一 亼ㄥ 人 ㄱ ㄱ ㄩ ㄩ

Diagram N° 2.

(10) In the ancient notation, the higher octave was denoted by the character *ch'ing* (2188) prefixed to any tone of the fundamental octave; for the lower octave, the character *cho* (2409), or sometimes *hsia* (4230) was employed.

(11) I do not quite understand for what reasons Mr. LEVIS, on p. 79, makes *F♯* (*kou*) precede *F♮* (*shang*).

(12) From LEVIS, *op. cit.*, p. 91.

(13) For the Sung notation (No. 7) see the researches published by HSIA CH'ÊNG-CH'OU in the *Yenching Journal of Chinese Studies*, No. 12, December, 1932.

3. It is important and interesting enough, though not yet generally recognized in modern works, that Chinese musical theoreticians not only distinguish between the pentatonic scale : *C, D, E, G, A,* and the two heptatonic scales : *C, D, E, F♯, G, A, B,* and *C, D, E, F, G, A, B,* but also consistently refer to the different *modes* of both the pentatonic and heptatonic scales. The modes of the latter are nothing but the well-known church-modes (14) which, in the West, were mentioned for the first time in the writings of FLACCUS ALCUIN (8th century) and AURELIANUS REOMENSIS (9th century). Thus there are 60 Chinese pentatonic modes, namely the five modes of diagram no. 3 (15) played in all the twelve keys of the chromatic scale, and 84 heptatonic modes, i.e. the seven modes of diagram no. 4 (16) in the twelve different keys. It is, therefore, a complete misunderstanding to consider the 60 or 84 scales to be "exaggeratedly large sets of tones", as has been recently done by Mr. JOSEPH YASSER (17).

Mode " C—c "
Ancient *kung tiao* 宮 詞

Mode " D—d "
Ancient *shang tiao* 商 詞

Mode " E—e "
Ancient *chiao tiao* 角 詞

Mode " G—g "
Ancient *chih tiao* 徵 詞

Mode " A—a "
Ancient *yü tiao* 羽 詞

Diagram N° 3.

(14) Extrinsically, the Chinese modes can, of course, also be paralleled with the Ancient Greek modes, but it must be remembered that the Greek scales theoretically are obtained in an entirely different way (by descending tetrachords), whereas there is a striking congruence between the Chinese and the church-modes, also as far as theory is concerned. In LEVIS' work, always read "church-modes" instead of "Ancient Greek modes". See note 9.

(15) From LEVIS, *op. cit.*, p. 72.

(16) From LEVIS, *op cit.*, p. 73.

(17) *A Theory of Evolving Tonality*, New York, 1932.

A method of indicating the shifting intervals of the five modes of the pentatonic scale given in Diagram no. 3 and of their eleven transpositions into all possible keys, was very conveniently obtained by combining the syllables of the ancient five tone notation : *kung, shang, chiao, chih, yü*, with the first syllables of the names of the twelve *lüs* : *huang, ta, t'ai, chia, ku, chung, jui, lin, i, nan, wu, ying.*

Thus, if *huang chung* (C) be the fundamental, the notation will run as follows :

Mode "*C-c*" : *kung-huang, shang-t'ai, chiao-ku* (18), *chih-lin, yü-nan.*
Mode "*D-d*" : *shang-huang, chiao-t'ai, chih-chung, yü-lin, kung-wu.*
Mode "*E-e*" : *chiao-huang, chih-chia, yu-chung, kung-i, shang-wu.*
Mode "*G-g*" : *chih-huang, yü-t'ai, kung-chung, shang-lin, chiao-nan.*
Mode "*A-a*" : *yü-huang, kung-chia, shang-chung, chiao-lin, chih-wu.*

For *ta lü (C♯)* as the fundamental, the notation would be :

Mode "*C♯-c♯*" : *kung-ta, shang-chia, chiao-chung, chih-i, yü-wu.*

Mode "*D♯-d♯*" : *shang-ta, chiao-chia, chih-jui, yü-i, kung-ying.*

For *i tsê (G♯)* as the fundamental, the first mode *(G♯-g♯)* would run :

Kung-i, shang-wu, chiao-huang, chih-chia, yü-chung, and so on.

In the modes of the heptatonic scale, the fourth and seventh degrees of the fundamental scale (*pien chih* and *pien kung* respectively) are, as pointed out before, mostly treated as passing tones of lesser importance; in diagram no. 4 they are indicated by smaller notes. The musical notation remains, of course, the same as in the case of the pentatonic scales (19).

---

(18) LEVIS, *op. cit.*, p. 72, l. 23, erroneously gives "*chiao-chung*" as denomination of the third degree.

(19) In this diagram, it appears to be somewhat confusing that Mr. LEVIS, in order to avoid *F♯*, has transposed the fundamental tone *C* of diagram n° 3, into *F*. It would have been more consistent and better for direct comparison, to start with the "Lydian" mode, with *C* as the fundamental :

as this is the heptatonic *kung-tiao* directly corresponding to the pentatonic of diagram no. 3. This would have been so much the more justified, as the "Lydian"

4. To this point we have been able to state that the system of Chinese music does at least not fundamentally disagree with that of the West; seen from the evolutionary viewpoint, there even exist striking similarities and parallels between the two : according to various ancient authors, as ARISTOXENOS, PLUTARCH, and others,

Ancient : *chih tiao*
  Mode C-c (Ionian)

Ancient : *yü tiao*
  Mode D-d (Dorian)

Ancient : *pien kung tiao*
  Mode E-e (Phrygian)

Ancient : *kung tiao*
  Mode F-f (Lydian)

Ancient : *shang tiao*
Mode G-g (Mixo-Lydian)

Ancient : *chiao tiao*
  Mode A-a (Aeolian)

Ancient : *pien chih tiao*
  Mode B-b (Locrian)

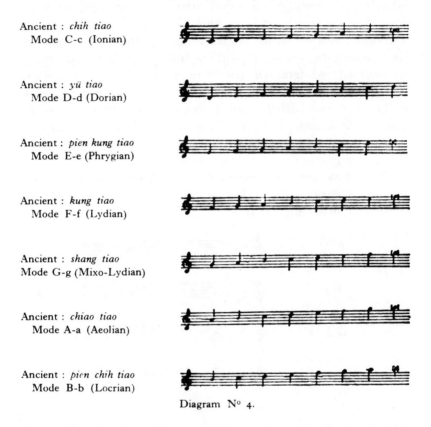

Diagram Nº 4.

the archaic Greek temple melodies were also based on a strictly anhemitonic, pentatonic scale from which the heptatonic scale was later developed; in the West as in the East, the chromatic gamut was known and subjected to theoretical discussions, but in

mode (*kung tiao*), up to the time of the Mongolian invasion, was the unrivalled chief mode of Chinese compositions.

The Chinese spatial notation of this scale would run : *kung-huang, shang-t'ai, chiao-ku, pien chih-jui, chih-lin, yü-nan, pien kung-ying*; whereas, in Mr. LEVIS' example, it would start with "*kung-chung, shang-lin*", the fundamental *lü* not being *huang-chung* (C), but *chung-lü* (F).

neither place it generally could play the rôle of a basis of musical composition. But, while in the West nobody even thought of the possibility of employing other tones than those of the chromatic gamut (19a), a Chinese composer of the Sung era, CHIANG K'UEI whose name we already mentioned, made the extremely interesting and novel attempt of introducing tones finer than a semitone into his musical compositions. For these fractional tones which he defines to be "a little higher than the neighbouring regular tone" without indicating their exact pitch, he invented the term *chê-tzŭ* (550, 12324), literally translated "turning note". In one and the same composition, there appear at the most two such fractional tones in addition to the seven tones of the heptatonic scale. Hence we are theoretically entitled to speak, as Mr. LEVIS does, of a nine tone scale (enneatonic scale) of which only seven tones belong to the normal chromatic division of the octave interval. From four of CHIANG K'UEI's songs, Mr. LEVIS derived the following four types of scales (referred to C as fundamental) (20) :

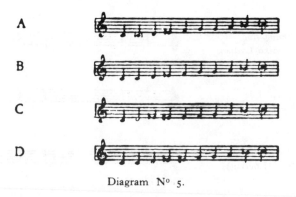

Diagram N° 5.

Here the larger notes represent the main tones of the scale, the smaller ones, auxiliary tones of lesser importance, and the diamond-shaped notes, the fractional notes or *chê-tzŭ* which sound a little higher than indicated by their position relative to the lines.

(19a) Only the Arabic *mathal* theory (see footnote 4) could, with some restrictions, be quoted as a parallel to CHIANG K'UEI's innovation.

(20) From LEVIS, *op. cit.*, p. 75.

As an illustration of this unique system, a facsimile of one of CHIANG K'UEI's compositions, together with Mr. LEVIS' transcription into modern notes, is given below (21). There the smaller characters of the first, third, fifth, and seventh columns (from right to left) accompanying the text of the poem (which is written in larger characters in the colums of even order), represent the musical notation of CHIANG K'UEI's composition.

The key is Lin-chung (eighth degree of the chromatic scale) and the mode, Yü (Dorian).

FACSIMILE OF A COMPOSITION BY CHIANG KUEI IN ANCIENT Lü NOTATION, WRITTEN BY THE HAND OF CHIANG KUEI (SUNG DYNASTY), (FROM THE *Pai Shih Tao Jin Ko Ch'u*, PAGE 6)

In this notation, the first syllables of the names of the twelve *lüs* are used (cf. p. 78); they indicate the absolute pitch of each tone, not only, as do the syllables of the ancient notation : *kung, shang, chiao, etc.*, relative tone intervals.

The *chê-tzŭ* note appears seven times : column 1, 11th and 14th characters; column 3, 9th; column 5, 7th and 13th; column 7, 5th

---

(21) " Music-poem VI. " From LEVIS, *op. cit.*, p. 174-75, from the *Pai Shih Tao Jên Ko Ch'u*, 8556, 9964, 10780, 5624, 6046, 3062), p. 6.

and 15th.   In each case, it stands *between two notes of the same pitch*, the sequences here being either : *ku, chê-tzŭ, ku*

or : *ying, chê-tzŭ, ying*

The same is true of all of the other three examples given by Mr. LEVIS (22), where the *chê-tzŭ* tone also consistently leads back to the preceding tone, never to any other.   To my mind this observation is utterly important for the appreciation of the true nature of the *chê-tzŭ*, and it seems to be a serious omission that Mr. LEVIS does not pay due attention to this fact.   There exists, of course, the possibility that other Sung compositions, known to Mr. LEVIS but unknown to me, are preserved in which the *chê-tzŭ* tone is given the same importance as the other natural tones, in other words, where it leads over from one tone to another, — but I must confess that I doubt it seriously.   In the four named examples the *chê-tzŭ* is obviously introduced in order to break the boring monotony of a three- or even four-fold repetition of one and the same tone.   It produces a kind of primitive *vibrato* effect on certain tones that are of special importance for the character of the music-piece, as the fundamental, the perfect fifth, or also the major third; it appears, therefore, to be not entirely justified to count the *chê-tzŭ* tones as independent additional tones to the heptatonic scale by which a true nine-tone scale is formed.

From the fact that the *chê-tzŭ* is found, in three out of four examples (A, B, and D of diagram no. 5), to stand together with the fundamental or the perfect fifth, or both, Mr. LEVIS derives the theory that the *chê-tzŭ* tones correspond to the 13th and 14th degrees of the circle of natural fifths, which would mean that the difference in pitch can be defined by the ditonic comma. The weakness of this theory is evident.   As Mr. LEVIS admits

---

(22) From LEVIS, *op. cit.* pp. 176-78, from the *Pai Shih Tao Jên Ko Ch'u.*, pp. 6 and 7; one of them is reproduced in facsimile on p. 87.

himself, it does not yield a satisfactory explanation for the fact
that in one example (D) the *chê-tzŭ* appears with the fundamental
and the major third, not the perfect fifth. There can hardly
be a doubt that it was not for theoretical but for purely aesthetical
reasons that the *chê-tzŭ* tones were introduced, and the discussion
of the probable exact pitch of those queer tones would appear
to be of no practical value whatsoever.

Pai Shih Tao Jên Ko Ch'ü, page 7. Ancient Chinese *lü* notation.

The key is *Nan-lü* (tenth degree of the chromatic scale) and the
mode, *Shang* (Mixo-Lydian*

5.♮ However, it is not only the *chê-tzŭ* tones that make Sung
music more interesting and fascinating than that of the other
periods of Chinese history altogether. Let us, before leaving
this question, have a look at the third of Mr. LEVIS' four Sung
music-poems (23) rendered above in facsimile and transcription.

---

(23) "Music-poem VIII", p. 177.

The key is *nan-lü* (*A*), the mode, *shang tiao* (Mixo-Lydian), it is asserted at the end of the piece, according to Mr. LEVIS. Thus the basic scale of our composition ought to run :

the smaller notes (*D♯* and *G♯*) indicating the "auxiliary or passing tones of lesser importance" in the ordinary Mixo-Lydian mode (cf. diagram no. 4). But evidently, the scale here in use is not the pure Mixo-Lydian (*shang tiao*) because it employs in several instances *D♮* instead of the scale tone *D♯*, not to speak of the *D♯-chê-tzŭ*. And this irregularity is not the only one. A brief statistical analysis will perhaps be useful.

The complete scale of the piece employs the following tones :

it corresponds to type "C" of diagram no. 5 (in which *C* is chosen ground tone instead of *B*). The total number of tones of which the piece consists, is 62. The frequency of their occurrence is the following :

B (ground tone, chief tone of the heptatonic scale) . . .	7	times
C♯ (chief tone)	1	»
D♮ (tone foreign to the scale)	4	»
D♯ (auxiliary tone)	14	»
D♯-chê-tzŭ	2	»
E (chief tone)	3	»
F♯ (chief tone)	12	»
G♯ (auxiliary tone)	14	»
A (chief tone)	5	»
TOTAL	62	»

The first thing to be observed is, that the total number of the chief tones of the ancient heptatonic scale (which form the basic pentatonic scale, *B*, *C♯*, *E*, *F♯*, *A*) amounts only to 28, which is less than one-half of the number of tones of the whole piece. The two "auxiliary" or "passing" tones, *D♯* and *G♯*, occur 14 times each. Thus the two auxiliary tones alone keep the balance against the five chief tones, which means that the ancient scale system is set completely upside-down, the auxiliary tones having a marked prevalence over the chief tones. This together with the occurrence of the foreign tone $D\sharp$ shows with sufficient strength that there is but little reason for calling the scale "Mixo-Lydian" (24). In reality, the composer of this music-poem has entirely broken the narrow frame of the ancient musical laws, and he clearly aims at creating a new style which, perhaps, we may call "chromatic".

We really must regret that this fascinating new creation was never continued but was condemned to fall into complete oblivion during the following centuries.

6. As concerns the rhythmical structure of Chinese music-pieces, it is a matter of fact that the Chinese lay a considerably lesser stress on this component than on that of melody. This is also evident from the fact that the notation of time values stands on a much lower level than that of melodic movement. In many musical scores, rhythm is not denoted at all (25); on the other hand, when rhythmic signs are employed, their significance is not always unequivocal. Without entering upon details, as this would carry us too far, we shall restrict ourselves to a few general statements.

Anything corresponding to our *bar* is unknown in Chinese musical scores. Therefore it is necessary to employ different rhythmic symbols for the different beats of a measure or musical phrase. The time which is most commonly used is $\frac{4}{4}$; next after it comes $\frac{2}{4}$. $\frac{3}{4}$ or $\frac{6}{4}$ time is nowadays rather unusual,

---

(24) LEVIS, in his discussion of these compositions, (p. 173) goes only half-way, when he says : "The actual tonality of the music agrees with that in which it is claimed to have been written. Several of these songs are not written upon a five tone basis but more truly upon a seven tone basis. This becomes evident not only through the effect upon the ear but also by observing the relative frequency of the different tones in each piece".

(25) A rhythmic notation is, of course, not yet to be found in the scores by CHIANG K'UEI. The first man to discuss the various kinds of rhythm in music was CHANG YEN (416, 13069) of the late Sung era (born, 1248). Later, the composer Wei Liang-fu (12567, 7017, 3627), about the year 1530, wrote on the subject.

though they seem to have been much more frequently employed during the T'ang and Sung eras. This is true particularly of popular music; in artistical compositions, all kinds of time are met with, such as $\dfrac{6}{4}$, $\dfrac{4\frac{1}{2}}{4}$, $\dfrac{5}{4}$, $\dfrac{7\frac{1}{2}}{4}$, etc.

The common principle of all classes of rhythmical notation may be expressed as follows : supposing, for the sake of simplicity, that a music-piece be written in $\dfrac{4}{4}$ time, then

> 1 rhythmic symbol standing with 1 note indicates 1 beat, or 1 crotchet;
> 1 rhythmic symbol standing with 2 notes indicates $\frac{2}{2}$ beats, or 2 quavers;
> 1 rhythmic symbol standing with 4 notes indicates $\frac{4}{4}$ beats, or 4 semi-quavers;
> 2 rhythmic symbols standing with 1 note indicate 2 beats, or 1 minim;
> 3 rhythmic symbols standing with 1 note indicate 3 beats, or 1 dotted minim;
> 4 rhythmic symbols standing with 1 note indicate 4 beats, or 1 semi-breve.

1 rhythmic symbol standing with 3 notes may indicate either 1 quaver triplet, or 1 quaver followed by 2 semiquavers, or 2 semiquavers followed by 1 quaver. It is in every case left to the personal taste of the musician or singer to choose between these three possibilities.

The following diagram no. 6 (26) will serve as an illustration of the above statements.

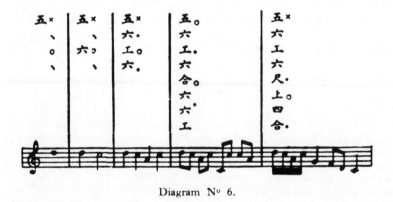

Diagram Nᵒ 6.

(26) Reproduced from LEVIS, *op. cit.*, p. 96.

For further information, the reader is referred to the discussion of the subject in LEVIS' work, pp. 94-96, and to an article by WANG GUANG KI, *Über die chinesischen Notenschriften* (27).

7. The foregoing paragraphs were mainly devoted to the discussions of the general principles governing Chinese musical art. They are applicable to instrumental as well as vocal music. In the following, we shall have to occupy ourselves with the special problem of the laws of vocal composition. As a matter of fact, this stands in the focus of interest, and it is entirely justified that by far the greater part of Mr. LEVIS' work treats of this subject.

One of the most striking features of the Chinese language are the so-called "tones" (shêng, 9883) by which, even in ordinary speech, a decidedly musical effect is produced. Of course, there are also other languages than the Chinese and those related to it, in which a characteristical word (not only phrase) intonation is met with, as for instance Norwegian, Swedish, Lithuanian, or Croatian. But in the latter the different intonation serves only in exceptional cases as a means of distinguishing between the two words which for the rest are identical in pronunciation [thus, in Norwegian *seter* (plural of *sete*, "seat") is pronounced with a compound (falling and then rising) intonation, but *seter*, "dairy cottage", with a simple rising intonation], while in Chinese this forms the rule. There can, however, hardly be a doubt that also the Indo-European languages once possessed a melodic movement or intonation inherent in the pronunciation of every word. The accents introduced by the Alexandrian grammarians into written Greek certainly indicate melodic, and also rhythmic movement; and a similar function must be attributed to the accents which are found, in Sanskrit, in the Vedic Chant (*udātta*, acute, *an-udātta*, grave, *svarita*, circumflex). As to the mediaeval Byzantine and Latin *neumes*, it must be remembered that the Byzantine neume notation, although in some way connected with the Greek accents, has not been directly developed from these. This is evident from the earliest preserved texts in which the neume notation always appears *together* with the accents, but

---

(27) Published in *Sinica*, Vol. III (1928), pp. 110-23 (in the preceding volume of *Sinica*, the same author's name is spelled "WANG KUANG KI", cf. footnote (6a)).

without a consistent congruence between the two; thus, for instance, the acute accent is not necessarily coupled with a rising melodic movement, nor the grave with a falling (28).

Mr. LEVIS, in his discussion of the matter, is clearly aware of this. If he is quoting the European neumes at all in his book he does it for the reason that they play a rôle which is strikingly similar to that of the "tones" in Chinese music. He therefore also suggests to abolish the equivocal and inadequate term "tone ", and to substitute for it the expression "neume". Both in China and in the West, there are three well defined elements of melodic movement : level, rising, and falling, and in both parts of the world these elements were used, intentionally or unconsciously, as the basis of melody in music. The whole Chapter V of Mr. LEVIS' work is devoted to the parallelism between the neumes of China and Europe; it shows clearly that the similarity is not only a superficial one. Here we can of course only consider the part played by the neumes in Chinese music.

The first man to classify the four neumes, was SHÊN YÜEH 9849, 13791) (A. D. 441-513). It is reported that, when asked by the Emperor T'AI WU TI (10573, 12744, 10942) about the nature of these neumes, he answered with a bon-mot : "They are T'ien-tzǔ shêng chê (29) (11208, 12317, 9892, 553), "The Son of Heaven is learned and wise", the four words consecutively illustrating the four neumes : t'ien (p'ing shêng, 1st neume), tzǔ (shang shêng, 2nd neume), shêng (ch'ü shêng, 3rd neume), chê (ju shêng, 4th neume). The discovery (of course not "invention") of these four neumes was certainly an excellent achievement, and due credit has been given for it to SHÊN YÜEH by the historians. But citing LEVIS literally, "the true greatness of SHÊN YÜEH lies in this, that he was the first to recognize, use and institute the neumes as the basis of musical composition in conjunction with poetry". If, possibly or even probably, before the time of SHÊN YÜEH, the melodic movement of songs was unconsciously formed according to the neumes of the words to be sung, this became through him the governing principle of composition for all times to come, and no composition could henceforth be called perfect if these

(28) Cf. HUGO RIEMANN, Musik-Lexikon, s.v. "Byzantinische Musik".
(29) LEVIS transliterates "chih" instead of "chê".

rules were neglected. Thus SHÊN YÜEH's name stands at the beginning of the vocal musical art in China.

In analyzing musical compositions, LEVIS distinguishes between the *immaterial* and the *material* form, the former of which he terms "structure", the latter, "tonal superstructure". The structure of a music-poem is defined by the neumes of the words

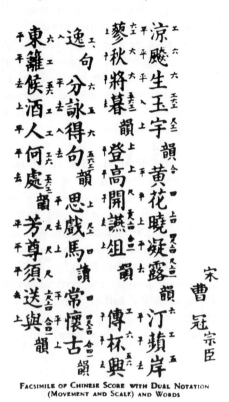

FACSIMILE OF CHINESE SCORE WITH DUAL NOTATION
(MOVEMENT AND SCALE) AND WORDS

of which it consists; it indicates the direction in which the melody has to move : where it has to poise, where to rise, where to fall. All that is left to the composer, is to invent an artistically satisfactory melody, to find a tonal superstructure which obeys the general laws of movement inherent in the structure of the poem. Thus the Chinese composer of a poem has considerably less freedom of invention than his European colleague.

The Chinese theoreticians divide the neumes into two groups : the first, *p'ing* (9310), "level" group, is made up of the *p'ing shêng* only, the second, *tsê* (11682), "oblique" group, comprises the *shang* (9729), *ch'ü* (3068) and *ju* (5690) *shêng*. The *p'ing* group, because poising, motionless, is considered negative, *yin*, the *tsê* group, because moving upwards or downwards, positive, *yang*. In the spoken language, the *ju shêng* sounds utterly short; therefore, in song, it is usually replaced by one of the other three neumes, not only those of its own group but also by the *p'ing shêng*. In poetry (not only in music), it is important to keep as perfect a balance as possible between the positive and negative elements, with regard to the frequency of their occurrence as well as to their distribution. In his analyses of music-poems, Mr. LEVIS makes the very interesting experiments to mark down the *yin* and *yang* neumes by white and black circles respectively; he thus obtains patterns which show a simply surprising regularity.

In many Chinese music-poems, both the structure and the tonal superstructure are denoted. As an illustration, the first of Mr. LEVIS' examples is reproduced on p. 93 (30).

The congruence between the neume structure and the tonal superstructure is good although it is not perfect. The entire piece consists of 48 units; in six of them (i.e. one-eighth of the whole), the melodic movement violates the rule prescribed by the neume structure — horribile dictu !

But no ! This offence against the venerable law of olden time shall fill our hearts with joy rather than horror : it shows that music in China, just as in Europe, is not a mere handicraft constrained by a hundred eternal iron laws, but a living and ever rejuvenescent art.

WILLY HARTNER

---

(30) *Op. cit.*, p. 104 [Music-poem I, untitled, text by TS'AO KUAN (11636, 6373) of the Sung era, composed by HSIEH YÜAN-HUAI (4432, 13744, 5034) in 1848]. The same music-poem was analyzed, from another point of view, in an interesting article by HEINZ TREFZGER ("Die Musik in China", in *Sinica*, XI (1936), pp. 171-97).

The Geometrical Basis of the Ancient Chinese Square-Root Method

By Lam Lay Yong*

INTRODUCTION

IN TRACING THE DEVELOPMENT of ancient and medieval Chinese mathematics, it is found that the Chinese mathematicians encountered no difficulty in root extractions and in the solving of higher numerical equations[1] as compared with their Western counterparts. The Western mathematicians, after their attempts in finding the square root and in solving the quadratic equation, struggled on very slowly to find the cube root and to solve the cubic and higher numerical equations, treating each problem separately and by a different procedure.[2] The Chinese, on the other hand, as early as the Han dynasty (206 B.C.–221 A.D.) had already evolved a basic method for root extraction, and this method was extended to solve first the quadratic equation and, later, higher numerical equations. It is now generally accepted by historians of mathematics that this method is essentially the same as that used by William Horner (1819) for solving higher numerical equations.

The procedures for root extraction are to be found in the *Chiu chang suan shu*.[3] In a paper on "Horner's Method in Chinese Mathematics,"[4] Wang and Needham have brought to light from the obscurity of this Chinese text the methods of root extraction and have illustrated them by solving two of the problems given in the text—

---

* University of Singapore.

[1] Problems on root extractions and numerical equations may be found in the following books: Anon., *Chiu chang suan shu*[a] (Nine Chapters on the Mathematical Art) c. 100 (*Szu pu ts'ung k'an*,[b] Shanghai: Commercial Press, 1919); Sun Tzu,[c] *Sun Tzu suan ching*[d] (Master Sun's Mathematical Manual), between 280 and 473 (*Ts'ung shu chi ch'eng*,[e] Shanghai: Commercial Press, 1939); Ch'in Chiu-shao,[f] *Shu shu chiu chang*[g] (Mathematical Treatise in Nine Sections), 1274 (*Ts'ung shu chi ch'eng*, Shanghai: Commercial Press, 1936); Yang Hui,[h] *T'ien mou pi lei ch'eng ch'u chieh fa*[i] (Practical Rules of Arithmetic for Surveying), 1275, in *Yang Hui suan fa*[j] (Yang

Hui's Methods of Computation) (*Ts'ung shu chi ch'eng*, Shanghai: Commercial Press, 1936).

At the end of this article is a list of the Chinese characters which represent the romanized Chinese words appearing herein. Superscript letters indicate the placement in the list of equivalent characters.

[2] D. E. Smith, *History of Mathematics*, 2 vols. (New York: Ginn, 1925), Vol. II, pp. 144 ff., 443 ff.

[3] *Op. cit.*, pp. 9a–11b, 12b–15a.

[4] Wang Ling and Joseph Needham, "Horner's Method in Chinese Mathematics; Its Origins in the Root-Extraction Procedures of the Han Dynasty," *T'oung Pao*, 1955, *43*:345–401.

56

finding the square root of 55225 and the cube root of 1860867. The importance of this paper cannot be overemphasized: it has established the similarities between the ancient root-extraction methods and Horner's method for solving numerical equations and has traced the origin of the methods used by Sung mathematicians for solving numerical equations, thereby showing the great potentialities and flexibility of the basic Chinese method of root extraction.

It is generally believed that the concept of the root-extraction methods in the *Chiu chang suan shu* was based on geometrical representations. This is vaguely hinted at in the original text and is supported by the meanings of the technical terms. Commentators on the *Chiu chang suan shu*—Liu Hui[k] of the third century and Li Shun-feng[l] of the seventh century—have indicated the geometrical basis of the square-root and cube-root methods. In the modern editions of the book there is a diagram, included by Tai Chen[m] (1724–1777), showing the square-root-extraction process.[5] However, the oldest surviving diagram is found in Yang Hui's *Hsiang chieh chiu chang suan fa*[n] (A Detailed Analysis of the Mathematical Methods in the "Nine Chapters") of 1261. In this book Yang Hui explained the methods of root extraction as found in the *Chiu chang suan shu* and illustrated the working of a problem in the *Chiu chang suan shu*—finding the square root of 71824. The procedures of this problem are given very lucidly, and the various stages on the counting board are shown in detail, accompanied by the diagram illustrating the geometrical origin of the method. Since the text of the *Chiu chang suan shu* is obscure and pithy, it is the purpose of the present paper to give a translation of the relevant passages in the *Hsiang chieh chiu chang suan fa* so that the reader can have a concrete picture of the square-root method which was common knowledge among ancient and medieval Chinese mathematicians.

TRANSLATION FROM THE *HSIANG CHIEH CHIU CHANG SUAN FA*

The following passages are taken from the *Yung lo ta tien*.[o,6] The existing editions of the *Hsiang chieh chiu chang suan fa* are incomplete and do not have these passages.

Find the square root of 71824 *pu*[p] (paces).
Answer: 268 *pu*.
Working: Put down the given amount to be the *shih*,[q] dividend < 71824 *pu* >.[7] Separately, put down one counting rod to be the *hsia fa*[r] (lit. lowest divisor), < this is originally the divisor >. From the extreme right end, shift this counting rod by steps of two places each to correspond with the *shih*. < If it is below the hundredth place of the *shih*, then this determines the tenth place of the root, and if it is below the ten thousandth place, this determines the hundredth place of the root. > From the *shih*, obtain the number for the first *shang*[s] (the root to the hundredth place) < 200 >. Above the *hsia fa* put down

---

[5] *Chiu chang suan shu*, p. 19a.

[6] (Great Encyclopedia of the Yung-lo Reign) 1407 (China: Chung Hwa Co., 1960), Ch. 16, 344, pp. 7a–9a. This book was compiled in the reign of Yung-lo in the Ming dynasty by the order of Emperor Chu Ti. Subjects dealing with astronomy, geography, the dual principle of Chinese philosophy, medicine, divination, Taoism, arts and crafts, and Chinese calligraphy in all forms were to be found in this encyclopedia. In short, it was to embrace all forms of human knowledge known at that time, compiled by over three thousand men from all ranks of literary skill under the leadership of Hsieh Chin. The completed encyclopedia had 11,095 volumes, only about 370 being still extant.

[7] Words within braces <> correspond to the smaller Chinese characters of the text, which are usually explanatory notes.

[the product of the *hsia fa* and] the first *shang* <200> and call it the *fang fa*[t] <200>. Subtract the product of this with the first *shang* from the *shih*. <The product is 40000 and the remaining *shih* is 31824.> Multiply the *fang fa* by 2 <to obtain 400 *pu*> and shift it one place toward the right and call it the *lien*[u] <400>. Also shift the *hsia fa* back-ward by a step of two places <so as to correspond with the hundredth place of the *shih* in order to determine the tenth place of the root>. Obtain the number for the second *shang* (the tenth place of the root) <which is 60, giving a root number of 260>. Next to the *lien*, place [the product of the *hsia fa* and] the second *shang* and call it the *yü*[v] <60>. Multiply the two divisors, the *lien* and the *yü*, by the second *shang* and sub-tract the product from the *shih*. <The product is 27600 and the remainder is 4224.> Multiply the divisor *yü* by 2 and add this to the *lien* <to obtain 520>. Shift this one place toward the right <520>. Shift also the *hsia fa* by a step of two places to reach the last place, <in order to determine the unit place of the root>. Next, obtain the number of the third *shang* (the unit place of the root). Next to 260 of the *shang* place 8 as the third *shang*. Above the *hsia fa* put down 8, [which is the product of the *hsia fa* and the third *shang*], as the *yü*. [Continue as above] to obtain no remainder for the *shih*. Hence the answer.

The square-root diagram

Diagrams on the process of extracting the square root. <With the method illustrated by diagrams, its use will become clear.>

I   To determine the various positions of the root

ten thousand	thousand	hundred	ten	*pu*
7	1	8	2	4
				1
When it is shifted by a further step of two places, this position determines the hundredth place of the root.		When it is shifted by a step of two places, this position determines the tenth place of the root.		Separately, put down one counting rod called *hsia fa*. This position determines the unit place of the root.

II   The hundredth place of the root—the first *shang*

		shang hundredth place		
		2		

200 times 200 gives a product 40000,
which is subtracted from 70000.

3	1	8	2	4

2
The first *shang* is
multiplied by the
hsia fa to obtain
2, which is called
the *fang*.

1
The *hsia fa* in this
position determines
the hundredth place
of the root.

When the first *shang* is obtained, this is multiplied by the *hsia fa* and the product is called the *p'ing fang*[w]. This is again multiplied by the first *shang*, and the product is subtracted from the *shih*.

III   The process in determining the tenth place of the root

		2		
3	1	8	2	4

The first *shang* is
multiplied by the
hsia fa and added to
the *fang*.

1
hsia fa

The first *shang* 2 is multiplied by the *hsia fa* and added to the *p'ing fang*.

IV   The process in determining the tenth place of the root

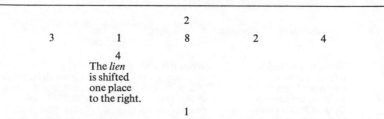

		2		
3	1	8	2	4
	4			

The *lien*
is shifted
one place
to the right.

1

The *fang fa* is shifted one place to the right and called the *lien*. The *hsia fa* is shifted two places to the right.

V   The tenth place of the root—the second *shang*

		shang tenth place

The *lien* and *yü* are multiplied by the second *shang* and the product 27600 is subtracted from the *shih*.

	2	6
4	2	2	4

The second *shang* is multiplied by the *hsia fa* to obtain 6, which is called *yü*.

lien

| 4 | | 6 |

1

The *hsia fa* in this position determines the tenth place of the root.

6 is obtained as the second *shang*. This is multiplied by the *hsia fa* and the product is called the *yü*. This [together with the *lien*] is multiplied by the second *shang* and the product is subtracted from the *shih*.

VI   The process in determining the unit place of the root

	2	6	
4	2	2	4
5	2		

The second *shang* is multiplied by the *hsia fa* to obtain 6 and this is added to form the *lien*.

1
hsia fa

Multiply the second *shang* by the *hsia fa* and this is added to the *yü* and *lien*.

VII   The process in determining the unit place of the root

	2		6	
4	2		2	4
	5		2	

Shift the *lien* one place to the right.

1
Shift the *hsia fa* two
places to the right.

Shift the *lien* one place and the *hsia fa* two places to the right.

VIII   The unit place of the root—the third *shang*

	hundred	ten	*pu*
	2	6	8

The *lien* and *yü* are multiplied by the
third *shang* and the product subtracted
from the *shih* leaves no remainder.
Hence the answer.

4	2	2	4
	5	2	8

The third *shang* is
multiplied by the
hsia fa and entered as
the *yü*.

1
The *hsia fa* here
determines the unit
place of the root.

Multiply the third *shang* by the *hsia fa* and call the product *yü*. Multiply all the *lien*
[including the *yü*] by the third *shang* and subtract the product from the *shih* to
leave no remainder.

COMMENTARY ON THE TRANSLATION

An explanation of the technical terms would throw much light on the geometrical
significance of the method. The close relations between division and root extraction
are evident from such terms as *shih* (dividend), *fa* (divisor), *shang* (quotient), and
ch'u[x] (to divide, though in ancient usage it means to cast out, and was subsequently
used for divisions in which what was wanted for further operations was the remainder
only—the integral result was discarded). The number 1 as the *hsia fa* (the lowest divisor)
is the coefficient of x^2 in our notation of the equation $x^2 = 71824$. The position of the
hsia fa below the *shih* or the number to be extracted is important, for if it is placed
below the ten thousandth figure, then this amounts to determining the hundredth value
of the root, and when it is shifted below the hundredth and unit figures of the *shih*, the

tenth and unit values respectively of the root are determined. When the *hsia fa* is placed below the ten thousandth figure of the *shih*, the root to the hundredth place 100*R*, called the first *shang*, is obtained by trial. In the problem, if 300 is taken as a root, then it would be found that the product 1 (*hsia fa*) × 300 (*shang*) × 300 (*shang*) exceeds the *shih*. Hence 300 is not a possible figure as the root, and the largest possible number is 200.

The product of the *hsia fa* and the first *shang* is called the *fang fa*, and when this is multiplied by the first *shang* the area of a square is obtained, since in the extraction of square roots the *hsia fa* is always 1. (Fig. 1, *ABCD*). The word *fang* means a square. When the root to the hundredth place is obtained, this square is removed and preparation is made to obtain the root to the tenth place from the remaining area. The handling of this intermediary step is important, for it is this and subsequent similar stages which allow the application of the technique to the solution of numerical equations. It is interesting to note that Yang Hui gave two versions of this intermediary stage. In his explanation he stated that the *fang fa* is multiplied by 2, and in his diagrammatic representation of the counting board III (see the translation above) he stated that the first *shang* is multiplied by the *hsia fa* and the product is added to the *fang fa*. Both produce the same result called the *lien*, but it is the latter technique which establishes a ladder system similar to that of Horner.

The last stage of the preparation amounts to shifting the number representing the *lien* on the counting board one place to the right and the *hsia fa* two places to the right so that it is below the hundredth digit of the remaining number to be extracted. The tenth digit of the root *r* is now obtained by trial and 10*r* is called the second *shang*. The ladder system of the working continues as before. The second *shang* is multiplied by the *hsia fa*, and the product is called the *yü*. This together with the *lien* is multiplied by the second *shang* to form the area *DCBIGED* (Fig. 1). *Lien* refers to the space at the two sides of the square, and *yü* means a corner. When this area is removed, preparation

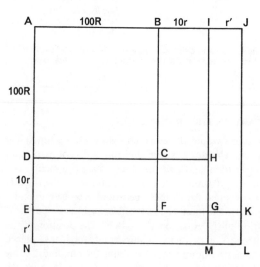

Figure 1

is made to find the root to the unit place from the remaining area. The correct procedure is shown in Yang Hui's diagram VI, where the second *shang* is multiplied by the *hsia fa* and added to the *lien* and *yü* to form a new *lien*. When this is shifted one place to the right and the *hsia fa* two places, the unit digit of the root *r′*, called the third *shang*, is obtained by trial. The procedure is the same as before and the area *EGIJLNE* (Fig. 1) is removed, leaving no remainder. Should there be a remainder, the above technique can be applied to obtain the first decimal place of the root and so on.

Yang Hui's diagrammatic

representations of the counting board (diagrams I to VIII) show the *shang* and *shih* occupying the first and second rows respectively, while the *hsia fa* is placed in the last row. If these rows are instead placed in columns from right to left as shown below, then it is interesting to note that the presentation of the working of the problem is analogous to the methods given in textbooks on the solution of numerical equations by Horner's method.[8]

hsia fa			*shih*	*shang*
1			71824	200
	fang fa	200 (1 × 200)	− 40000 (200 × 200)	
		+ 200 (1 × 200)	31824	60
	lien	400	− 27600 (460 × 60)	
	yü	+ 60 (1 × 60)	4224	8
		460	− 4224 (528 × 8)	
		+ 60 (1 × 60)	....	
	lien	520		
	yü	+ 8 (1 × 8)		
		528		

The author of the *Chiu chang suan shu* did not confine himself to the extraction of roots. In the twentieth problem of the last chapter there is a problem involving the quadratic equation

$$x^2 + 34x - 71000 = 0.$$

Though the method of solution is not given, we are told to use the *ts'ung fa*[y] to solve the problem. The significance of the technical term *ts'ung* has already been pointed out by Y. Mikami,[9] Ch'ien Pao-tsung,[z, 10] and Li Yen[ab, 11] as a forward step toward the solving of higher numerical equations from the root-extraction method. In solving a quadratic equation of the form $ax^2 + bx = c$, *ts'ung* refers to the coefficient b. The general meaning of *ts'ung* is "to follow," and as its name implies, it is preceded by another term. In the case of the quadratics, the *fang fa* is generally "followed" by the *ts'ung*.

After the *Chiu chang suan shu*, later mathematicians not only adopted its methods of root extractions, but modified and extended them. Sun Tzu in *Sun Tzu suan ching*[12] changed the names of the technical terms in his solution of finding the square root of 234567, and these same terms were adopted by Yang Hui. After the achievements of the *Chiu chang suan shu*, the next step was to extend the method to the solutions of more

[8] *Cf.* W. S. Burnside and A. W Panton, *The Theory of Equations*, 2 vols. (Dublin: Dublin Univ. Press, 1928), Vol. I, pp. 227 ff.

[9] *The Development of Mathematics in China and Japan* (New York: Chelsea, 1913), pp. 24–25.

[10] *Chung kuo shu hsieh shih*[aa] (A History of Chinese Mathematics) (Peking: Science Pub., 1964), p. 51.

[11] *Chung kuo shu hsieh ta kang*[ac] (Outline of the History of Chinese Mathematics), 2 vols. (Shanghai: Commercial Press, 1931), Vol. I, p. 159.

[12] *Op. cit.*, pp. 14–15.

general quadratic equations, and this was accomplished by Liu I[ad] (1080) in his book *I ku kên y üan*[ae] (Discussions on the Old Sources). This book is now lost, but his methods were appreciated by Yang Hui, who credited the solutions of the quadratic equations found in *T'ien mou pi lei ch'êng ch'u chieh fa*[13] to Liu I. Yang Hui gave an exhaustive and detailed account of the various methods of solving the equation $ax^2 + bx = c$, where c is positive and a,b could either be positive or negative. Each method is accompanied by a geometrical diagram which illustrates the various areas discussed in the working.[14] In one of Yang Hui's earlier books, *Hsiang chieh suan fa*[af] (A Detailed Analysis of the Method of Computation), there is a solution on the extraction of the fourth root, namely $x^4 = 1336336$.[15] Once the ladder system of the basic method of root extraction was established, there was little difficulty in adopting it for generalized purposes, and the end of the Sung dynasty saw a number of mathematicians—in particular, Ch'in Chiu-shao, Chu Shih-chieh,[ag] and Li Chih[ah]—applying it to solve numerical equations of higher order.

COMPARISON WITH THEON OF ALEXANDRIA'S METHOD OF EXTRACTING A SQUARE ROOT

The first European to use a figure identical to that displayed by Yang Hui was Theon of Alexandria[16] (*c.* 390) when he explained Ptolemy's method of extracting square roots with sexagesimal fractions. It is interesting to compare this method and the Chinese method. Theon based his on Euclid's *Elements*, Book II, Proposition 4, which states, "If a straight line be divided at any point, the square of the whole line is equal to the squares of both the segments together with twice the rectangle contained by the segments."[17] Theon illustrated this with the simple case of finding the square root of 144. Taking the square on the straight line *AB* as 144 (see Fig. 2), the line is divided such that the square on *AC* is 100. The remainder 44 must be equal to the square on *CB* and two rectangles formed by *AC* and *CB*. The length of *AC* is 10, and this is multiplied by 2. The number 44 is next divided by 20, and the remainder 4 is the square on *CB*, which must be 2.

A   C   B

Figure 2

Theon extended this principle to find the square root of 4500 degrees, which is approximately 67° 4' 55". First, a square of 4500 degrees is drawn, and the largest square number less than this is 4489, whose square root is 67. The square root is expressed in terms of sexagesimal fractions of the form $\sqrt{4500} = 67 + x/60 + y/60^2$, where x and y are integers to be found. Take *ABCD* (see Fig. 3) as the square of side 67° so that the remaining area *DCBJLND* contains 11°. This is divided by 2*AB*—

[13] *Op. cit.*, Ch. 2, pp. 34 ff.

[14] See Lam Lay Yong, *A Critical Study of the Yang Hui suan fa* (Singapore: Univ. Malaya Press, in press).

[15] See Lam Lay Yong, "On the Existing Fragments of Yang Hui's *Hsiang chieh suan fa*," *Archive for History of Exact Sciences*, 1969, 6

(No.1): 82–88.

[16] J. Gow, *A Short History of Greek Mathematics* (New York: Stechert, 1923), pp. 55 ff. T. L. Heath, *A History of Greek Mathematics*, 2 vols. (Oxford: Clarendon Press, 1921), Vol. I, pp. 60 ff.

[17] See T. L. Heath, *The Thirteen Books of Euclid's Elements*, 3 vols. (Cambridge: Cambridge Univ. Press, 1926), Vol. I, pp. 376-378.

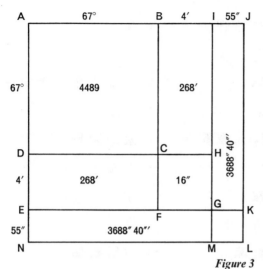

that is, 134—and the quotient must be of the form $x/60$; x is found to be 4. Take $BI = DE = 4/60$ so that the two rectangles $BIHC$ and $DEFC$ have a total area of $2 \times 67 \times 4/60$, and the area of the square $CHGF$ is $(4/60)^2$. The next sexagesimal fraction $y/60^2$ is obtained by a similar procedure from the remainder $11 - 2 \times 67 \times 4/60 - (4/60)^2$. This is divided by $2IG$—that is, $2(67 + 4/60)$—and the quotient is $55/60^2$. After a similar deduction of the areas of two rectangles and a square of side $55/60^2$, there is still a remainder, but Theon did not proceed further.

Figure 3

CONCLUSIONS

We see that the extraction of the square root was known in both East and West in very early times. The derivation of the methods was basically the same: it was from the Euclidean concept $(a + b)^2 = a^2 + 2ab + b^2$. However, the Greek method was purely geometrical and depended solely on the Euclidean figure to extract the root. This figure is two-dimensional, and no attempts were made to construct a three-dimensional figure in order to extend the square-root method to find the cube root. In fact, little was known about the extraction of the cube root in the West until the sixteenth century.[18] In contrast, the Chinese method of the square root was more sophisticated and abstract. By introducing the coefficient of the highest power x^2 as the *hsia fa*, later mathematicians were able to generalize the square-root method and apply it to solve any numerical equation. In the cube-root method, commentators on the *Chiu chang suan shu* Liu Hui and Li Shun-feng both hinted at the use of cubical blocks to demonstrate the process.[19]

[18] Smith, *History of Mathematics*, p. 148.

[19] Wang and Needham, "Horner's Method," pp. 395–396.

GLOSSARY

a 九章算術	b 四部叢刊	c 孫子
d 孫子算經	e 叢書集成	f 秦九韶
g 數書九章	h 楊輝	i 田畝比類乘除捷法
j 楊輝算法	k 劉徽	l 李涼風
m 戴震	n 詳解九章算法	o 永樂大典
p 步	q 實	r 下法
s 商	t 方法	u 廉
v 隅	w 平方	x 除
y 從法	z 錢寶琮	aa 中國數學史
ab 李儼	ac 中國數學大綱	ad 劉益
ae 議古根源	af 詳解算法	ag 朱世傑
ah 李治		

The *Jih yung suan fa:* An Elementary Arithmetic Textbook of the Thirteenth Century

By Lam Lay Yong*

INTRODUCTION

The *Jih yung suan fa*[a] (Arithmetical Methods for Daily Use), written in 1261, is one of the earlier works of Yang Hui.[b,1] It is a simple arithmetic textbook, and its purpose as stated in the preface of the book is "to assist the reader on the numerous matters of daily use and also to instruct the young in observation and practice." The book does not claim to go beyond the basic operations of multiplication and division, but its historical value lies in the detailed presentation and rich variety of these methods. Multiplication and division were of course known centuries prior to Yang Hui's time, and their methods were taken for granted in the mathematical books as far back as the *Chiu chang suan shu*[c] (Nine Chapters on the Mathematical Art, *ca.*100). The earliest existing book to have some explanations of these methods is the *Sun Tzu suan ching*[d] (Master Sun's Mathematical Manual, between 280 and 473), but the most extensive collection of the various methods is to be found in Yang Hui's *Jih yung suan fa, Hsiang chieh suan fa*[e] (A Detailed Analysis of the Methods of Computation, 1261)[2] and *Ch'eng ch'u t'ung pien pen mo*[f] (Alpha and Omega of Variations on Multiplication and Division, 1274).

The *Jih yung suan fa* consists of two chapters containing thirteen headings below which are presented sixty-six problems accompanied by solutions and diagrams. Among the different methods of multiplication and division discussed are the additive (*chia*[g]) method of multiplication and the subtractive (*chien*[h]) method of division. There are also problems on conversion into decimal fractions of weight and length measurements, and simple linear equations. Though the book is no longer extant, the few existing fragments have been collected by Li Yen[i] and printed in his book *Chung suan shih lun ts'ung*[j] (A Discussion on the History of Chinese Mathematics).[3] There are altogether ten problems with solutions, nine of them taken from *Chu chia suan fa*[k]

---

* Department of Mathematics, University of Singapore, Singapore 10.

[1] At the end of this article is a glossary of the Chinese characters which represent the romanized Chinese words appearing herein. Superscript letters indicate the placement in the list of equivalent characters.

[2] See Lam Lay Yong, "On the Existing Fragments of Yang Hui's *Hsiang Chieh Suan Fa,*" *Archive for History of Exact Sciences*, 1969, 6(No. 1):82–88.

[3] Li Yen, *Chung suan shih lun ts'ung* (A Discussion on the History of Chinese Mathematics), 5 vols. (Peking:Science Publishing House, 1954), Vol. II, pp. 60–68.

68

(Mathematics for Everyone)[4] and one from *Yung lo ta tien*° (Great Encyclopedia of the Yung-lo Reign, 1407).[5] These ten problems and their methods of solutions are partially translated and explained in the following discussion.

THE ADDITIVE METHOD OF MULTIPLICATION AND THE SUBTRACTIVE METHOD OF DIVISION

The additive method of multiplication and the subtractive method of division were very popular during the Sung dynasty. They probably appealed to the Chinese, who had already acquired a high sense of the place values of numbers. These methods are confined only to multipliers and divisors whose first digit from the left is 1; for example, 16 and 127. The additive method of multiplication is very similar to our present multiplication method of adding the products of the multiplicand and each digit of the multiplier. However, in the former method, since the first digit of the multiplier is 1, steps are considerably shortened by omitting writing down the product of the multiplicand and this digit. Instead, the digits of the multiplicand are added to their products with the other digits of the multiplier, taking into consideration the place value of the numbers. To illustrate the method an example is given below.

734 multiplied by 126.

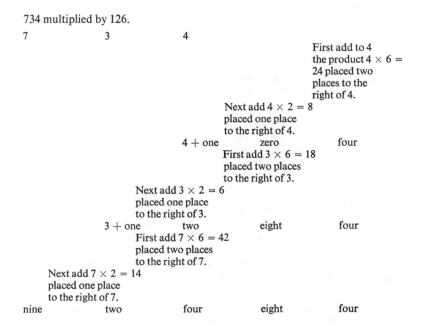

7   3   4

First add to 4 the product $4 \times 6 = 24$ placed two places to the right of 4.

Next add $4 \times 2 = 8$ placed one place to the right of 4.

4 + one   zero   four

First add $3 \times 6 = 18$ placed two places to the right of 3.

Next add $3 \times 2 = 6$ placed one place to the right of 3.

3 + one   two   eight   four

First add $7 \times 6 = 42$ placed two places to the right of 7.

Next add $7 \times 2 = 14$ placed one place to the right of 7.

nine   two   four   eight   four

[4] Li Yen possessed a copy of the *Chu chia suan fa* which was copied from the original work found in the collection of Sheng Sun,¹ the son of Mo Yu-chih ᵐ (1811–1871). See *Chung suan shih lun ts'ung*, Vol. II, p. 54. See also Li Yen, *Chung kuo suan hsieh shih lun ts'ung*ⁿ (A Discussion on the History of Mathematics in China) (Hong Kong: Cheng Chung Press, 1954), pp. 433–434.

[5] *Yung lo ta tien* (Great Encyclopedia of the Yung-lo Reign), (China: Chung Hwa Co., 1960), Ch. 16,343, pp. 19b–20b. This book was compiled in the reign of Yung-lo in the Ming dynasty by the order of Emperor Chu Ti. Subjects dealing with astronomy, geography, the dual principle of Chinese philosophy, medicine, divination, Taoism, arts and crafts, and Chinese calligraphy in

The subtractive method of division is the reverse of the additive method of multiplication.[6] However, it is not so straightforward. Comparing the two methods, Yang Hui says, "In the additive method of multiplication the number is increased, while in the subtractive method of division a certain number is deducted." "One who learns the subtractive method of division," he continues, "should test the result by applying the additive method of multiplication to the answer of the problem. This will enable one to understand its very source."[7] As in the additive method of multiplication, manipulation with the unit digit of the divisor is omitted in the subtractive method of division. However, whereas the working of the additive method of multiplication starts from the right, that of the subtractive method of division starts from the left. A comparison of the subtractive method of division and the present long-division method shows that the steps are very similar. An example is given below to illustrate the subtractive method of division.

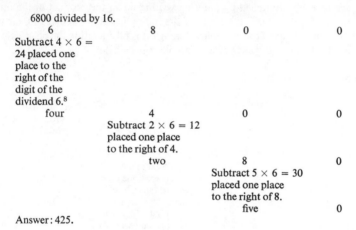

6800 divided by 16.

6	8	0	0
Subtract 4 × 6 = 24 placed one place to the right of the digit of the dividend 6.[8]			
four	4	0	0
	Subtract 2 × 6 = 12 placed one place to the right of 4.		
	two	8	0
		Subtract 5 × 6 = 30 placed one place to the right of 8.	
		five	0

Answer: 425.

THE PROBLEMS

Below is a list of the ten existing problems and an explanation of their methods of solution. A full translation of the solutions for problems 3, 6, and 10 is also given to enable the reader to have an idea of Yang Hui's presentation.

1.   The correct weight of certain articles is 112 *chin*[q] (pounds); find the nominal weight.
       Answer: 140 *chin*.

---

all forms were to be found in this encyclopedia. In short, it was to embrace all forms of human knowledge known at that time, compiled by over three thousand men from all ranks of literary skill under the leadership of Hsieh Chin. The completed encyclopedia had 11,095 volumes, only about 370 being still extant.
 [6] A more detailed discussion of these methods can be found in Lam Lay Yong, *A Critical Study of the Yang Hui Suan Fa* (Singapore: Univ. of Singapore Press, in press).

[7] Yang Hui, *Ch'eng ch'u t'ung pien pen mo*, 1274 (*I chia t'ang ts'ung shu*,[p] 1842), Ch. 1, p. 2a.

[8] 4 × 6 = 24; 4 is arrived at from inspection, and 6 is the unit digit of the divisor 16. The tenth digit of the product 24 is placed directly below 6, the first digit of the dividend from the left. On subtraction, the remainder 4 corresponds to the number which multiplied by the unit digit 6 of the divisor gives the product 24. "Four" now forms the first digit of the quotient.

The nominal weight[9] is assumed to be 125 per cent of the correct weight. 112 *chin* is multiplied by 125 by the additive method of multiplication, and the unit place in *chin* of the product is fixed in the same column as the unit place of the multiplicand. An alternative solution is given, where the unit digit of the correct weight, 112 *chin*, is fixed as the tenth digit (i.e., 112 *chin* becomes 1120 *chin*), and this amount is halved thrice to give the result.

2.   The correct weight of certain articles is 313 *chin*; find the nominal weight.
      Answer: 391 *chin* 4 *liang*[r] (ounces).

1 *chin* = 16 *liang*. As in the previous problem, 313 *chin* is multiplied by 125 by the additive method of multiplication to give a product of 391.25 *chin*. The fractional value of this amount in *chin* is called $2\frac{1}{2}$ *fen*,[s] and this is multiplied by 16 by the additive method of multiplication to convert it into 4 *liang*. The Chinese since ancient times have acquired a highly developed sense of decimal values. The names given for the first, second, and third decimal places are *fen*, *li*,[t] and *hao*,[u] respectively.

Below this problem, the following tables converting decimal fractions of 1 *chin* into *liang* and *ch'ien*[v] (1 *liang* = 10 *ch'ien*) and vice versa are given:

1 *fen* (i.e., 0.1 *chin*) =	1 *liang* 6 *ch'ien*	
2 *fen*	=	3 *liang* 2 *ch'ien*
3 *fen*	=	4 *liang* 8 *ch'ien*
4 *fen*	=	6 *liang* 4 *ch'ien*
5 *fen*	=	8 *liang*
6 *fen*	=	9 *liang* 6 *ch'ien*
7 *fen*	=	11 *liang* 2 *ch'ien*
8 *fen*	=	12 *liang* 8 *ch'ien*
9 *fen*	=	14 *liang* 4 *ch'ien*
10 *fen*	=	16 *liang*

1 *liang* = 6 *li* $2\frac{1}{2}$ *hao* (i.e., 0.0625 *chin*)
2 *liang* = 1 *fen* $2\frac{1}{2}$ *li*
3 *liang* = 1 *fen* 8 *li* $7\frac{1}{2}$ *hao*
4 *liang* = $2\frac{1}{2}$ *fen*
5 *liang* = 3 *fen* 1 *li* $2\frac{1}{2}$ *hao*
6 *liang* = 3 *fen* $7\frac{1}{2}$ *li*
7 *liang* = 4 *fen* 3 *li* $7\frac{1}{2}$ *hao*
8 *liang* = 5 *fen*
9 *liang* = 5 *fen* 6 *li* $2\frac{1}{2}$ *hao*
10 *liang* = 6 *fen* $2\frac{1}{2}$ *li*
11 *liang* = 6 *fen* 8 *li* $7\frac{1}{2}$ *hao*
12 *liang* = $7\frac{1}{2}$ *fen*
13 *liang* = 8 *fen* 1 *li* $2\frac{1}{2}$ *hao*
14 *liang* = 8 *fen* $7\frac{1}{2}$ *li*
15 *liang* = 9 *fen* 3 *li* $7\frac{1}{2}$ *hao*
16 *liang* = 10 *fen*

---

[9] The "nominal weight" is measured by means of a special steelyard which is adjusted to give 25% more in weight than an accurate steelyard. See *Ch'eng ch'u t'ung pien pen mo*, Ch. 3, p. 7a.

3.   The correct weight of certain articles is 123 *chin* 5 *liang*; find the nominal weight.
Answer: 154 *chin* 2 *liang* 2½ *ch'ien.*

Explanation: < In this problem the given weight in *chin* produces a fractional product which has to be converted to *liang*. The given weight in *liang* gives a product in *liang* and the two quantities are added together.> [10]

Method: Put down separately the quantities in *chin* and *liang* of the given weight. < These quantities are placed separately because they are different.> Multiply each quantity by 125 using the additive method of multiplication, and as for the fractional value of the product in *chin*, multiply this by 16 by the additive method of multiplication to convert it to *liang*, < as in the previous problem>.

Working: Put down separately the quantities in *chin* and *liang* of the given weight. < Place 123 *chin* in the upper row and 5 *liang* in the lower row.> Multiply each by 125 using the additive method of multiplication. < On multiplication, 123 *chin* becomes 153 *chin* 7½ *fen* and 5 *liang* becomes 6 *liang* 2½ *ch'ien.*> Next, multiply the fractional value of the quantity in *chin* by 16 using the additive method of multiplication to convert it to *liang*; < 7½ *fen* multiplied by 16 by the additive method of multiplication becomes 12 *liang*. When the products are added, the answer, 154 *chin* 2 *liang* 2½ *ch'ien*, is obtained>.

Another method: Put down the quantity in *chin* and halve the quantity in *liang* four times. < To halve it four times is equivalent to dividing it by 16, to convert it to a decimal fraction.> On obtaining the fractional value, add it to the quantity in *chin*. < On obtaining the fractional value, attach it to the end of the quantity in *chin*.> Use the original method of multiplying by 125 and then multiplying the fractional value in *fen* by 16 to convert it to *liang*. < The explanation is in the previous problem.>

Working: Put down the quantity in *chin*, < 123 *chin*>, and halve the quantity in *liang* four times to convert it to a decimal fraction; < 5 *liang* is equivalent to 3 *fen* 1 *li* 2½ *hao*>. Add this < to obtain 123 *chin* 3 *fen* 1 *li* 2½ *hao*>. Multiply this by 125 using the additive method of multiplication. < Fix the unit place in *chin* of the product in the same column as the unit digit in *chin* of the multiplicand. Multiply by 125 using the additive method of multiplication to give 154 *chin* 1 *fen* 4 *li* 6 *ssu*ʷ 2½ *hu* ˣ,[11]>. Multiply the fractional value by 16 using the additive method of multiplication to convert it to *liang*. < Fix the unit place in *liang* in the same column as *fen* and multiply 1 *fen* 4 *li* 6 *ssu* 2½ *hu* by 16 to give 2 *liang* 2½ *ch'ien*. Add this to 154 chin.> Hence the answer.

Another method: Convert the quantity in *chin* into *liang* and add this to the remaining quantity in *liang*. Multiply the sum by 125 using the additive method of multiplication. < This method gives the nominal weight in *liang*.> This will still have to be divided by 16 to be converted back into *chin*. < In the question the weight is in *chin* so the quantity has to be converted to its original form in *chin*.>

Working: Convert the quantity in *chin* into *liang*. < Put down 123 *chin* which is equivalent to 1968 *liang*, since 1 *chin* equals 16 *liang*. > Add the given quantity in *liang* to this. < Add 5 *liang* to obtain a total of 1973 *liang*.> Multiply by 125 using the additive method of multiplication < to give 2466 *liang* 2 *ch'ien* 5 *fen*>, and then divide it by 16 *liang* to convert it back to *chin* < to obtain 154 *chin* 2 *liang* 2½ *ch'ien*>.

The above three methods can be described briefly as follows:

I.   123 *chin* × 1.25 = 153.75 *chin*
                      = 153 *chin* 12 *liang*, ... (1)
since 0.75 *chin* × 16 = 12 *liang*.
5 *liang* × 1.25 = 6 *liang* 2½ *ch'ien* ... (2)
Add (1) and (2) to give 154 *chin* 2 *liang* 2½ *ch'ien*.

---

[10] Words within pointed brackets < > correspond to the smaller Chinese characters of the text, which are usually explanatory notes.

[11] *Ssu* and *hu* are the names of the fourth and fifth decimal places respectively.

II. 123 *chin* 5 *liang* = 123.3125 *chin*,
   since $\frac{5}{16}$ *chin* = 0.3125 *chin*.
   123.3125 *chin* × 1.25 = 154.14025 *chin*
   $\qquad\qquad\qquad\qquad$ = 154 *chin* 2 *liang* 2$\frac{1}{2}$ *ch'ien*,
   since 0.14025 *chin* × 16 = 2 *liang* 2$\frac{1}{2}$ *ch'ien*.

III. 123 *chin* 5 *liang* = 1973 *liang*,
   since 123 *chin* × 16 = 1968 *liang*.
   1973 *liang* × 1.25 = 2466 *liang* 2$\frac{1}{2}$ *ch'ien*
   $\qquad\qquad\qquad\qquad$ = 154 *chin* 2 *liang* 2$\frac{1}{2}$ *ch'ien*,
   since 2466 *liang* ÷ 16 = 154 *chin* 2 *liang*.

4. The nominal weight of certain articles is 140 *chin*; find the correct weight.
   Answer: 112 *chin*.

140 *chin* is multiplied by 8, and the column where the tenth digit lies is fixed as the unit column for the product.

5. The nominal weight of certain articles is 391 *chin* 4 *liang*; find the correct weight.
   Answer: 313 *chin*.

The quantity in *chin*—that is, 391 *chin*—is multiplied by 8 and, as in the previous problem, the decimal place of the product is moved one place to the left. The product becomes 312.8 *chin*. Next, 4 *liang* is now halved to obtain 0.2 *chin*. The two quantities are added to obtain the answer, 313 *chin*.

6. If 1 *chin* of goods costs 6 *kuan*[y] 800 *wen*,[z,12] find the cost of 1 *liang*.
   Answer: 425 *wen*.
   Explanation: < In this problem it is required to find the price of one *liang* from that of one *chin*, but on examination it is found that the method is generally applicable.> There are five methods.
   Method I. Let the price of 1 *chin* be the dividend and halve it four times. < Above the unit place in *kuan*, fix the unit place in *kuan* for the result; there are 16 parts.>
   Working: Let the price of one *chin* be the dividend, < that is, put down 6 *kuan* 800 *wen*> and halve it four times. < Halving the quantity four times gives 425 *wen*.> Hence the answer.
   Method II. Study the method, < which is to memorize the table below>, and work the problem from the right.
   To calculate 1, place 625 two places to its right, [i.e., 1/16 = 0.0625]. < If 1 *chin* costs 1 *kuan* then 1 *liang* costs 62$\frac{1}{2}$ *wen* which is two places to the right.>
   To calculate 2, place 125 one place to its right, [i.e., 2/16 = 0.125]. < If 1 *chin* costs 2 *kuan* then 1 *liang* costs 125 *wen* which is one place to the right.>
   To calculate 3, place 1875 one place to its right, [i.e. 3/16 = 0.1875]. < If 1 *chin* costs 3 *kuan* then 1 *liang* costs 187 *wen* 5 *fen*.>
   To calculate 4, place 25 one place to its right, [i.e., 4/16 = 0.25]. < If 1 *chin* costs 4 *kuan* then 1 *liang* costs 250 *wen*.>
   To calculate 5, place 3125 one place to its right, [i.e., 5/16 = 0.3125]. < If 1 *chin* costs 5 *kuan* then 1 *liang* costs 312$\frac{1}{2}$ *wen*.>

---

[12] *Wen* means "cash" or "coins," and *kuan* means "a string of 1,000 cash." Hence 1,000 *wen* = 1 *kuan*.

To calculate 6, place 375 one place to its right, [i.e., 6/16 = 0.375]. < If 1 *chin* costs 6 *kuan* then 1 *liang* costs 375 *wen*.>

To calculate 7, place 4375 one place to its right, [i.e., 7/16 = 0.4375]. < If 1 *chin* costs 700 *wen* then 1 *liang* costs 43 *wen* 7½ *fen*.>

To calculate 8, place 5 one place to its right, [i.e., 8/16 = 0.5]. < If 1 *chin* costs 800 *wen* then 1 *liang* costs 50 *wen*.>

Conversely, to find the price of 1 *chin* from 1 *liang*, multiply by 16 using the additive method of multiplication < with the tenth column fixed as the hundredth column>.

Working: Put down the price of 1 *chin* as the dividend, < that is, put down 6 *kuan* 800 *wen*>. Start working from the right with the help of the above table. < Fix the tenth place above the tenth place of the dividend. Firstly, for 800 *wen*, the result is 50 *wen* and next, for 6 *kuan*, the result is 375 *wen*. The total is 425 *wen*.> Hence the answer.

Method III. Let the price of 1 *chin* be the dividend and halve it < to obtain the price of 8 *liang* instead of 16 *liang*>. Divide by 8 using the *kuei*[aa] method < to obtain the price of 1 *liang* instead of 8 *liang*>. The working is shown in the diagrams below.

(i)   Halve the price of 1 *chin* and above the unit place in *kuan* fix the hundredth place in *wen*.

                         3   4

       This becomes 340 *wen*.

(ii) To find the first place.
                         3   4 + 6

       On encountering
       3 add 6 [placed
       one place to the
       right].

(iii) To find the second place.
                         4   2   4

       When it is 10
       subtract 8.

                                       On encountering
                                       2 put down 4 [one
                                       place to the right]

(iv) To find the third place.
                         4   2   5
                                                          On encountering
                                                          4 make it 5.

       The answer 425 is obtained.

Method IV. Let the price of 1 *chin* be the dividend and divide it by 16 using the subtractive method of division. < The price of 1 *chin* is divided by 16 *liang* and the idea is to retain 10 *liang* and subtract 6 *liang*. Above the unit place in *kuan*, fix the hundredth place in *wen* for the price of 1 *liang*>. The diagrams below illustrate the working.

(i)   To subtract the first place.
       Above the unit
       place in *kuan* fix
       the hundredth
       place in *wen*.
                         6   8

       This becomes 680 *wen*. Keep 400
       and subtract 240.
                         4   4

       400 is kept as
       quotient.

                                       40 has to be
                                       further subtracted.

(ii) To subtract the second place.
        The hundredth
        place is fixed
        here.

4		*4*

        400 is kept.

                        Keep 20 and
                        subtract 12.

4	*2*	8

                        Keep 420.

                                        This place has
                                        to be further
                                        subtracted.

(iii) To subtract the third place.
        The hundredth
        place is fixed
        here.

4	*2*	8

                        Keep 420.

                                        Keep 5 and
                                        subtract 30.

4	*2*	*5*

        Each *liang* costs 425 *wen*.

Method V. Let the price of 1 *chin* be the dividend, 16 *liang* the divisor and divide;
< that is, the price of 1 *chin* is divided into 16 parts in order to find the price of 1 *liang*>.
The working is shown in the diagrams below.

(i) The first place of the quotient.
        Fix the hundredth
        place for the
        price of 1 *liang*
        here.

Quotient		
Dividend	6	8
Divisor	1	6

        400 is obtained for the quotient.

Quotient	4	
Dividend	0	4
Divisor	1	6

        6 *kuan* 400 *wen* is subtracted from the dividend.

(ii) The second place of the quotient.
                        The next place
                        of the quotient
                        is 2.

Quotient	4	2	
Dividend	0	0	8
Divisor		1	6

320 *wen* is subtracted from the dividend.
(iii) The third place of the quotient.

		The next place of the quotient is 5.	
Quotient	4	2	5
Dividend	0	0	0

When the dividend is subtracted there is no remainder.

Divisor		1	6

80 *wen* is subtracted from the dividend to give no remainder. Hence the answer, 425 *wen*.

The different methods shown in this problem illustrate the various types of division known prior to and during Yang Hui's time. In the first method the divisor 16 is treated as 2^4. The second method applies the conversion table of *liang* into *chin*. This table is apparently well known to the student, and since the measures of *chin* and *liang* are in daily use, the table must have been readily available. This method and others show that the ancient and medieval Chinese could handle the place values of numbers with confidence and ease.

The third method illustrates the *kuei* method of division. This method is confined to divisors of unit digit only, with a table written out for each divisor. For example, when 8 is the divisor, the table is as follows[13]:

> On encountering 1 put down 2 below one place to its right. [Here 2 represents the re-mainder whose place value is one place lower than the given digit 1. From the above statement, since there is no number directly below the given digit 1, it is then implied that 1 is a digit of the quotient. This digit is written in full Chinese character below it, and its place value is relative to the dividend. Thus, if the dividend is 10, then the quotient is 1 and the remainder is 2, but if the dividend is 100, the quotient is 10 and the remainder is 20. Similar meanings apply to the statements below.]
> On encountering 2 put down 4 below one place to its right.
> On encountering 3 put down 6 below one place to its right.
> On encountering 4, then 5 is a digit of the quotient whose position is directly below 4. There is no remainder.
> On encountering 8, then 1 is a digit of the quotient whose position is one place to its left.

On applying the division table one should always begin with the first digit from the left of the dividend. The remainder, which is placed one place to its right, should be added to the remaining digits of the dividend (that is, the digits of the dividend excluding the first one). The place value of the remainder is the same as the digit of the dividend immediately above it. After addition, the division table is then applied to the first digit from the left of the sum obtained, and the same procedure is carried on with this sum. The above division table is sufficient for working out a division problem with 8 as divisor. Should a digit of the dividend represent more than half of the divisor 8, that is, 5, 6, or 7, then the method is to subtract 4 from it and the other remaining digits, with 4 placed directly below that particular digit. After subtraction, the procedure described above is continued with the remainder, while the digit of the quotient 5 is added to the quotient of the first digit of the remainder directly below 4.

---

[13] A complete list of division tables for the *kuei*   *t'ung pien pen mo*, Ch. 2, pp. 9a–9b. method of division can be found in *Ch'eng ch'u*

Example: 3400 divided by 8.
Fix the hundredth
place here.

	3	4

For 3 add 6
placed one place
to the right.

		+ 6	
	three	4 + 6	
		For 10 subtract	
		8 (or add 2).	
	three + one	2	
		For 2 add 4	
		placed one place	
		to the right.	

			+ 4
	four	two	4
			For 4 the quotient
			is 5.
	four	two	five

Answer: 425.

The fourth method of division is the subtractive method of division, and the last method given is the common long-division method.[14]

7. Find the price of 37 *chin* 9 *liang* of goods when 1 *chin* costs 1 *kuan* 120 *wen*.
    Answer: 42 *kuan* 70 *wen*.

37 *chin* 9 *liang* is converted to 601 *liang*. 1 *kuan* 120 *wen* is multiplied by 601 and divided by 16.

8. Find the price of 601 *liang* of goods when 1 *chin* costs 1 *kuan* 120 *wen*.
    Answer: 42 *kuan* 70 *wen*.

This problem is exactly the same as the previous one except that the given quantity is now in *liang*. The slight variation in the working is that 1 *kuan* 120 *wen* is now halved four times and then multiplied by 601.

9. 42 *kuan* 350 *wen* can buy 37 *chin* 13 *liang* of goods. Find the price of 1 *chin*.
    Answer: 1 *kuan* 120 *wen*.

37 *chin* 13 *liang* is converted to 605 *liang*. 42 *kuan* 350 *wen* is multiplied by 16 and divided by 605.

---

[14] For an account of this method see Lam Lay Yong, "On the Chinese Origin of the Galley Method of Arithmetical Division," *British Journal for History of Science*, 1966, *3* (No. 9): 66–69.

10. Each *shih*[ab] of pulse costs 785 *wen* and each *shih* of wheat costs 1 *kuan* 160 *wen*.
　　If 297 *kuan* can buy 300 *shih* of pulse and wheat, find the amount of each type bought.
　　Answer: 136 *shih* of pulse and 164 *shih* of wheat.

Explanation: <In this problem on pulse and wheat, use the method of division in a group.>

Method of division into parts: Multiply the total amount by the cheaper price <and the product gives the cheaper cost>. Subtract this from the total amount of money and let the remainder be the dividend. <This is the excess of the more expensive quantity.> Subtract the cheaper price from the more expensive one and let the remainder, which is the difference in price of the two quantities, be the divisor. On division, the quotient gives the more expensive quantity. Subtract this from the total amount and the remainder is the cheaper quantity.

	136 *shih* of pulse	164 *shih* of wheat	
1 *shih* of pulse costs 785 *wen*	Total cost of pulse is 106 *kuan* 760 *wen*	Total cost of wheat is 128 *kuan* 740 *wen*	1 *shih* of wheat costs 1 *kuan* 160 *wen*
Difference in prices is 375 *wen*	Area of 61 *kuan* 500 *wen*		

The total amount of pulse and wheat is 300 *shih* and the total cost is 297 *kuan*.

Working: Multiply the total amount, <300 *shih* of pulse and wheat>, by the cheaper price. <1 *shih* of pulse costs 785 *wen* and on multiplying, 235 *kuan* 500 *wen* is obtained.> Subtract this from the total amount of money, <297 *kuan*>, and the remainder, <61 *kuan* 500 *wen*>, forms the dividend of the more expensive quantity. Subtract the two prices and let the difference be the divisor. <1 *shih* of pulse costs 785 *wen* and 1 *shih* of wheat costs 1 *kuan* 160 *wen*. The difference is 375 *wen* which is the divisor.> Divide to obtain the more expensive quantity. <Divide 61 *kuan* 500 *wen* by the divisor to obtain 164 *shih* of wheat.> Subtract the more expensive quantity, <that is, the wheat> from the total amount <of pulse and wheat> to obtain the amount of the cheaper quantity, <136 *shih* of pulse>. Hence the answer.[15]

[15] This problem is recorded in *Yung lo ta tien*.

The setting out of the solution of a problem is usually in two parts, the first being the outline of the method to be used and the second the working of the problem in numerical values. A concise basic set of mathematical notations is lacking, and hence the solution is always presented in a long-winded manner. The solution of the above problem is the solving of a pair of simultaneous linear equations in two unknowns. Using the present algebraic notations, the solution may be condensed as follows:

Let x be the amount in *shih* of pulse and y the amount in *shih* of wheat.

$$x + y = 300 \tag{1}$$
$$785x + 1160y = 297000 \tag{2}$$

Multiply (1) by 785 and subtract from (2) to give

$$(1160 - 785)y = 297000 - (300 \times 785)$$
$$y = \frac{61500}{375}$$

The diagram given is quite an interesting one. The area of *ACEFGH* represents the total cost of pulse and wheat, 297000 *wen*, since *AC* denotes the total quantity of pulse and wheat, 300 *shih*. The area of rectangle *ABGH* represents the total cost of pulse while the area of rectangle *BCEF* represents the total cost of wheat, since *AB* = x, *BC* = y, and *AH* represents the cost of 1 *shih* of pulse, *CE* represents the cost of 1 *shih* of wheat. The unknown side y of the rectangle *GDEF* can be easily found from its area, 61500, and the known side, 375.

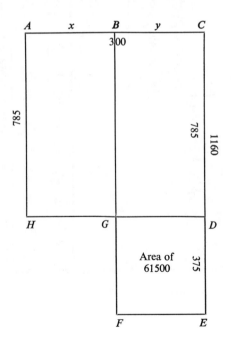

麥為問。分身為法。

分率術曰共物為實。以賤率乘之。俱為賤價。

錢餘為貴實。貴物所多之繫貴賤二率相減。餘為法求見一價。

左除之先見貴物以貴物減總數。餘為賤也。

麥直
六十　積一百二十八貫七百四十文

四石

穀直
二十　積一百六貫七百六十文

六石

一貫一百六十文麥價

七百八十五文穀價

多穀三百七十五

積六十一貫五百

穀麥共三百石。共錢

二百九十七貫文。

物為實。穀麥共三百石。以賤率乘之。穀賤每石七百八十五文。

百三十五貫五百文以減總錢二百九十七貫。餘為貴實。六十

百。貴賤二率相減餘為法。穀石價七百八十五。麥石價一貫一

The original Chinese diagrams, from Yung lo ta tien ([China] : Chung
Hwa Co., 1960), Ch. 16, 343, p. 20a.

CONCLUSION

It is unfortunate that the larger portion of the *Jih yung suan fa* is missing, the contents of which can only be left to conjecture. However, apart from the methods of multiplication, conversion into decimal fractions, and simultaneous linear equations, the above ten problems depict varied alternatives in division. These deal with the method of "halving" when the divisor is a power of 2, the special reference to the table of converting *liang* into *chin* when the divisor is 16, the *kuei* method when the divisor is a unit digit, the subtractive method of division when the divisor has 1 as the first digit on the left, and lastly, the general method of division which corresponds to our present long-division method.

GLOSSARY

a 日用算法   b 楊輝

c 九章算術   d 孫子算經

e 詳解算法   f 乘除通變本末

g 加   h 減

i 李儼   j 中算史論叢

k 諸家算法   l 繩孫

m 莫友芝   n 中國算學史論叢

o 永樂大典   p 宜稼堂叢書

q 斤   r 兩

s 分   t 厘

u 毫   v 錢

w 絲   x 忽

y 貫   z 文

aa 歸   ab 石

Qualitative
Sciences

A Contribution to the History of Chinese Dietetics [§]

BY LU GWEI-DJEN [‡] AND JOSEPH NEEDHAM [*]

M ODERN knowledge of nutrition and disease has brought the realisation that vitamins are essentials of a complete diet as much as the proteins, carbo- hydrates and fats, and that each has a function peculiar to itself, being essential for the maintenance of some normal function or functions of the body. Deficiency of any of the vitamins in the diet may result in ill health and even death. The League of Nations formed a special committee for the investigation of such prob- lems, and its work is now continued by the Food and Agricultural Organization of the United Nations. Considerable progress has been made each year though it is to be admitted that our approach is still somewhat slow in view of the importance of dietetics for human welfare. Nevertheless it is a striking fact in the history of science that the first pioneer of vitamin studies, Sir Frederick Gowland Hopkins, has lived to see the establishment of the chemical constitution, and even the synthesis, of many of the most important vitamins.

The present contribution arises from the fact that one of us (G.D.L.), while en- gaged in experimental work [1] on the physiology of vitamin B_1, became interested in the question of the antiquity of human knowledge of beri-beri as a deficiency disease. Such knowledge has certainly existed at least since the 5th century A.D. in China, as may be found from writings of that period which we possess. The fact that the Chinese knowledge of dietetics has been so largely overlooked is due partly to the lack of any proper index to Chinese literature, and partly to the extreme divergence of the Chinese language from Western alphabetical languages, which has sealed off the history of thought and knowledge in China from the scholars of the west. Even the standard history of Chinese medicine, however, (Wong & Wu) has practically nothing to say regarding Chinese dietetics.

A detailed review of the development of our present knowledge of nutrition and disease which led in the end to the discovery of vitamin B_1 and its relation to the disease beri-beri would of course include an account of the isolation of the vitamin in crystalline form, the various theories proposed to account for its action, its rôle in cell oxidations, etc. But we shall simply refer the reader to reviews (Peters; Harris; Williams & Spies).

By the end of the last century, it was clear that diseases like scurvy, beri-beri and rickets, could be cured empirically by the addition of suitable foods to the diet, although there was no knowledge of the chemistry or the nature of the deficient substances. In connection with beri-beri, a disease which we now know to be due to vitamin B_1- deficiency, Harris writes:

This disease, known in China as early as 2600 B.C., was first conclusively shown to have a dietary origin in 1880 by Takaki, the Director- General of the Medical Department of the Japanese Navy. By simply increasing slightly the allowance of vegetables, fish, meat and barley in a diet still consisting predominantly of rice, he was able to reduce the incidence of disease to virtually vanishing point.

[§] Contributed to *Isis* in 1939, printing inter- rupted by the war.
[‡] The Henry Lester Institute for Medical Research, Shanghai, China.
[*] The Biochemical Laboratory, University of Cambridge, England.
[1] Lu; Platt & Lu; Needham & Lu.

This recognition of a dietary cure, followed by Eijkman's discovery of experimental beri-beri (1890), led to attempts at the chemical isolation of the potent substance in rice polishings which gave origin to the vitamin view of beri-beri, (Funk, 1912). We might say that the recognition of the dietary cure was the initial step in the late development of the relation between food and health in the West.

Beri-beri, a disease known in China as "Chiao Ch'i" (1362, 1064) [2] has been cured by dietary means for many centuries but in modern times it took a new form in society, owing to the industrialisation of the great cities, the introduction of the milling machine and the appallingly poor conditions (heat, humidity, lack of ventilation and long hours of working) in some factories. Wong & Wu (p. 88) believe that the disease "Chüeh Ch'i" (3198, 918) mentioned in the oldest Chinese medical classic, the *Huang Ti Su Wên Nei Ching* (5124, 10942, 10348, 12650, 8177, 2122), 3rd-2nd cent. B.C., was in fact beri-beri.

We can hardly avoid a preliminary statement concerning the different basis on which the organization of Chinese society was built, if a clear understanding of the facts to follow is desired.

The official class of China, fundamentally an agricultural country, has always had an interest in the food of the people. In fact, Shên Nung (9820, 8408) one of the earliest legendary sages of Chinese history, who is supposed to have introduced the cultivation of the five sorts of grain, and invented the plough, is also pictured as an experimentalist who tasted all kinds of plants and classified them according to their nature and their effects on the individual, into various groups for ordinary and medicinal uses. Although the book, *Shên Nung Pên Ts'ao* (9820, 8408, 8846, 11634), which contains three volumes, with 365 varieties of plants for medicinal use, has no real connection with the legendary culture-hero, it cannot be later than the beginning of the Christian era since in A.D. 4 the emperor convoked an assembly of experts in *Pên Ts'ao*,[2a] and it is referred to in the biography of an important physician [2b] of the same period.[2c] The oldest version probably existed before the 1st century B.C. It may be significant that the *Pên Ts'ao* divides plants into three categories, superior, medium, and inferior, according as they possess "rejuvenating," tonic or curative, properties. Hence even at this early date there was a suggestion that prevention was more important than cure.

The existence of a knowledge of dietary treatment of various diseases in China can be traced back as early as the time of the Warring States. The *Chou Li*, (2450, 6949), or *Record of the Rites of the Chou dynasty*, which was probably put together between the 4th and 1st century B.C. includes in the list of the four imperial medical officers a Shih I, (9971, 5380) i.e. an Imperial Dietetician, as well as an Imperial Physician, an Imperial Surgeon, and a Regius Professor of Medicine.[2d]

A survey of the earlier medical literature, including the *Pên Ts'ao* editions, together with the *Chin Kuei Yao Lüeh* (2032, 6465, 12889, 7564) of Chang Chi,[3] (416, 787) (3rd century A.D.) shows that the latter is the first of the ancient medical books to give an account of the causes of the diseases and hence a theory of therapy. It contains many vivid accounts of deficiency diseases in their various stages, and describes preparations which, as we know today, would be rich in the various vitamins.

It is interesting that the famous writer Han Yü (3827, 13572) (762-824 A.D.) states in one of his essays that the disease beri-beri was particularly prevalent south

[2] The numbers following Chinese words refer to the second edition of Herbert A. .Giles, *Chinese-English dictionary* (Shanghai, 1912).
[2a] See *Chhien Han Shu* (1737, 3836, 10024), ch. 12, p. 92.
[2b] Lon Hu (7343, 4979).
[2c] See *Chhien Han Shu* (1737, 3836, 1024), ch. 92, p. 7B.
[2d] Cf. E. Biot's translation, vol. 1, pp. 8, 93.
[3] Chang Chung-Ching (416, 2876, 2143), "the Chinese Hippocrates."

of the river, (the Yang-Tze). This was still true in 1935 A.D. when Hou noted a similar difference, and referred it to the difference in diet between the wheat-eating north and the rice-eating south. In the Sung dynasty there appeared a monograph especially devoted to beri-beri, the *Chiao Ch'i Chih Fa Tsung Yao* (1362, 1064, 1845, 3366, 12010, 12889) by Tung Chi (12259, 844) of about +1078.

When we reach the 14th century A.D., we find a very interesting book which hitherto has been little known, even among Chinese historians of science. Hu Ssu-Hui (4927, 10271, 5193), who during the reign of Jên Tsung (5627, 11976) of the Yuan dynasty, occupied the post of Imperial Dietetician from 1314–1330 A.D., wrote a book called *Yin Shan Chêng Yao* (13269, 9715, 687, 12889), or, *The Principles of Correct Diet*. He says in his preface that his book is a collection of material selected from the *Pên Ts'ao* and other well-known medical books of his day, as well as the data collected by himself during his 15 years of office. He set out only to describe plants which had a beneficial action on man; but he was more than a botanist for he dealt with methods of cooking and serving food. It would be difficult indeed today to find the manuscript material of Hu Ssu-Hui's sources, since so much of it must have been lost.

It may be significant that actually during his period of office there were many public calamities such as might lead to deficiency diseases. In 1319 all the roads in Shantung and Huainan were flooded. In 1320 there was famine in Honan, and civil war because of Achakpa the Mongol, a brother of the reigning prince of the dynasty. Earthquake troubles occurred in 1322, and five years later there was drought, locust plagues, famine, landslides and earthquakes again. Finally in 1330 there was another famine in Honan. It is clear, therefore, that considerable opportunity existed for the practice of dietetic knowledge. Hu's book was presented to the Emperor.

One hundred and fifty years later, Ching Tsung (2143, 11976) Emperor of the Ming dynasty, after a further succession of public calamities, issued a royal edict that the book should be published for the general benefit. This is the version which we now have.

The book is divided into three volumes, of which the second is devoted to the cure of diseases by diet. A differentiation was made between two types of cases in beri-beri, known today as the "wet form" and the "dry form." The acute, or wet, form, was regarded as due to "fiery ch'i"; and the chronic, atrophic, or dry, form, was regarded as due to "cold ch'i." Among the sixty-two diets given for the cure of various diseases, the following are of special interest as applying to beri-beri: —

(1) For the cure of "wet" beri-beri,
    (a) Make a soup of rice and horse-tooth-vegetable, and let the patient drink it on an empty stomach early in the morning.
    (b) Cook 16 ozs. pork with one handful of onion, 3 dried grass-seeds, pepper, fermented beans and rice (½ lb.) and let the patient eat it in the morning.
(2) For the cure of "dry" beri-beri,

    (a) Cook 1 big fish (carp) with ½ lb. small red beans, two-tenths of an oz. of ch'eng fruit skin, two-tenths of an oz. of small dried peppers, and two-tenths of an oz. of dried grass-seed. Let the patient eat it.
(3) For cases with oedema.
    (a) Stuff 1 lb. small red beans and 5 grass-seeds into a duck, and make a soup. Let the patient eat it in the morning.

It will be seen from these recipes that a good intake of vitamin B_1 was assured, but other vitamins would be present in addition, an important fact since many or most of the deficiency cases would be lacking in more than one vitamin. The plants shown in figs. 1–6, prepared from Hu Ssu-Hui's book, include the constituents of the above recipes; all of them were recommended by him for use in cases of beri-beri. In particular an infusion of the Hu lu (gourd) in wine was recommended for mild cases in the *Ch'ien Chin Fang* (1725, 2032, 3435), the *Thousand Golden Remedies* compiled by Sun Ssu-Mo (10431, 10271, 7998) in the 7th cent. A.D.

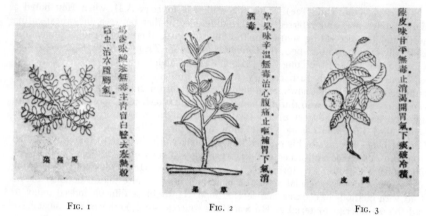

FIG. 1   FIG. 2   FIG. 3

FIG. 1: MA CH'IH TS'AI, *horse tooth vegetable*, tastes sour, cold, non-poisonous, for clearing eye-troubles, gets rid of the hot and the cold, disinfectant, cures beri-beri. (Unidentifiable.)

FIG. 2: TS'AO KUO, *grass seed*, tastes mildly hot, non-poisonous, cures abdominal & epigastric pain, good against nausea, a tonic driving down the ch'i, dissolves wine poison . (Bretschneider, III, No. 58; *Amomum globosum*, a cardamon.)

FIG. 3: CH'EN P'I, *cheng fruit skin*, tastes slightly sweet, quenches thirst, stimulates the stomach ch'i, dissolves mucus, disperses accumulated cold. (Bretschneider, III, No. 281; ripe orange peel, spp.)

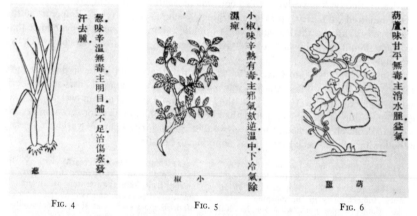

FIG. 4   FIG. 5   FIG. 6

FIG. 4: TS'UNG, *spring onion*, tastes mildly hot, non-poisonous, brightens the eye, supplies what is not enough, cures injuries by cold, causes perspiration & removes oedema. (Bretschneider, II, No. 357; *Allium fistulosum*, the Chinese onion.)

FIG. 5: HSIAO CHIAO, *little pepper*, tastes very hot, non-poisonous, cures rising evil ch'i, warms the viscera, drives down cold ch'i, banishes diseases of wet ch'i. (Laufer, p. 374.)

FIG. 6: HU LU, *hu lu gourd*, tastes flat, non-poisonous, cures water oedema, beneficial for the ch'i. (Unidentifiable.)

Figure 7 shows the frontispiece of the book. A consultation of two dieticians is seen in progress in the presence of the patient and assistants. At the top right-hand corner is placed the motto: "Food cures various diseases."

At this point it is necessary to say a word or two about the term "ch'i" (1064)

要 正 膳 飲

撰 慈 思 忽

FIG. 7

Principles of Correct Diet

Yin Shan Cheng Yao

by

HU SSU-HUI

"she liao chu bing"

"food cures various diseases"

which has been used in the foregoing paragraphs and which will be found in what follows.[4] It is usually translated "spirit" or "quintessence" and the ancient Chinese medical books attribute all diseases to disturbances in the normal state of the ch'i characteristic of the part of the body in question. Thus we have: —

p'i ch'i (9078, 1064) spleen spirit
kan ch'i (5822, 1064) liver spirit, and
chiao ch'i (1362, 1064) foot spirit, or beri-beri.

These essences may be thought of as analogous to the "archaei" or vital forces regarded by sixteenth-century European medicine (cf. Paracelsus) as residing in the various organs. But in addition to these there are also influences from without, for example: —

huo ch'i (5326, 1064) fiery or hot spirit
han ch'i (3825, 1064) cold spirit.

These were also involved in the causation of disease. But there is also another meaning of ch'i. It will be remembered that in the Aristotelian doctrine of "souls," plants possess a vegetative principle ($\psi v \chi \grave{\eta} \; \theta \rho \epsilon \pi \tau \iota \kappa \acute{\eta}$), animals a vegetative principle and a sensitive principle ($\psi v \chi \grave{\eta} \; a \grave{\iota} \sigma \theta \eta \tau \iota \kappa \acute{\eta}$) while man alone possesses a rational principle ($\psi v \chi \grave{\eta} \; \delta \iota a v o \eta \tau \iota \kappa \acute{\eta}$) as well as the vegetative and sensitive principles. Aristotle thus held

---

[4] Cf. also I. A. Richards' remarkable commentary on the psychological doctrine of Mencius.

a doctrine of levels which had only to be translated into terms of time to become evolutionary. Now in ancient Chinese thought, the "ch'i" is the basis of everything; animals have a nature of their own, the "shou hsing" (10022, 4600) as well as ch'i; and man alone has the true "hsing" (4600) to which the adjective good, "shan" (9710), can be applied. A scheme like that of Aristotle could therefore be drawn up as follows: —

PLANTS: .... ch'i (1064)
ANIMALS: .... Shou hsing (10022, 4600) + ch'i (1064)
MAN: .... Shan hsing (9710, 460) + Shou hsing (10022, 4600) + ch'i (1064)

But it is doubtful whether the ancient scholars looked at the matter in this way, for they sometimes spoke of a "hsing" in plants; sometimes it was beneficial to man, sometimes it was poisonous. On the other hand it is certainly right to say that they attributed "shou hsing" to man, for they identified it with the "o ch'i" (8452, 1064) i.e. tendencies to evil.

As an illustration of this, we may quote from the writings of Hsün Ch'ing (4875, 2198) [Hsün Tzu (4875, 2198)] the well-known Confucian thinker (298–238 B.C.). In the book which bears his name, he says (ch. 5, p. 13a):

Water and fire have essences [ch'i (1064)] but not life [sheng (9865)]; plants and trees have life [sheng] but not knowledge [chih (1783)]; birds and animals have knowledge [chih] but not a sense of justice [i (5454)]. Man has essence, life, and knowledge, and in addition the sense of justice; therefore he is the noblest of earthly beings. In strength he does not equal the ox, nor in power of running the horse, and yet they are used by him; how can this be? Because man is able to form social organizations [ch'ün (3304)] and they are not. How is it that men can do this? Because they can cooperatively play their parts and receive their portions. How is it that they can carry this out? Because of justice [i] which unites the parts into a harmony, and therefore a unity, and leads to strength, and in the end to triumph.

As an illustration of the solution of the famous controversy about the goodness or badness of human nature by recognizing higher and lower elements in man's constitution, we may refer to the book called *The Rat Jade, Shu P'o* (10072, 9443), written by Tai Chih (10567, 9976) in the early 13th century A.D. at the end of the Sung dynasty, in which he reconciles the views of Mencius that man's nature is good with that of Hsün Tzu that it is bad, by pointing to the different components in this "Aristotelian" ladder of souls.

As regards Hu Ssu-Hui's book it only remains to add that he gave special dietaries for children and for pregnant and lactating women, and that he emphasized seasonal factors. His book is not the only one in the literature dealing with nutrition. There is, for example, the *T'ai i yüan chi chiu liang fang chai yao* (10573, 5380, 13752, 892, 2281, 7017, 3435, 228, 12889). Unfortunately, this book is not dated. It is a "selection of prescriptions employed by the Imperial Medical College for saving life in cases of extreme peril." It contains many nutritional recipes, among which there is one which recommends the use of fermented food in general for beri-beri cases.

Another pamphlet dealing with nutrition was written by Ch'ên Chih (658, 1846) [4a]; the *Yang Lao Fêng Ch'in Shu* (12856, 6783, 3574, 2081, 10024) a treatise on the care and feeding of old people. Later, in the Yüan dynasty, Tsou Hung (11811) added four further chapters and retitled it *Shou Ch'in Yang Lao Hsin Shu* (10019, 2081, 12856, 6783, 4574, 10024).

So deeply imbedded in the Chinese people was the ancient dietetic knowledge, that to this day the phrase "grandmothers' cures," is used of traditional dieting and, even in present-day Shanghai, both groceries and drugs are sold in the same store. The traditional knowledge sufficed till modern times, and for the following reason. Anciently rice was ground in every family on the day it was intended for use; this

---

[4a] About 1200 A.D.

preserved in all probability a thin fatty layer on the grain and a large portion of the embryo which contained sufficient vitamin B_1. With modern industrialisation there grew up the practice of milling the rice in factories and then storing it in the polished form so that bacterial action and growth of moulds necessitated the subsequent washing which removes all the vitamin B_1 of the product. To combat this severe deterioration of the diet, together with bad working conditions, research into beri-beri along modern lines was needed. In unpublished work carried out in Shanghai the Medical Department of the Henry Lester Institute in collaboration with the Industrial Department of the Shanghai Municipal Council found, in a large-scale feeding experiment on eighty factory workers, one third of whom had in the previous year shown signs of beri-beri, that when rice polished on the day of use (not washed) was fed, together with various vegetables cooked exactly the same as the previous year, only one single individual showed slight signs of beri-beri. It is, therefore, possible to continue to use highly-milled rice for feeding adults without giving rise to beri-beri, even when climatic and other stresses are existent, if the grain is not washed before it is cooked.

In China it has been found that in the absence of flood and war, the country folk know by experience what to eat and how to eat to maintain a satisfactory state of well-being at low economic levels.

It has already been mentioned that Han Yü, writing in the 8th century A.D. stated that beri-beri was particularly prevalent south of the river (the Yang-tze). A systematic survey of the diet of representative communities of different parts of China with data on the daily intake of vitamin B_1 in terms of international units per adult male per day has recently been calculated, by Platt (unpublished) using the data of W. Y. Swen and J. Lossing Buck collected by the department of Agricultural Economics of the University of Nanking. It is now generally agreed that the daily requirement of vitamin B_1 for an adult is about 300–350 I.U. per day. The calculation showed that the vitamin B_1 consumption of the people in the south (e.g. Kuangtung or Hunan) is below or just on the lower limit (250–322 I.U. per day), while that of the people in the north (e.g. Hopeh or Shansi) is well above it (450–690 I.U. per day). It is interesting that the geographical difference in the incidence of the disease is as true today as it was in Han Yü's time.

Should the rice be prepared and stored in the milled form, the vitamin B_1 content falls to sub-optimal levels and beri-beri results. Thus the conviction of Hu Ssu-Hui that proper feeding may eliminate certain diseases was vindicated after six hundred years, and modern science was able to redress one of the evils of modern industrialism.

In conclusion, it seems clear that the empirical knowledge of diet, especially in relation to certain deficiency diseases, is much older than commonly supposed. Credit for this, as for certain other valuable discoveries, must be given to ancient Chinese civilisation. At the same time it was only by the analytical methods of Western science that the relations of the food-constituents to health and disease could be incorporated into a logical system. Finally, it may be suggested that an intensified study of Chinese literature might bring to light some useful indications and clues for modern physiologists and pathologists.

REFERENCES

There are numerous editions of the *Pên Ts'ao* printed in China in various versions. Hu Ssu-Hui's book was reissued in printed form by the Shanghai Commercial Press in the series *Kuo Hsüeh chi pên ts'ung shu* (6609, 4839, 850, 8846, 12039, 10024), (1935).

Bretschneider, E.: *Botanicon Sinicum* (Shanghai, 1892).

Eijkman, E. J.: See Bibliography in Harris' book.

Funk, W.: *Journ. Physiol.* (1912), vol. 45.

Harris, L. J.: *Vitamins* (Cambridge, 1935), also *Vitamins and Vitamin Deficiencies; vol. 1, B₁* (London, 1938).

Hou Hsiang-Chuang: *Nutrition Notes* (China, 1935), No. 5.

Laufer, B.: *Sino-Iranica*, (Chicago, 1919).

Lu Gwei-Djen: *Biochem. Journ.* (1939), vol. 33.

Needham, D. M. & Lu, G. D.: *Biochem. Journ.* (1939), vol. 33.

Peters, R. A.: *Trans. Roy. Soc. Trop. Med. & Hyg.* (1938), vol. 31.

Platt, B. & Lu, G. D.: *Quart. Journ. Med.* (1936), vol. 5.

Richards, I. A.: *Mencius on the Mind* (London, 1932).

Williams, R. R. & Spies, T. W.: *Vitamin B₁ and its uses in Medicine* (New York, 1938).

Wong, K. C. & Wu, L. T.: *History of Chinese Medicine* (Tientsin, 1932; 2nd ed. Shanghai, 1936).

Characteristics of Chinese Astrology

By Shigeru Nakayama *

ALTHOUGH THE TRADITIONAL Chinese approach to astrology is unique, it will not be misleading to begin by classifying its components with respect to the distinction Otto Neugebauer successfully applied to Western astrology — namely that between judicial astrology and genethliacal astrology.[1] Within the following discussion the term " portent astrology " will be used in place of " judicial astrology " in order to avoid confusion with the familiar distinction between judicial and natural astrology.[2] While in the West " genethliacal astrology " is synonymous with " horoscopic astrology," it will be shown that the Chinese art of fate calculation is not properly called " astrology "; Chinese horoscopes as such are founded on ideas imported from the West.

Chinese court astrology consists purely in the accumulation of portents in the form of celestial, meteorological, and seismological phenomena — supernovae, planetary conjunctions, comets, hailstorms, earthquakes — and their empirical correlation with events in human society which are relevant to the success of the Imperial rule. Here is a typical interpretation from the *Shih chi* [A,3] (Records of the Grand Astrologer-Historian; *c.* 90 B.C.): " When Mercury appears in company with Venus to the east, and when they are both red and shoot forth rays, then foreign kingdoms will be vanquished and the soldiers of China will be victorious." [4]

The Chinese term *t'ien-wen* [B] is now used simply to mean " astronomy," but this is a decided shift in denotation; in classical writings it is ordinarily used in the sense of " portent astrology." The major part of the " T'ien-wen chih " (Treatise on *t'ien-wen*) generally included in the Standard Histories is devoted to observational records of celestial and other natural anomalies and their correlation with political events. As a result, the Histories preserve

* University of Tokyo.

[1] Otto Neugebauer, " The History of Ancient Astronomy: Problems and Methods," *Publications of the Astronomical Society of the Pacific*, 1946, *46*:39; and A. Sachs, " Babylonian Horoscope," *Journal of Cuneiform Studies*, 1952, 6:51.

[2] E.g., the *Oxford English Dictionary*, s. v. " astrology."

[3] Appended to the end of this article is a list of the Chinese characters which represent the romanized Japanese and Chinese words appearing in the text and footnotes. Superscript capital letters — i.e., *Shih chi* [A] — indicate the placement in the list of the equivalent characters.

[4] Joseph Needham, *Science and Civilisation in China* (Cambridge: Cambridge Univ. Press, 1954–), Vol. 3, p. 353.

a voluminous and detailed collection of such observations extending over two millennia — a collection which is even today proving of immense value to seismologists, astronomers, and other scientists.[5] But what were the motives behind it? How was it possible for the interest to be sustained if the correlations on which the system was based had no real predictive power? This problem arises equally in other cultures — there were, for instance, similar concepts in Babylonian astrology — but nowhere else did portent astrology withstand so many historical vicissitudes, spanning without substantial modification the two millennia between the formative period of Chinese natural philosophy and its final rejection in favor of modern science.

Celestial phenomena were matters of great concern to the throne, implying grave political consequences. According to Chinese theories, the monarch is a man of transcendent virtue, whose title to the throne is bestowed by heaven.[6] In other words, he is the agent of the natural order which heaven stands for, and he rules under its auspices. If, then, his conduct is contrary to the natural order, he is no longer qualified for the throne. The *Han shu* [C] (Standard History of the Former Han) quotes an Imperial edict:

> When the prince of men is not virtuous, a reproach appears in Heaven or Earth, and visitations and prodigies happen frequently, in order to inform him that he is not governing rightly. Our experience in governing has been (only) for a brief time, so that (we) have not been correct in (our) acts, hence on (the day) *mou-shen* (i.e. 5th Jan. B.C. 29) there was an eclipse of the sun and an earthquake. We are greatly dismayed. . . .[7]

In order to detect these celestial admonitions and to take countermeasures as quickly as possible, the ruler appointed court astrologers (who were, of course, also the astronomers of the time) and required them to engage assiduously in sky-gazing. They supplied data which played a part in the ruler's political decisions, and in return enjoyed high positions in the court bureaucracy.[8]

Chinese astrology shares three essential features with that of ancient Babylon: [9]

1) *Empirical collection of data.* The point of departure in portent

---

[5] To cite one of very many instances, statistics on meteor showers have been compiled from ancient Chinese, Japanese, and Korean sources. See Susumu Imoto [AB] and Ichirō Hasagawa,[AC] " Chūgoku, Chōsen oyobi Nihon no ryūseiu kokiroku " [AD] (Ancient Records of Chinese, Korean, and Japanese Meteor Showers). *Kagakushi Kenkyū,*[AE] 1956, No. 37:7–15. An English translation was published as " Historical Records of Meteor Showers in China, Korea, and Japan," *Smithsonian Contributions to Astrophysics,* 1958, 2:131 ff.

[6] Kiyoshi Yabuuchi,[AF] *Shina no tenmongaku* [AG] (Chinese Astronomy) (Tokyo: Kōseisha, 1943), pp. 63–66.

[7] Homer H. Dubs (trans.), *History of the Former Han Dynasty* (Baltimore: Waverly, 1938–), Vol. 2, p. 382. Interpolations are Dubs'.

[8] It is interesting in this connection that Hans Bielenstein has interpreted the records of portents in the early Standard Histories as deliberately manipulated by metropolitan officials as a means of indirect criticism of rulers. See " An Interpretation of the Portents in the Ts'ien-Han shu," *Bulletin of the Museum of Far Eastern Antiquities,* 1950, 22: 127–143.

[9] Reginald Thompson, *Reports of the Magicians and Astrologers of Nineveh and Babylon* (2 vols. London: Luzac, 1900), has been consulted for Babylonian sources.

astrology was day-to-day observations. In theory, at least, no ideological or other dogmatic bias was allowed to affect their impartiality. While it is impossible to completely reconstruct the foundations upon which the interpretation of portents was based, however, it is clear that these foundations were in the main deductively derived from cosmological principles.

2) *Official character.* Under the Chinese and Babylonian types of "oriental despotism" the ruler's administrative and personal concerns always took precedence over private interests, and thus made up almost the entire subject matter of astrological interpretations. The clientele of astrologers was limited to the ruler and to members of his household. The institutional framework of astrology was tightly incorporated into the bureaucratic machinery, responsible only to the ruler. The astrological records were included in the official archives. The heads of the astrological profession were royal advisors, men of high rank and position. Under the Grand Astrologer were a number of officials who on certain occasions gathered and addressed the ruler.[10]

3) *Secrecy.* Astrological information was directly relevant to national security. As a means of keeping the powers possessed by astrologers under official control, the office of Astrologer was hereditary in Babylonia as well as in Japan.[11] In China, although the post was not formally hereditary, elements of secrecy were also present, and astrological knowledge, books, and tools were not freely available outside the court.

On the basis of these similarities, we may be tempted to postulate an early interchange of ideas between the Near East and the Far East in the pre-Christian period. This is a matter of classic controversy among sinologists,[12] not confined to astrological topics, but extending to astronomy and culture in general. There are too many arguments on both sides to allow this question to be settled at present; for instance, against the similarities cited above may be poised the considerable differences in the demarcation and naming of constellations in the two areas.

THE EMPIRICAL DEVELOPMENT

As knowledge is accumulated and systematized we can distinguish in astrology two lines of development, empirical and dogmatical. While celestial and terrestrial phenomena are the subject matter of more-or-less objective description, thus providing data for scientific analysis, their interpretations are subjective, based on rules which are not likely to be altered if they are shown to lack predictive value. We shall first look at the empirical aspects.

Scientific astronomy seeks regularity and periodicity, while only irregular phenomena invite astrological attention. As the recognition of celestial

---

[10] Morris Jastrow, *Aspects of Religious Belief and Practice in Babylonia and Assyria* (New York/London: G. P. Putnam's Sons, 1911), p. 117.

[11] Thompson, *op. cit.*, Vol. 2, p. xvi.

[12] Shigeru Nakayama, "Japanese Studies in the History of Astronomy," *Japanese Studies in the History of Science.* 1962, No. 1:14–22, especially p. 14.

periodicity is established, how and when are phenomena transferred from the realm of astrology to that of mathematical astronomy?

We can classify natural portents into two classes, according to whether or not their recurrence was known to be periodic. All sublunary (meteorological) portents, such as lightning, earthquakes, and meteors, were considered nonperiodic. The periodicity of comets was not known, and they were considered the worst portents of all. The only way to detect nonperiodic phenomena was constant observation. Their effect on superstitious minds made them omens; they were regarded either as unwelcome or as unusually auspicious signs. On the other hand, the study of periodic phenomena was identical with the subject which we nowadays call celestial mechanics. The best example of the transfer of a type of phenomenon from astrology to astronomy through the discovery of its periodicity may be found in the case of eclipses.

Eclipses.

It is interesting that in the " Treatise on Astrology " (T'ien kuan shu),[D] of the *Shih chi,* lunar eclipses are described as periodic occurrences but solar eclipses are not. Solar eclipses were therefore unfavorable omens, while lunar eclipses were trivial from the astrological point of view. Attempts to predict solar eclipses, increasingly sophisticated as the level of mathematical astronomy rose, establish that Chinese scientists were willing to pursue the hypothesis of periodicity. Since these attempts must have begun early, it is only natural to expect that the astrological importance of eclipses of the sun should have been lost.[13] On the contrary, it was maintained to

---

[13] The following remark has been prepared by Nathan Sivin, of the Massachusetts Institute of Technology:

Homer H. Dubs suggests (in " Solar Eclipses during the Former Han Period," *Osiris,* 1938, 5:518–519) that the earliest solar eclipse which was computed (by a simple linear interval) but could not have been observed in Asia took place on 16 February of A.D. 26. This remark must refer to the eclipse of Julian 6 February (J.D. 1730591), the correct Gregorian equivalent of which is 4 February, not 16 February. The central annular eclipse of 6 February was in fact visible in Asia, as T. R. von Oppolzer and F. K. Ginzel (*Spezieller Kanon der Sonnen- und Mondfinsternisse* [Berlin, 1899], pp. 30–31) assert. Dr. Nakayama has confirmed this by the method Dubs himself used — recalculation using Oppolzer's elements as directed by P. V. Neugebauer in *Astronomische Chronologie* (Berlin: W. de Gruyter, 1929). He finds that at longitude 115° east and latitude 35° north, in the vicinity of the Royal Observatory at Yang-ch'eng, the maximum phase was 4.0/12 at 16^h 56^m, 16 minutes before sunset. It would of course have been still greater for an observer further north.

It is also untrue that this eclipse is recorded only in *Ku-chin chu.*[AH] It is cited at the beginning of Ch. 6 of the " Treatise on the Five Elements " in the History of the Later Han (*Hou Han shu chi shieh,*[AJ] Wan yu wen k'u [AK] ed., p. 3777), where it is noted that it " took place 8 *tu* in the mansion Wei," [AL] i.e., at $22\frac{1}{2}^h$ right ascension, approximately.

While it is clear that an even moderately successful method for predicting solar eclipses comes late, the first attempts at prediction almost certainly predate the *Ching-ch'u* [AM] calendrical treatise of the period 226–239. By how long remains to be rigorously established. The best account in a Western language of Chinese eclipse prediction methods is given in K. Yabuuti (Kiyoshi Yabuuchi), " Astronomical Tables in China from the Han to the T'ang Dynasties," in Kiyoshi Yabuuchi (ed.), *Chūgoku chūsei kagaku gijutsu shi no kenkyū* [AN] (Studies in the History of Science and Technology in Medieval China) (Tokyo: Kadokawa, 1963), English section, pp. 460–489.

modern times. It is true that observation of solar eclipses provided an excellent method for checking the accuracy of the calendar, but they kept their astrological significance for quite other reasons.

First, the periodicity of lunar eclipses had been discovered before the court ritual had attained a fixed pattern, whereas the periodicity of solar eclipses was not accepted until after they had come to acquire a definite ritual significance. Thus, as a matter of conventional inertia, the custom of closing offices on the day of a solar eclipse was observed even during and after the T'ang period. Second, the prediction of solar eclipses requires considerably more sophisticated techniques than that of lunar eclipses. For this reason early attempts at prediction could not have attained a high rate of confirmation. Inaccuracies were attributed not necessarily to imperfections in scientific technique, but often to the indeterminacy of celestial motions — or, to put it more accurately, to their susceptibility to at least some control by human desires operating through ritual and magic.

It is even doubtful to what extent the recognition of periodicity freed people's minds from fear of portents. Failure of a predicted eclipse to happen was regarded as the result of exorcism, and those responsible were rewarded. On another level, ascribing the nonappearance of a predicted eclipse to the virtue of the Emperor, his courtiers made a custom of congratulating him at such times.[14]

Many Chinese solar eclipse records seem not to have been based on observation.[15] This is also true in the case of Japan, at the time a cultural satellite of China. Of the 576 solar eclipses recorded there before A.D. 1600, less than 20 per cent were clearly listed as having been observed, and only half of those, according to Oppolzer's *Canon der Finsternisse*, were visible in Japan. A quarter of the lunar eclipses recorded could not have been visible in Japan. It is clear, therefore, that most Japanese records of eclipses were based on computation rather than observation.[16]

Why, then, did astrologers predict an excessive number of eclipses while their duty was to make accurate prognostications? When a prediction failed, the relieved Emperor did not punish them, but considered the failure of the eclipse to occur as a happy omen. Conversely, when an eclipse occurred unexpectedly, the Emperor, defenseless against calamity, was painfully embarrassed. As a result, the astrologer might be removed from office or even executed. Small wonder that he considered an occasional unconfirmed prediction to be merely good insurance!

Planetary motions.

In ancient China, as we see in the " Treatise on Astrology " of *Shih chi*, the positions of planets, particularly their meetings (conjunctions, occulta-

---

14 Tsutomu Saitō,[AP] *ōchō jidai no in'yō dō*[AQ] (The Yin-Yang Art of the ōchō Era) (Tokyo: Kōin Sōsho, 1915), p. 86.
15 Chu Wen-hsin,[AR] *Li-tai jih-shih k'ao*[AS] (A Study of Eclipses in Chinese History)

(Shanghai: Commercial Press, 1935), p. 62.
16 Takanobu Suzuki,[AT] " Honpō kodai no nisshoku ni tsuite "[AU] (On Ancient Japanese Eclipse Records), *Nihon tenmongakkai yōhō*,[AV] 1910, 6:143–169.

tions, and apparent contiguities), also had remarkable astrological significance. The Babylonian idea of geographical association was paralleled in the identification of twelve divisions of the sky where meetings could occur with twelve states of the empire, so that the terrestrial location affected by a celestial event was immediately specified. On the other hand, the " Treatise on Harmonics and Calendrical Astronomy " (Lü-li chih) [E] of the Former Han History furnishes us with fairly precise figures for the synodic periods of the five planets, reflecting the state of knowledge at the end of the first century B.C.[17] On the basis of these, the Chinese proceeded to establish a " Grand Conjunction " period in which the initial conditions of the sky (and thus of time) were to recur, and to adopt the beginning of this great cycle as the epoch for computation of their ephemerides. Oddly enough, however, a very detailed understanding of the periodicities in planetary motions did not reduce the importance of astrology any more than it was reduced by scientific attempts to predict eclipses of the sun. Regularity was one thing, astrological significance another.

Once cyclical behavior was established, astral influence was simply considered to operate in regular sequences. This idea was one of the basic factors which, in the Hellenistic period in the West, gave rise to the horoscopic art. It did not have the same effect in China, as we note in the next section.

THE THEORETICAL DEVELOPMENT

Two metaphysical principles underlie Chinese astrology as well as medicine, alchemy, and many other interrelated aspects of the Chinese intellectual framework: the yin-yang [F] and " five elements " principles. The yin-yang principle explains all phenomena in the universe in terms of a fundamental dichotomy which corresponds to that between heaven and earth, male and female, and so on. The five-elements principle was used to systematize the relations of things by placing them in the constellation of natural agents — wood, fire, earth, metal, and water. When and how these principles came into being still challenges the historian. One careful assessment has it that they were first formulated and put together during the early Warring States period (c. 4th century B.C.) and became established during the Former Han dynasty (2nd and 1st centuries B.C.).[18]

It is undeniable that these two principles were from their inception closely related to astronomical and cosmological thinking; according to Shinzō Shinjō, about 360 B.C. it was recognized that the presence of five and only five planets in the heavens points to the primacy of the number five over the others. The yin-yang dualism was elaborated into cosmological theories based on a celestial-terrestrial dichotomy.[19] The application

[17] Needham, op. cit., Vol. 3, pp. 398 ff.

[18] Nobuyuki Kobayashi,[AW] Chūgoku jōdai in'yō gogyō shisō no kenkyū [AX] (A Study of Yin-Yang and Five-Elements Thought in Ancient China) (Tokyo, 1956), pp. 9–13.

[19] Unlike the Aristotelian dichotomy, heaven (yang) and earth (yin) are not foreign to each other; the interaction between them provides an explanation of various phenomena in the universe.

of these conceptions specifically to astrological forecasting must have come early.[20]

In the " Treatise on Astrology " of the *Shih chi*, although much attention is given to the establishment of correlations on the basis of historical events, it is clear that the substance of the correlations was determined deductively. An analogy is drawn between the celestial realm and the Imperial court. Each asterism and star is considered analogous to a government bureau and its officials. On the basis of this analogy various interpretations were made. That the twenty-eight lunar mansions are divided into twelve groups and associated with twelve feudal states was mentioned earlier. The five planets are related to the seasons, directions, and so on, as shown in Table 1. On the basis of these and other analogies, the positions of planets against their background constellations, their relative positions, and changes in their magnitudes were interpreted and correlated with terrestrial events.

TABLE 1. Associations of the five planets

Planet	Direction	Season	Element
Jupiter	east	spring	wood
Mars	south	summer	fire
Venus	west	autumn	metal
Mercury	north	winter	water
Saturn	center	end of summer	earth

The notion of individual fate, *ming*,[G] is found very early in the history of Chinese thought. The problem of fate often entered the discussions of Han philosophers. Its close association with the calendar was established during the Han period and developed thereafter.[21] The five-elements principle played an indispensable role in fate calculation. It was combined in practice with the sexagenary cycle, the pairing of ten " stems " (*kan*)[H] and twelve " branches " (*chih*)[J] to count off days or years in groups of sixty, which began in the Former Han dynasty as part of the development of calendar making. The use of these symbols for fate calculation is not, however, found in any Han dynasty book. The earliest work in which they are so used — in which, that is, we can see traces of a system of prognostication based on elements independent of celestial motions as such — was, so far as we know, written by Kuan Lu [K] in the Three Kingdoms period (A.D. 221–265).[22] The further development of the art of fate calculation was a complex process. It was augmented by Taoist elements, particularly during the Six Dynasties period (A.D. 3rd to 6th centuries). It probably reached its zenith in the T'ang,[23] when Western cultural influence was prominent.

[20] Shinzō Shinjō,[AY] *Tōyō tenmongaku shi kenkyū* [AZ] (Researches in the History of Astronomy in the Far East) (Tokyo: Kōbundō, 1928), pp. 640–643.

[21] Tadashi Sugimoto,[BA] " Shin'isetsu no kigen oyobi hattatsu " [BB] (The Origin and Development of ch'an-wei Thought), *Shigaku*,[BC] 1934,

13:637–661.

[22] Chao Wei-pang, " The Chinese Science of Fate-Calculation," *Folklore Studies*, 1946, 5: 280–283. The author places the possible origin of this sort of prognostication in the final years of the Later Han (end of the 2nd century).

[23] See particularly *Li Hsu-chung ming-shu* [BD]

We must take care, however, not to confuse this Chinese art with the genethliacal astrology of the Hellenistic tradition. Although the basic idea of foretelling the individual's fate is the same, the method employed in each case is distinctly different; the Chinese art, from its outset, was not directly concerned with celestial motions as was Ptolemaic astrology. So far as the State was concerned, the goal of astronomy was the production of the official calendar; Chinese astrologers relied heavily on calendrical indications rather than directly upon astronomical computations or observations. Counting cycles based on planetary periodicities could be replaced with much simpler abstract cycles. The "stem-branch" system was in this sense purely mathematical.

Interpretation depended upon a combination of the five elements (with their auxiliary correlates, such as geographical directions and seasons) and the sexagenary calendrical indices of the year, month, day, and even hour of the individual's birth.[24] The direct influence of celestial phenomena could be dealt with only by complex and laborious calculation, as anyone acquainted with the writings of Ptolemy is only too aware of. If astrological calculations were to be drastically simplified for widespread use in a culture where advanced mathematical training was not common, the replacement of an intricate system of celestial periods by a simple numerical cycle is hardly to be wondered at. Thus, Chinese fate calculation is not astrology in any literal sense, but an application of calendrical (i.e., time-numbering) elements to mundane personal affairs.

Closely related to developments in calendrical science was the notion of lucky and unlucky days. As early as the Han dynasty, hemerology was practiced to determine the propitious moment for undertaking any act of daily life. Illustrations and notes in the margins of the official calendar, printed and distributed to the public after the tenth century, were used widely as guides for scheduling personal conduct.

Once the yin-yang and five-elements hypostases were taken to be metaphysically prior to such sensible phenomena as the five planets, nature could be understood more directly in terms of their much more basic and much simpler rhythms. But at the same time they were too abstract — more so than similar principles in Western philosophy — to be influenced in any way by the accumulation of empirical knowledge.

From then on traffic was one-way, from principle to explanation of particular phenomena. The later development of Chinese Naturphilosophie ran toward superstition and was increasingly divorced from the Chinese scientific tradition. When, during the Ch'ing period, news came that the earth was regarded in the Copernican theory as another planet, and that the discovery of Uranus had delivered yet another blow to the primacy of the number five, it did not occur to Confucian philosophers to examine seriously the bases of their cosmological thought; the planets are only one manifestation of reality.[25]

(Book of Fate Calculation of Li Hsu-chung; c. 8th century).

[24] Needham, op. cit., Vol. 2, pp. 358–359.

[25] Shigeru Nakayama, Senseijutsu, sono kagakushi jō no ichi (Astrology, Its Place

THE TRANSMISSION OF HOROSCOPIC ASTROLOGY

While portent astrology appears to be widely distributed in early times, the horoscopic art is unique; there is no sign of parallel developments outside the Mediterranean area. Traces of horoscopic astrology in China come late, and are clearly transmitted through the contact of cultures.

The first translation of foreign astrological knowledge was in the *Mo-teng-chieh ching* [L] of Chu Lu-yen [M] and Chih-ch'ien [N] (A.D. 230). This work introduces an astrology based on the *nakshatra*, the twenty-eight lunar mansions of India. There are in the Chinese version materials and concepts (such as the week) of Iranian or even Babylonian origin not found in the Sanskrit original, the *Śārdūlakarṇāvadāna sūtra*, which dates from before the middle of the second century.[26] All these influences played an increasing role in the T'ang period, when Sogdian astrology was especially popular, and Iranian and Tantric Buddhist elements contributed to the elaboration and spread of hemerology.[27] A number of sutras translated at this time were devoted to Indian methods of casting horoscopes. While these methods can hardly be considered an integral part of Buddhist theology, their transmission was not wholly adventitious. The fatalistic tendencies of Buddhism and its emphasis on spiritual enlightenment for the individual were certainly congenial to genethliacal astrology, just as the Confucian concern for socio-political institutions provided the ideological basis of portent astrology.

Perhaps the best known of the T'ang astrological sutras, the *Hsiu-yao ching*,[P] was translated from an Indian language by Pu-k'ung [Q] in 759. In it the Western horoscope is explicitly expounded, but it is superimposed on the *nakshatra* and the week. The Western zodiac is found even earlier in the *Ta-chi ching* [R] (late 6th century). The *Tu-li-yu-ssu ching*,[S] translated early in the ninth century, seems to be purely Greek or even early Islamic rather than Indian, so far as can be judged from extant passages in a Japanese horoscope entitled *Shukuyō unmei kanroku* [T] (1113) and the Chinese compendium *Hsing tsung* [U] (c. 14th century). These works are, however, general in nature, and ignore the problems of determining precise planetary positions. A treatise which could be used for actually computing horoscopes is the *Ch'i-yao jang tsai chueh* [V] (Formulas for Avoiding Calamities According to the Seven Luminaries; 9th century) which provides planetary ephemerides from 794 on. Kiyoshi Yabuuchi has noted that the motions are described in terms of synodic periods, clearly a concession to Chinese usage.[28]

The Sung scholar Wang Ying-lin [W] (1223–1296) has noted that the Fu-t'ien [X] calendrical treatise was based on Indian astronomical methods, but

in the History of Science) (Tokyo: Kino-kuniya, 1964), p. 57.

[26] Makoto Zenba,[BG] " Matoga kyō no tenmon rekisū ni tsuite "[BH] (On Astronomical and Calendrical Problems in the *Śārdūlakarṇāva-*

dāna sūtra), in *Tōyōgaku ronsō* [BI] (Orientalist Essays) (Kyoto: Heirakuji, 1952), pp. 171–213.

[27] Yabuuchi, *Shina no tenmongaku*, p. 143.

[28] *Chūgoku chūsei* . . . , p. 164.

no further details were known until very recently. According to the "Treatise on Calendrical Astronomy" in the Standard History of the Five Dynasties period (907–960), the Fu-t'ien treatise was an unofficial compilation, completed between 780 and 793 in China. Its calendrical epoch, instead of the usual Grand Conjunction ages earlier, was the time of conjunction of the sun and moon in yu-shui [Y] (equivalent to 330° longitude) in A.D. 660. This day (J.D. 1962169) was a Sunday, the religious significance of which was brought to T'ang China by Manicheans. The Tiao-yuan [Z] calendrical treatise, adopted for official use in 939, copied the Fu-t'ien treatise and the characteristics of its epoch.

We are now able to fill in this sketchy outline with the aid of Japanese works. It was recently proved that the Fu-t'ien calendar was brought to Japan in 957 by a Buddhist monk. So far we have found that two extant Japanese horoscopes, of 1112 and 1269 respectively (the earlier is reproduced in Fig. 1, and will be described below) were calculated by the Fu-t'ien method, since the epoch of calculation for both is yu-shui of A.D. 660.[29]

We have moreover discovered a fragmentary manuscript copy of the solar table of the Fu-t'ien calendar in the Tenri Library. By analyzing it, we have shown that it employed a parabolic function for expressing the solar equation of center as follows:

$$y = \frac{2}{3,300} \, x \, (182 - x),$$

where x is the solar mean anomaly and y the equation of center, both expressed in Chinese degrees (tu; in this case, for calculating convenience, 364 tu rather than the usual $365\frac{1}{4}$ tu equals 360°).

This expression resembles neither Indian trigonometric functions as translated in the Chiu-chih [AA] (Navagrāha) treatise (718)[30] nor the complicated Chinese interpolation formulae of the period.[31] Functions of this sort were often employed by Chinese astronomers, who lacked trigonometry, to express cyclic changes in such cases as parallactic elements in eclipse prediction, but this is the earliest known application of the function to the solar equation of center.

A similar expression was employed for the same purpose in the Uighur calendar, as reported in Islamic sources recently studied by Professor E. S. Kennedy:

$$y = \frac{2}{90,000} \, x \, (182 - x).$$

The solar equation of center y is here expressed in terms of the correction from mean to true syzygies, and hence should be multiplied by 12.19° (the

[29] Hiroyuki Momo,[BK] "Futenreki ni tsuite"[BL] (On the Fu-t'ien Calendrical Treatise), Kag. Ken., 1964, No. 71:118–119. The Fu-t'ien treatise was also employed at times in competition with the official Hsuan-ming [BM] treatise for the usual functions demanded of a Chinese luni-solar ephemeris, such as eclipse predictions.

[30] An English translation of the Chiu-chih treatise is available in Yabuuchi, Chūgoku chūsei . . . , pp. 493–538.

[31] Ibid., pp. 466–467. For details, see Yabuuchi, Zuitō rekihō shi no kenkyū [BN] (Researches in the History of Calendrical Science during the Sui and T'ang Periods) (Tokyo: Sanseidō, 1944), pp. 70–73 and 77–79.

difference between the mean daily motions of the sun and moon) in order
to obtain the angular equation of center.[32] Therefore, although the *Fu-t'ien*
calendar chronologically precedes this Uighur calendar, there is some possibility that a unique astronomical method originated in Central Asia.[33]

On the other hand, the horoscopes, like all others of Far Eastern origin,
unmistakably exhibit the influence of Indian and Hellenistic astrology. The
earliest Japanese nativity, earlier than any in China known to us, is reproduced as Figure 1. The innermost circle represents the twelve directions
marked out by the Chinese " branch " cycle. The next circle represents
the twelve signs of the zodiac; their associations with the seven luminaries
as illustrated in Ptolemy's *Tetrabiblos* are also indicated. The third circle
indicates the positions of the nine luminaries (the sun, the moon, the five

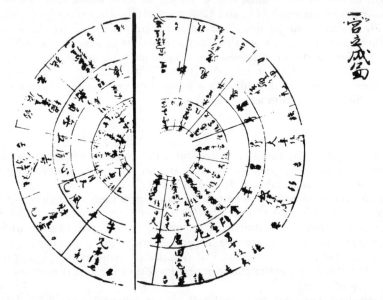

FIGURE 1. The earliest known Japanese horoscope, for the year 1112
(copied *c.* 17th century).

planets, and the two Indian pseudo-planetary nodes of the moon, Rāhu and
Ketu) at the time of birth. The fourth circle contains the twenty-eight
lunar mansions. The outermost circle is the twelve houses.

The most strikingly apparent difference between Japanese and Western
horoscopes lies in their shapes. While Japanese nativities are arranged in
a circle, those of the West (Indian as well as Arabic) are invariably square
or rectangular.[34] The earliest known Chinese horoscope, which may date

[32] E. S. Kennedy, "The Chinese-Uighur
Calendar as Described in the Islamic Sources,"
Isis, 1964, 55:435–443.
[33] Shigeru Nakayama, "Futenreki no tenmongaku shi teki ichi" [BP] (The Significance

of the *Fu-t'ien* Calendar in the History of
Astronomy), *Kag. Ken.*, 1964, No. 71:120–122.
[34] This was confirmed in personal communications from David Pingree on Indian horoscopes and E. S. Kennedy on those of Islam.

from the fourteenth century, is not quite circular, but a dodecagon.[35] At any rate, neither in China nor in Japan is any special preference given to the inner houses (I, IV, VII, and X). It appears that horoscopic diagrams were considerably modified in the course of their transmission through China to Japan.

By the time the Western horoscopic art was introduced into China, the Chinese had already developed their own techniques for determining the fates of individuals; it was welcomed only into the "underworld" of Tantrism and hemerology, and played no part in the development of astronomy.

APPENDIX. THE EQUIVALENT CHARACTERS FOR THE JAPANESE AND CHINESE WORDS APPEARING IN THE TEXT AND FOOTNOTES

A、史記　　　B、天文　　　C、漢書　　　D、天官書

E、律暦志　　F、陰陽　　　G、命　　　　H、干

J、支　　　　K、管輅　　　L、摩登伽經　M、竺律炎

N、支謙　　　P、宿曜經　　Q、不空　　　R、大集經

S、都利聿斯經　　　　　　　T、宿曜運命勘録

U、星宗　　　V、七曜攘災決　　　　　　W、王応麟

X、符天　　　Y、雨水　　　Z、調元　　　A A、九執

A B、井本進　　A C、長谷川一郎

A D、中国、朝鮮、及び日本の流星古記録　　A E、科学史研究

A F、藪内清　　A G、支那の天文学　　　A H、古今注

A J、後漢書集解　　　　　　A K、萬有文庫

A L、危　　　A M、景初　　A N、中国中世科学技術史の研究

A P、齋藤勵　　A Q、王朝時代の陰陽道　　A R、朱文鑫

A S、歴代日食考　　　　　　A T、鈴木敬信

A U、本邦古代の日食について

A V、日本天文学會要報　　A W、小林信行

A X、中国上代陰陽五行思想の研究　　　　A Y、新城新蔵

A Z、東洋天文学史研究　　B A、杉本忠

[35] Reproduced in Needham, Vol. 2, Plate XVII (facing p. 352).

ＢＢ、讖緯説の起原及び発達　　ＢＣ、史学　　　ＢＤ、李虚中命書

ＢＥ、中山茂　　ＢＦ、占星術、その科学史上の位置

ＢＧ、善波周　　ＢＨ、摩登伽經の天文暦数について

ＢＪ、東洋学論叢　　　　　　　ＢＫ、桃裕行

ＢＬ、符天暦について　　　　　ＢＭ、宣明

ＢＮ、隋唐暦法史の研究　　　　ＢＰ、符天暦の天文学史的位置

A Chinese version of the alchemical Great Work, portrayed as an array of correspondences. The adept is surrounded by correlates of yin and yang (the moon containing a hare pounding the elixir in a jade mortar; the sun containing a three-legged raven; tiger and dragon), the Five Phases, and the trigrams of the Book of Changes. Below the tiger in the circle is the elixir of immortality, and issuing from it the essences characteristic of the Five Phases and four of the visceral spheres of function within the body. From a text of internal alchemy (Chen-yuan miao tao hsiu tan li-yen ch'ao,[au] probably between the seventh and tenth centuries, in Yun chi ch'i ch'ien), 72:34a.

Chinese Alchemy and the Manipulation of Time

By N. Sivin*

TOWARD WHAT ENDS did the Chinese alchemists master chemical manipulations? On what principles did they shape their processes? Their goals, I claim, were not in any significant sense chemical—a point easily overlooked if we restrict our attention, as we usually do, to those isolated accomplishments that entitle the alchemist to credit in the light of modern chemistry.

Alchemy (wai tan)[a,1] was the art of making elixirs of immortality, perfected substances which brought about personal transcendence and eternal life—which, in other words, communicated their own perfection to human beings when ingested (or, as I will show, even without being ingested). Their power could also be applied to the "maturation" of metals into silver and gold, to the cure of sickness, and other matters of practical benefit. But almost every alchemical treatise used its manipulations as concrete metaphors for cosmic and spiritual processes.

Even so routine and preliminary a process as the extraction of silver from lead by cupellation was assimilated in many-layered imagery reflecting the separation of the positive and negative macrocosmic energies and the emergence of the perfected self when a human is metamorphosed into an immortal:

> Refine silver within lead
> And the spiritual being is born of itself.
> In the ash reservoir, melting in the flame,

Invited paper.

*Technology Studies, Massachusetts Institute of Technology, Cambridge, Massachusetts 02139.

This essay summarizes as concisely as possible parts of an extensive and considerably more technical study of the theoretical foundations of laboratory alchemy, completed in 1970 and due for publication c. 1978 in Joseph Needham, Ho Peng Yoke, Lu Gwei-djen, and N. Sivin, Science and Civilisation in China, Vol. V, Pt. 4 (Cambridge: Cambridge University Press). Many more examples and citations are provided there for each step in my argument, but this version incorporates five years' further reflection. I intend only to exhibit the general character of alchemy over the period in which it flourished (second to eleventh century A.D.). Because few texts can yet be dated precisely, I am not prepared to trace its origin or evolution without excessive resort to guesswork. The reader should be aware that given the enormous limitations of understanding and control of the literature, even a rough sketch of the sort I offer can have little value except to encourage further study. For additional orientation and bibliography, see Needham and Lu, Science and Civilisation, Vol. V, Pt. 2 (1974), and Sivin, Chinese Alchemy: Preliminary Studies (Harvard Monographs in the History of Science, 1) (Cambridge, Mass.: Harvard University Press, 1968).

All translations below are my own. I acknowledge with gratitude support from the National Science Foundation and the National Library of Medicine and hospitality from the Sinologisch Instituut, Leiden, Gonville and Caius College, Cambridge, and the Research Institute for Humanistic Studies, Kyoto.

[1] Superscript letters refer to the list of Chinese characters at the end of the article.

Lead sinks down, silver floats up.
Pristine white the Treasure appears
With which to make the Golden Sprout.[2]

A necessary first step toward understanding the character of Chinese alchemical thought is to look at how alchemists drew on concepts of natural change,
especially temporal change, that practically all Chinese thinkers shared.

TIME IN NATURAL PHILOSOPHY

Scientific thought began, in China as elsewhere, with attempts to comprehend
how it is that although individual things are constantly changing, always coming
to be and perishing, nature as a coherent order not only endures but remains
conformable to itself. In the West the earliest such attempts identified the
unchanging reality with some basic stuff out of which all the things around
us, despite their apparent diversity, are formed.

In China the earliest and in the long run the most influential scientific
explanations were in terms of time. They made sense of the momentary event
by fitting it into the cyclical rhythms of natural process. The life cycle of
every individual organism—its birth, growth, maturity, decay, and death—
followed essentially the same progression as those more general cycles which
went on eternally and in regular order: the cycle of day and night, which
regulated the changes of light and darkness, and the cycle of the year, which
regulated heat and cold, activity and quiescence, growth and stasis. It was
the nested and intermeshed cycles of the celestial bodies that governed the
seasonal rhythms and, through them, the vast symphony of individual life
courses.

This pattern held for minerals as well as for flora and fauna. The Chinese
shared what seems to have been a traditional belief among miners everywhere
that earths matured within the terrestrial womb. The metallurgist was merely
accelerating a natural progress of metals toward perfection—a perfection that,
except in the cases of silver and gold, nature's corrosion would eventually
undo.[3]

The concepts that molded this organismic world-view were above all functional
and relational. They specified what was significant about each thing (its "way"
or *tao*)[b] in terms of what it did in relation to what others did, its role in
the cosmic system (the greater *Tao*) rather than its static qualities. Yin and

---

[2] *Chin pi ching*[t] (Gold and Cerulean Jade Canon, c. A.D. 200?), cited in *Tan lun chueh chih
hsin chien*[u] (Heart-Mirror of Mnemonics and Explanations from Writings on the Elixir, probably
prior to the tenth century; in *Cheng-t'ung Tao tsang*[v] [hereafter TT], Vol. 598), p. 7a; excerpted
from an unpublished complete translation of the Heart-Mirror, based on a text edited from
three printed versions. Although this quotation refers to a chemical operation, the Heart-Mirror
contains a great deal of material on the physiological analogue of alchemy (*nei tan*),[w] which uses
alchemical language to describe the pursuit of immortality through breathing or meditative
disciplines. This "internal" (*nei*) form and "external" (*wai*) alchemy were so closely complementary,
and were so regularly practiced in conjunction before the eleventh century (when external alchemy
began to die out), that it is often impossible to distinguish which sort of operation a text is
concerned with. Their theoretical basis was largely identical. The Golden Sprout is an intermediary
in the preparation of the elixir.

[3] This view of the miner, the smith, and the metallurgist in all the major civilizations has been
documented by Mircea Eliade in *Forgerons et alchemistes* (Paris: Flammarion, 1956), translated by
Stephen Corrin as *The Forge and the Crucible* (New York: Harper, 1962); for more recent references
see Mircea Eliade, "The Forge and the Crucible: A Postscript," *History of Religions*, 1968, 8:74–88.

yang were the passive and active phases through which any cycle must pass. Although used analytically, to break down change into its parts, the aim of these concepts was as often as not synthetic, to explain the relation of parts in the whole. Yin and yang could also be used to distinguish aspects of configurations in space—down and up, back and front, inside and outside. Here too the *Oxford English Dictionary* sense 2 of "phase" ("any one aspect of a thing of varying aspects; a state or stage of change or development") is not out of place; yin and yang were almost always applied to spatial relations that are either changing or in dynamic equilibrium. The Five Phases (*wu hsing*,[c] often mistranslated as Five Elements) were a very closely analogous division of cycles or configurations into five functionally distinct parts (not, in scientific explanations, into five types of ultimate constituents). The phases Fire and Wood, for instance, made possible a finer analysis of the yang, or active, aspect of change. Wood is the name of the phase of growth and increase, and Fire is the maximal flourishing phase of activity, when the yang is about to begin declining and yin must once again reassert itself. In the cycle of the year Wood and Fire thus correspond to spring and summer respectively.[4] A third set of concepts of the same kind, much used in alchemy, were the eight trigrams and a set of twelve hexagrams from the Book of Changes, each defined with a function that played a due part in the preparation of elixirs.

All these concepts belong to the most general level of early Chinese thought about nature. The various fields of science, such as medicine and alchemy, simply applied them to different classes of phenomena, redefining and supplementing them with technical concepts as necessary.

THE NATURAL ELIXIR

Gold was an obvious endpoint for the evolution of minerals, for it was exempt from decay. In China a second line of development, more important in alchemy, led toward cinnabar (HgS). Chinese gave its large, translucent, nearly vitreous crystals, the color of fresh blood, a special place among semi-precious stones. Cinnabar was, like gold, associated with vitality and immortality from very early times. Alchemists associated it with the maximal phase of yang, Fire, and adapted *tan*[d] (cinnabar) as a term for elixirs whether or not mercuric sulfide was among their ingredients.[5]

The alchemist's elixir, although certainly the product of artifice, reproduced the work of nature. Here is a theoretical text, so far undatable but before

---

[4] The aspects of yang that correspond to Wood and Fire are conventionally distinguished as young or minor yang (*shao yang*)[x] and mature, old, or major yang (*t'ai yang*).[y] Metal and Water are the corresponding subphases of yin, and Earth is the neutral point of balance between yin and yang. I capitalize the names of the Five Phases to avoid confusion with ordinary wood, fire, and so on. The first philologically adequate explication of yin-yang and the Five Phases in any language is that of Manfred Porkert, *The Theoretical Foundations of Chinese Medicine* (MIT East Asian Science Series, 3) (Cambridge, Mass.: MIT Press, 1974), pp. 9–54. Porkert is concerned only with their use in certain aspects of medicine.

[5] A typical scheme of cinnabar evolution is given in *Huang ti chiu ting shen tan ching chueh*[z] (The Yellow Emperor's Canon of the Nine-Vessel Spiritual Elixir, with Explanations, Chs. 2–20 apparently of the late tenth century; TT584–585), 14:1a. Some early writers, e.g. those cited c. 320 by Ko Hung,[aa] make cinnabar a stage in the evolution of gold; see *Pao p'u tzu nei p'ien*[ab] (The Inner Chapters of the Philosopher Pao p'u tzu; *P'ing chin kuan ts'ung-shu*[ac] ed.), 16:5a, translated by James R. Ware in *Alchemy, Medicine, and Religion in the China of A.D. 320. The Nei P'ien of Ko Hung (Pao-p'u tzu)* (Cambridge, Mass.: MIT Press, 1966), p. 268.

A.D. 900, on the elixir that the cosmic order produces in its own time. It corresponds to the alchemist's two-ingredient process in which silver extracted from lead and mercury recovered from cinnabar are conjugated to bring about a perfect balance of yin and yang function. The natural elixir, unattainable by mere mortals, is called "cyclically transformed" because it is matured by the alternations of the seasons, which we will see the alchemist modeling by alternations of his fire:

> Natural cyclically transformed elixir is formed when flowing mercury, embracing Squire Metal[e] [i.e., lead], becomes pregnant. Wherever there is cinnabar there are also lead and silver. In 4320 years the elixir is finished. . . . It embraces the ch'i[f] of sun and moon, yin and yang, for 4320 years; thus, upon repletion of its own ch'i, it becomes a cyclically transformed elixir for immortals of the highest grade and for celestial beings. When in the world below lead and mercury are subjected to the alchemical process for purposes of immortality, [the elixir] is finished in one year. . . . What the alchemist prepares succeeds because of its correspondence on a scale of thousandths.[g,6]

The number 4320 implies the "scale of thousandths" that relates the natural and artificial elixirs. There were twelve "hours" (shih)[h] in a Chinese day and 4320 in a round year of 360 days. An hour in the laboratory thus recapitulated a year in the terrestrial matrix.

Just as the diversity in color and other physical properties of native gold suggested that there could be red and purple varieties which excelled the normal, cinnabar was believed to constitute a range of substances. The rarest and most evolved kinds (again unattainable by mortals) were in effect natural elixirs. Ch'en Shao-wei's[i] great monograph Arcane Teachings on the Alchemical Preparation of Numinous Cinnabar (written perhaps c. 712) first discusses the best cinnabar at the alchemist's disposal (similar to that shown in Fig. 1), then its eventual maturation into an elixir, and finally the endpoint of cinnabaric evolution which the alchemist tries to approximate:

> Now the highest grade, lustrous cinnabar,[j] occurs in the mountains of Ch'en-chou and Chin-chou [in modern Hunan] upon beds of white toothy mineral. Twelve pieces of cinnabar make up one throne. Its color is like that of an unopened red lotus blossom, and its luster is as dazzling as the sun. There are also thrones of 9, 7, 5, or 3 pieces or of one piece. . . . In the center of each throne is a large pearl [of cinnabar], ten ounces (liang)[k] or so in weight, which is the "monarch." Around it are smaller ones . . . ; they are the "ministers." They surround and do obeisance to the great one in the center. About the throne are a peck (tou)[l] or two of various kinds of cinnabar, encircling the "jade throne and cinnabar bed." . . . If lustrous cinnabar is further taken in the sevenfold-recycled or ninefold-cyclically-transformed state, then without ado the yin soul is transformed and the outer body destroyed, the spirit made harmonious and the constitution purified. The yin ch'i [i.e., the material configuration organized by energy of yin type] dissolves and the persona floats up, maintaining its shape, to spend eternity as a flying immortal of the highest grade of realization . . . .

---

[6] Tan lun chueh chih hsin chien, p. 12b. Ch'i refers to the active energy (in the colloquial, purely qualitative sense of the word) that organizes matter into configurations, causing change, or that maintains the organization of configurations and thus resists change. The word is also applied to configurations of matter so organized. Such configurations are generally defined by their functions rather than by their constituents. See Porkert, Theoretical Foundations, pp. 167–168. The earliest senses of the term are surveyed in Kuroda Genji,[ad] "Ki," Tōhō shūkyō, 1953, No.3:1–40; 1955, No.7:16–44.

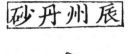

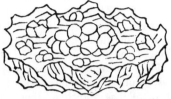

Figure 1. "Throne" formations of large cinnabar crystals, growing on beds of what is apparently drusy quartz. From a 1957 Peking reproduction of the Chinese Materia Medica printed in 1249 (Ch'ung hsiu Cheng-ho ching shih cheng lei pei yung pen-ts'ao),[ar] 3:2b.

After imbibing the active energy of the cosmos for twenty-two thousand years, lustrous cinnabar is spontaneously

> . . . transformed into celestial throne cinnabar, in which the throne is jade-green. There are nine pieces in the center, growing in layers, pressed closely about by 72 [smaller] pieces. It floats in the midst of the Grand Void, constantly watched over by one of the spirits of the Supreme Unity. On a Superior Epoch day[m] [the full moon of the first lunar month] the Immortal Officials descend to collect it. The mountain [on which it is found] suddenly lights up as if illuminated by fire. This celestial throne cinnabar is collected [only] by Immortal Officials; people of the world can have no opportunity to gather it.[7]

ALCHEMICAL PROCESSES

The alchemist's simulation of the hidden operations of nature was founded upon a much more ancient simulation, that which made possible the accomplishments of the metalworking artisan. In extracting a metal from its ore or in making strong steel from brittle cast iron, the metallurgist was demonstrating that humans can imitate natural processes, can stand in the place of nature and bring about natural changes at a rate immensely faster than nature's own time. People in traditional societies did not know of any way to speed up the growth cycle of plants or animals. Only the life rhythms of minerals could be manipulated.

Upon the craftsman's mastery of metalline growth alchemists built processes to fit their own purpose. They constructed in their laboratories working models of the cosmic order as manifested in time. They not only shrank the dimensions of the universe to fit their four walls but also compressed time to make the

---

[7] Ta tung lien chen pao ching hsiu fu ling sha miao chueh,[ae] (TT586), pp. 5b–7b; excerpted from a complete draft translation based on collation of two printed texts. The first six characters of the title indicate that this book (like the one cited in n. 11) provides supplementary instructions for the process given in the Great Void Canon on Making the Realized Treasure, an otherwise unknown work. "Realization" and "realized" throughout this essay are my rather literal translations of chen,[af] which denotes the authenticity and perfection attained by the elixir, or by the adept who attains a high grade of immortality. The text mentions Immortal Officials and grades because immortals were given appointments in the hierarchy of the gods, bureaucratically organized like the imperial government.

duration of their manipulations feasible. Thus through point-by-point correspondence the artificial circumstances of the laboratory were made profoundly
natural and responsive to the operation of the cosmic *Tao*, to the rhythms
of the great order outside: "When earth mixes with water to form mud, and
is kneaded [by subterranean forces] below a mountain, gold will be formed,
and generally cinnabar above it. When this [cinnabar] is ceaselessly metamorphosed and cyclically treated [by the alchemist], and once again forms gold,
this is merely a reversion to the root substance, and not something to be
wondered at."[8] How could the alchemists be sure that what went on within
their reaction vessels was identical with the work of nature? If they erred
they would be overruled by the *Tao*, and their elixirs would wither.

They depended upon numerological correlations, correspondences, and
resonances—in a word, upon analogy.[9] There were three points at which
analogies could be applied to cosmic process: materials, apparatus and other
aspects of spatial arrangement, and control of time and temperature in
combustion. Space permits examination only of the last, but it is important
to keep in mind that all were used in conjunction.

FIRE PHASING

Since the heat of the flame stood for the active cosmic forces, to recreate
those forces in the laboratory required that fire be bound by time. "Fire phasing"
(*huo hou*)[n] was the gradual increase and decrease of fire intensity by using
precisely weighed increments of fuel. It is true that a constant increment in
the weight of charcoal burnt in the furnace does not cause a constant increase
in temperature. But it was well known by craftsmen that each weight of fuel
burnt in the same way yields a fixed heat and a predictable product. What
the alchemists did was to make this concept of fire phasing dynamic, varying
the weight of fuel in a regular way over a very long period of heating. They
were bringing their processes under the control of one of the few exact measuring
instruments at their disposal, the balance. The profile of the flame's intensity
was thus governed by correspondences that tied it to the seasonal cycles. An
anonymous early author calls into play yin and yang phases and other divisions
of the annual cycle and cites two of the most canonic sources of alchemical
theory: "The amounts of fuel to be weighed out are increased and decreased
in cyclical progression according to the proper order of yin and yang. They
must conform with the signs of the Book of Changes and the Threefold
Concordance, tally with the 4, 8, 24, and 72 seasonal divisions of the year,
and agree with the implicit correspondences and *ch'i* phases of the year, month,
day and hour—all without a jot or tittle of divergence."[10]

---

[8] *Huang ti chiu ting shen tan ching chueh*, 13:2a.

[9] On correlative thinking see Joseph Needham and Wang Ling, *Science and Civilisation in China*,
Vol. II (1956), pp. 279–291 and elsewhere. The most important source on numerology in China
is Marcel Granet, *La pensée chinoise* (L'évolution de l'humanité, 5.3) (Paris: La Renaissance du
Livre, 1934), pp. 149–299.

[10] Cited in *Chu chia shen p'in tan fa*[ag] (Wonderful Elixir Formulas of the Masters, tenth century
or later; *TT* 594), 4:1b. On this collection, see Sivin, *Chinese Alchemy*, pp. 75–76. The Threefold
Concordance is the *Chou i ts'an t'ung ch'i*,[ah] traditionally dated c. A.D. 140, which would make
it the earliest extant alchemical text. Even though the present version is probably much later,
its putative antiquity gave it great status in the eyes of later writers. See *Chinese Alchemy*, pp.
36–40. There is a very rough English translation, Wu Lu-ch'iang, "An Ancient Chinese Treatise

Figure 2. Furnaces similar to the one described by Ch'en
Shao-wei, from Wu Wu,[as] Tan fang hsu chih[at] (Indispensable
Knowledge for the Alchemical Laboratory, 1163; in TT 588). Ch'en
specified nine openings in the top level, twelve in the middle,
and eight in the lower. His furnace was fired only through the
median doors, each of which was equipped with tuyeres. In Wu's
book these structures were no longer used as furnaces; they
merely supported a small stove which enclosed the reaction
vessel.

Consider a single fire-phasing scheme, from Arcane Teachings on the Ninefold
Cyclically Transformed Gold Elixir (c. 712?), a companion work to the book
by Ch'en Shao-wei cited earlier. Ch'en repeats a sixty-day cycle six times over
a round year to make his elixir. The furnace is built in three tiers (see Fig.
2), which correspond to sky and earth and man centered between them. The
central tier has twelve doors, which correspond to hours in the day and months
in the year. Ch'en fires through each door for five days to complete a cycle.

> As for the time of firing the furnace, the fire should be applied at a midnight
> which is also the first hour of a sixty-hour cycle, on the first day of a sixty-day
> cycle, in the eleventh month [i.e., the month which contains the winter solstice].
> Begin by firing through door A for 5 days, using 3 oz of charcoal. There must
> always be 3 oz of well-coked charcoal, neither more nor less, in the furnace. Then
> open door B and start the fire, firing for 5 days, using 4 oz of charcoal. Then
> open door C and start the fire, firing for 5 days, using 5 oz of charcoal.

This scheme proceeds until 8 oz of charcoal have been fed through the sixth
door for 5 days. Then the alchemist is told to move on to what Ch'en calls
the yin doors, reducing the fuel by 1 oz per day at each door, until at the
twelfth the charge has been reduced to 4 oz. The cycle is not quite closed,
for the weight of fuel has not been reduced to the original amount. The
second sixty-day cycle begins with 5 oz of fuel and goes up to 10 oz. The
third cycle begins with 7 oz and goes up to 12 oz. The sixth cycle, which
completes the elixir, begins at 17 oz, rises to 22 oz, and falls again to 16
oz.[11]

---

on Alchemy Entitled 'Ts'an T'ung Ch'i,' Written by Wei Po-Yang about 142 A.D. with an Introduction
and Notes by Tenney L. Davis," Isis, 1932, 18:210–289.

  "Ch'i phases" are phases of the regular temporal variations in energy and activity (see n. 6).
In other words, the cycles which yin-yang and the Five Phases analyze are cycles of ch'i, whether
manifested as heat, light, or some other factor that brings about change or maintains a configuration.

  [11] Ta tung lien chen pao ching chiu huan chin tan miao chueh[ai] (TT 586), pp. 12a–13a, excerpted
from a complete translation based on two texts. The first sentence quoted places the epoch of
the process at the moment when calendrical cycles begin. For certain purposes connected with
imperial ritual—including astronomical computation—the year was considered to begin in the
eleventh month, the month which contained the winter solstice. See N. Sivin, "Cosmos and

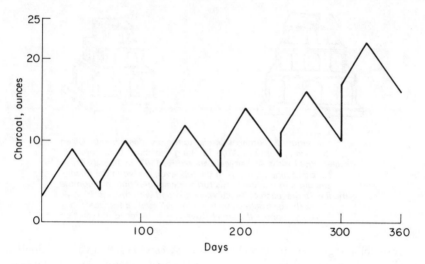

Figure 3. The cyclical pattern of precisely controlled "fire phasing."

As graphed in Figure 3, the zigzag time-weight profile represents most reasonably the notion of a change in the alchemical ingredients at once cyclic and progressive. If, mentally, we were to superimpose the six cycles, which in principle repeat each other, it would be very tempting to call this a helical fire-phasing schema.

Some fire-phasing schemata are a good deal simpler than this one, and some are more elaborate—for instance, a thirteenth-century "two-variable" procedure in which the weight of charcoal below the reaction vessel and that of water in a cooling apparatus in its upper part are varied regularly in such a way that the sum of this yin-yang pair remains constant.[12]

THE SIGNIFICANCE OF FIRE PHASING

From the documents I have quoted and many others like them, one can see that number and measure were being used in a way not much like their

---

Computation in Early Chinese Mathematical Astronomy," *T'oung Pao* (Leiden), 1969, *55*:10–11. There is an excellent discussion of furnaces and other apparatus in Ho Ping-yü and Joseph Needham, "The Laboratory Equipment of the Early Mediaeval Chinese Alchemists," *Ambix*, 1959, 7:57–115.

[12] A system in which the increase in weight of charcoal is linear and repeated in each cycle is given in *T'ai shang pa ching ssu jui tzu chiang wu chu chiang sheng shen tan fang*[aj] (Method for Making the Eight-Radiance Four-Stamens Purple-Fluid Five-Pearl Incarnate Numinous Elixir, a Most High Scripture, probably late fourth century; in *Yun chi ch'i ch'ien*,[ak] The Seven Bamboo Tablets of the Cloudy Satchel, c. 1023; *TT*691), 68:4b–5a. The two-variable system appears in *Chin hua ch'ung pi tan ching pi chih*[al] (Confidential Instructions on the Canon of the Heaven-Piercing [?] Golden Flower Elixir, preface dated 1225; *TT*592), 2:20a–21b. Although Meng Hsun,[am] who recorded this treatise, claims that he carried out the process, its great complexity suggests a thought experiment. On the origins of this treatise, see Needham and Lu, *Science and Civilisation*, Vol. V, Pt. 2, p. 316. Note that Needham and Lu translate the last word of the title as "manual." It is true that although *ching* usually designated books of philosophic or religious doctrine, it

use in modern science, where time-temperature specifications also control chemical preparations. What makes alchemical operations resemble those of modern chemistry is that the quantities that controlled the process were derived, by way of theory, from observations and measurements. While they served as models of cosmic change, the alchemical processes also had to transform one substance into another, and so at some point certain chemical and physical conditions had to be satisfied. Ch'en Shao-wei was well aware, for instance, that 1 lb of cinnabar would yield 14 oz of mercury (a figure he almost certainly learned from mercury smelters).

Alchemy and modern chemistry differ in the ways number served theory, as well as in the links between theory and the observations on which it was based. The foundation on which chemical theories rest is essentially mathematical; Chinese alchemical theories were essentially numerological. The latter used numbers, not as measures, but as a means of ranking phenomena into a qualitative order, an order that reflects their subjective values. It is not at all unusual for measured quantities to be combined with symbolic numbers in such schemes, nor for the results of numerological manipulations to be treated as though they were quantitative. (These complexities may be seen in the grading system prevalent in American universities, the most influential survival of numerology in modern society.)

Fire phasing was one of many means of constructing in the laboratory a system analogous part by part to another system—the cosmic order. Alchemists drew upon all the numerological and emblematic correspondences available to describe the activity of the cosmos, especially those correspondences related to the Five Phases, yin-yang, and the symbols of the Book of Changes. But because these concepts had already been extended to so many areas of thought, their connotations were too rich and diverse to be entirely integrated into alchemical theory (or any other sort of early scientific theory). They were quite adequate for some kinds of explanation, but not for others. By weaving a new fabric of associations to supplement conventional natural concepts adapted to their own concerns, the alchemists were in effect picking and choosing what meanings would be included in the overtones of their discussions. Fire phasing was a way of creating special structures embodying number—patterns of times and weights—which conveyed no more and no less than what alchemists wanted to teach their disciples.

SPATIAL AND MATERIAL CORRELATIONS

The alchemist's model could be complete only if it were aligned with the universal order in space as well as time. In spatial correlations an enormous variety of verbal and numerological associations, mostly analogous in kind to those I have already mentioned, were called into play. The laboratory was oriented to the cardinal points of the compass, the furnace centered in it, and the reaction vessel centered in the furnace to make it the axial point of change. The designs of furnaces and vessels were precisely specified to

was occasionally applied to what moderns would consider technical manuals; but this was done precisely to assert their status as canons of craft traditions, and there is no reason not to translate *ching* literally.

make them concrete metaphors for sky and earth, with the work of man centered between the two and in perfect concord with both.

The more theoretically inclined alchemists tended to use processes in which two ingredients represented the alternating conjugation and separation of yin and yang. One of the two most usual processes began with cinnabar and lead, representing yang and yin, from which were extracted mercury, representing yin emergent from yang, and silver, representing yang emergent from yin. The other combined mercury and sulfur to form vermilion (artificial cinnabar, finely divided), which could be repeatedly decomposed and resynthesized. As Ch'en Ta-shih° (Sung period) put it, "That cinnabar should come out of mercury and again be killed by mercury: this is the mystery within the mystery."[13]

Despite the many types of modeling that it incorporated, alchemy was not model-making for its own sake, nor was it the pursuit of chemical knowledge for its own sake, nor even an alternative way of doing natural philosophy. It had its own goals, and we are now ready to look at them.

THE GOALS OF ALCHEMY

I have already mentioned that in a certain sense alchemy grew out of the metalworking artisan's ability to accelerate mineral maturation. The alchemist's discovery was that the life courses of minerals could be accelerated by man in the interest not only of metal production but of understanding and even of wisdom. No one could wait 4,320 years to experience nature's production of an elixir, and in any case her womb was not open to contemplation. An alchemist who set out to fabricate an elixir in a few months or a year was creating an opportunity to witness the cyclical sweep of universal change.

The alchemists' enterprise, as they themselves defined it, was not chemistry. They were not primarily motivated by curiosity about the properties and reactions of specific substances (which they generally learned about from craftsmen). Those properties and reactions fascinated some alchemists, but they were no more intrinsically important than the characteristics of pigments that painters must know about in order to set down images. In fact the response of alchemists to their materials was at least as esthetic as it was manipulative. "Charming" $(k'o\text{-}ai)^p$ is one of the most common adjectives in alchemical writing for mineral specimens.

Chemical knowledge and concepts of a chemical kind, in other words, were to some extent indispensable means and to some extent byproducts (as were physical relations and ritual patterns). Alchemists recorded a great deal of their chemical knowledge as well as their attempts to make sense of it. Their documents are chemically far richer and more explicit than those of the West before the Renaissance.[14] But the aims of Chinese alchemy were of a different sort than those of modern science.

---

[13] *Pi yü chu sha han lin yü shu kuei*[an] (On the Cerulean Jade and Cinnabar Jade-Tree-In-a-Cold-Forest Process, eleventh to fourteenth century?; *TT*587), p. 1a.

[14] The chemical knowledge reflected in alchemical practice is being surveyed systematically in Needham and Lu, *Science and Civilisation*, Vol. V, Pts. 2 (1974) and 3 (1976), and I discuss chemical aspects of theory in Vol. V, Pt. 4 (see n. *). Chinese alchemy, no less than the traditions of other civilizations, depended on recondite language and symbolic images (see the Frontispiece) both to multiply levels of meaning and to exclude people unprepared spiritually and morally to use its teachings responsibly. But the degree of mystification has been badly overestimated

The old books reflect many different aims and many kinds of curiosity. Some interests were on the whole pragmatic; it was the product, the elixir, that mattered, and one way of making it was in principle as good as another. The emphasis was on benefits elixirs yielded: health, wealth, above all immortality. This was especially true of certain medical men such as Sun Ssu-mo[q] (fl. 673), who described his alchemical processes in much the same way as his techniques for preparing drugs.[15] But the aim that runs through most of the roughly one hundred treatises on laboratory alchemy that still exist,[16] the aim that makes them a coherent literature, that conditioned every step in the design of the processes, was to construct a model of the *Tao*, to reproduce in a limited space on a shortened time scale the cyclical energetics of the cosmos.

This goal values contemplating the process over using the product. In the Threefold Concordance, the most influential of the early alchemical canons, the product was practically ignored. There are no instructions for compounding or ingestion. There are a mere couple of cursory descriptions of that immortal beatitude which to pragmatic alchemists was the highest reward of their art.[17] Among the posterity of the Threefold Concordance we find such a concern with gnostic rapture over the process that the steps between understanding and becoming an immortal are occasionally skipped altogether. This gap is apparent in the Most Ancient Canon of Earth and *Tui* (between sixth and ninth centuries?): "If [the devotee] attains a clear and penetrating understanding of these Five Phases, one can proceed to a discussion of fire-subduing, and can then talk to him about the *tao* [i.e., the art] of projection. When he has comprehended every aspect of the Five Phases, he will be a man of balanced Realization, and the Three Worms[r] will leave his body."[18] In other words, one can understand the preparation of elixirs by rendering volatile minerals invulnerable to the fire, and the use of these elixirs in transforming lesser substances into gold, only after comprehending exactly how the Five-Phases theory governs these changes. The adept who thus has mastered the alchemical process is invulnerable to the Three Corpseworms (demonic presences who

---

by a few historians who, because they could not trouble themselves to understand the conceptual language of yin-yang and the Five Phases, assumed that it was also occult.

[15] His *T'ai ch'ing tan ching yao chueh*[ao] (Essential Formulas from the Alchemical Classics, a Most Pure Scripture; in *Yun chi ch'i ch'ien*, Ch. 71) was edited and translated with commentary in Sivin, *Chinese Alchemy*.

[16] Needham and Lu, *Science and Civilisation*, Vol. V, Pt. 2, Bibliography A and Concordance for *Tao tsang* Books and Tractates, lists over two hundred alchemical texts. One can only estimate how many are concerned with external alchemy (see n. 2). For a list of thirteen published translations, see Sivin, *Chinese Alchemy*, pp. 322–324.

[17] The extent to which the Threefold Concordance (see n. 10) was originally concerned with laboratory procedures remains problematic. It contains elements of external, internal, and sexual alchemy (i.e., disciplines for immortality through the most literal union of yin and yang), but was interpreted by external alchemists as primarily concerned with their discipline. See R. H. Van Gulik, *Sexual Life in Ancient China. A Preliminary Survey of Chinese Sex and Society from ca. 1500 B.C. till 1644 A.D.* (Leiden: E. J. Brill, 1961), pp. 80–84.

[18] *T'ai ku t'u tui ching*[av] (*TT*600), 1:4b. *Tu* refers to the phase Earth (in which opposed tendencies reach balance), and *tui* to one of the eight trigrams of the Book of Changes, but there is no indication in the text of what association of the latter the author meant to invoke. "Balanced Realization" describes the perfect complementarity of yin and yang energies that characterizes immortals; see n. 7.

normally live in the body and are responsible for natural death) and is therefore immortal.

The content, tone, and balance of the evidence strongly suggest that the dominant goal of Chinese alchemy was contemplative, and even ecstatic. A Taoist book of the early sixth century has this to say about the use of a small amount of elixir to turn lead to gold:

> You may also first place the lead in a vessel, heat it until it is liquefied, and then add one spatula of the Scarlet Medicine [i.e., the elixir] to the vessel. As you look on, you will see every color flying and flowering, purple clouds reflecting at random, luxuriant as the colors of Nature—it will be as though you were gazing upward at a gathering of sunlit clouds. It is called Purple Gold, and it is a marvel of the *Tao*.[19]

This is a splendid description of what a metallurgist sees on a lead button as it oxidizes and the oxide is moved by surface tension. The richness and vividness of the details suggest a state of heightened awareness. That is perhaps not surprising, since meditative practices were normally part of the alchemist's discipline. If Purple Gold was that fugitive iridescence on the surface of molten lead, as the text clearly says it was, it was not a gold for selling or spending, only for seeing and wondering.

The alchemists constructed their intricate art, made the cycles of the cosmic process accessible, and undertook to contemplate them because they believed that to encompass the *Tao* with their minds—or, as they put it, with their hearts and minds (comprised in one word, *hsin*)[s]—would make them one with it. That belief (by no means confined to Taoists) was not at all unlike one of the central convictions that underlay the birth of physics in the West before the time of Plato: the idea that to grasp the unchanging reality that underlies the chaos of experience is to rise above that chaos, to be freed at least for the moment from the limits of personal mortality.

CONCLUSION

In the great metaphysical and religious significance given to alchemical processes the Chinese art was remarkably similar to that of the Hellenistic tradition, although the dominant metaphors were constructed out of very different world-views and beliefs. The idea of immortality did not enter European alchemy until the twelfth century and may have been received from China (predominantly via Islam), but it would be arbitrary to deny that the tradition of Zosimos was true alchemy because it lacked this notion.[20] The Stoic and Gnostic scheme of spiritual death and rebirth played much the same role

---

[19] *T'ai ch'ing chin yeh* (or *i*) *shen tan ching*[ap] (Classic of Liquefied Gold and Divine Elixir, a Most Pure Scripture; in *Yün chi ch'i ch'ien*), 65:15b. A more extended version appears in *TT*582. The part of the text cited almost certainly dates from the early fifth century. See Henri Maspero, "Une texte taoïste sur l'Orient Romain," pp. 97–98 in *Études historiques (Mélanges posthumes sur les religions et l'histoire de la Chine*, 3) (Paris: Civilisations du Sud, 1950).

[20] Needham has taken the position that for this reason "the Hellenistic proto-chemists ought not to be called 'alchemists'" (*Science and Civilisation*, Vol. V, Pt. 2, p. 12); for early criticism see the review by Robert P. Multhauf in *Ambix*, 1975, 22:218–220. A view of Alexandrian alchemy broadly similar to my understanding of the Chinese tradition is taken by Allen G. Debus, e.g., in "The Significance of the History of Early Chemistry," *Journal of World History*, 1965, 9:39–58, and by H. J. Sheppard in "The Origin of the Gnostic-Alchemical Relationship," *Scientia*, 1962, 97:146–149 and other writings.

in thought about transcendence as the Chinese belief that the adept's perishable body would be left behind as a lifeless husk when his new, imperishable self (embodying his old but "purified" personality) was ready to hatch out and join the ranks of the immortals. The religious depth as well as the chemical ingenuity of alchemists at both ends of Asia deserves serious consideration. There is evidence that the arts of both civilizations were means toward self-perfection through disciplined contemplation of manual operations (sometimes the patterns of manipulation were so idealized that they could not possibly have been carried out). It misses the point to assume that either tradition was dedicated to the search for abstract knowledge.

In China the operative alchemy of the laboratory, no less than the physiological and introspective disciplines that borrowed its language and symbols, was a form for self-cultivation, a means toward transcendence.

> Necessary that the maturing come within man,
> Due to the maturing of his heart and mind.
> If heart and mind have reached divinity, so will
>   the Medicine;
> If heart and mind are confused the Medicine will be
>   unpredictable.
> The Perfect *Tao* is a perfect emptying of heart
>   and mind.
> Within the darkness, unknowable wonders.
> When the wise man has attained the August Source
> In time he will truly reach the clouds.[21]

CHINESE CHARACTERS

a. 外丹
b. 道
c. 五行
d. 丹
e. 金公 = 鉛
f. 氣
g. 象而成之大千之數
h. 時
i. 陳少微
j. 光明砂
k. 兩
l. 斗
m. 上元
n. 火侯

o. 陳大師
p. 可愛
q. 孫思邈
r. 三尸
s. 心
t. 金碧經
u. 丹論訣旨心鑑
v. 政統道藏
w. 內丹
x. 少陽
y. 太陽
z. 黃帝九鼎神丹經訣
aa. 葛洪
ab. 抱朴子內篇

---

[21] From the Arcane Memorandum of the Red Pine Master (*Ch'ih sung tzu hsuan chi*,[aq] probably T'ang or earlier), cited in *Tan lun chueh chih hsin chien* (see n. 6), p. 14a. "Medicine" is a conventional synonym for the elixir.

ac. 平津館叢書

ad. 黑田源次

ae. 大洞鍊真寶經修伏靈砂妙訣

af. 真

ag. 諸家神品丹法

ah. 周易參同契

ai. 九還金丹妙訣

aj. 太上八景四蘂紫漿五珠降生神丹方

ak. 雲笈七籤

al. 金華沖碧丹經祕旨

am. 孟煦

an. 碧玉朱砂寒林玉樹匱

ao. 太清丹經要訣

ap. 太清金液神丹經

aq. 赤松子玄記

ar. 重修政和經史證類備用本草

as. 吳悮

at. 丹房須知

au. 真元妙道修丹歷驗抄

av. 太古土兌經

Technology

ANSWER TO QUERY NO. 105 (*Isis*, 35, 177). NOTE ON A FEW EARLY CHINESE BOMBARDS.[1]

On October 5, 1931, the undersigned visited the Shansi Provincial Museum while on an excursion to Shansi province with Mr. CARL W. BISHOP. In my diary for that day I wrote of seeing "iron cannon with an inscription dated Ming Hung-wu 丁巳, or 1377." Examination of any table of comparative dates will show that this is the tenth year, and that Mr. BISHOP erred in placing it one year later. I took two photographs of the cannon (reproduced herewith), one of which was used by THOMAS T. READ in his paper, "The early casting of iron," *The Geog. Review* XXIV, 4, Oct. 1934, 548. It is to be regretted that no rubbing of the inscription is available. It should be noted that there used to be other smaller cannon preserved in this same museum, in the Peking Historical Museum, and in the Tsinan (Shantung) Provincial Library, but no dates have been found on any of them. According to KURODA GENJI 黒田源次, the cannon in the Peking Museum number several hundred, some coming from the period of Hung-wu (1368–98) and others from the ensuing periods of Yung-lo through Chêng-t'ung (1403–1449), all made of copper. (See *Man-chou hsüeh-pao* 滿洲學報 4[1935].[2])

The only cannon of earlier date reported are from a large find of 500 of assorted sizes said to have been manufactured in the years of the anti-Mongol rebel, CHANG SHIH-CH'ÊNG (d. 1367),[3]

who set himself up as emperor in 1354–1357 under the reign title of CHOU T'IEN-YU 周天祐.[4] These were unearthed at Nanking in the Hsien-fêng period (1851–1861), and two at least were given a home in the museum of Nan-t'ung, Kiangsu. Both of these are made of iron, weigh respectively 500 and 350 catties (approximately 666 and 466 lbs. avoirdupois), and were reportedly made in the years Chou 3 and 4, or 1356 and 1357. They are illustrated in an account of this rebel entitled *Wu wang Chang Shih-ch'êng tsai-chi* 吳王張士誠載紀, *ts'ê* 1 (publ. 1932) by HAN KUO-CHÜN 韓國鈞.[5] It is only right to add that CHANG WÊN-HU 張文虎 (1808–1885), who does not doubt that they come from the time of CHANG SHIH-CH'ÊNG, raises one or two pertinent questions concerning these weapons. Why, he writes (*Shu-i shih shih ts'un* 舒藝室詩存 2/22a) were they discovered in Nanking, which was not the territory of CHANG SHIH-CH'ÊNG: why too are they inscribed merely Chou such and such a year and month, and not T'ien-yu? Oh well, he concludes, what does it matter.

Iron cannon of a somewhat later date are illustrated in *Dai nihon shi kōza* 大日本史講座 ch. 5, by NAGANUMA KENKAI 長沼賢海 (Tokyo, 1929). The one made in China is said to be dated Ming Hsüan-tê 1.11, or the end of 1426, and that made in Korea may even, it is thought, be of earlier manufacture. We know from *Koryu sa* 高麗史 81/646–647 that a general bureau of gunpowder artillery (?) 火桶都監 was set up in Korea in 1377, and we learn that the gunpowder 火藥 in this bureau and the weapons in the general bureau of defense were both formally inspected in 1381.

L. CARRINGTON GOODRICH

---

[1] The Chinese type for this note was kindly loaned to *Isis* from the fonts of the Harvard-Yenching Institute.

[2] This illustrated article, entitled *Shên chi huo-p'ao lun* 神機火炮論, is resumed in the *Shih hsüeh hsiao-hsi* 史學消息 1:4, 24, March 1, 1937. I have not seen the original, and no copy appears to have reached this country.

[3] See GILES, Chinese Biographical Dictionary No. 103.

[4] Correct TCHANG, *Synchronismes Chinois* 409.

[5] I owe these two references to Mr. CH'ÊN HUNG-SHUN, of the Columbia University Libraries.

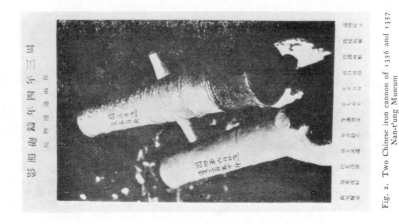

Fig. 2. Two Chinese iron cannon of 1356 and 1357
Nan-t'ung Museum

Fig. 1. Two illustrations of Chinese iron bombard of 1377

126

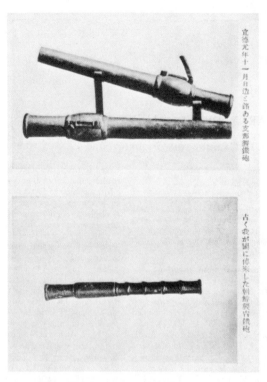

宣德元年十一月日造之銘ある支那製鐵砲

古く我が國に傳來した朝鮮製古鐵砲

Fig. 3. Top: Iron cannon made in China, dated 1426
Bottom: Iron cannon made in Korea, earlier (?)

THE EARLY DEVELOPMENT OF FIREARMS IN CHINA

By L. Carrington Goodrich and Fêng Chia-shêng

"The iron explosive" 鐵礮行

The black dragon lays eggs big as a peck.
Crack goes the egg, the dragon soars away, the spirit of
thunder departs.
First it leaps up; light follows; the lightning flash reddens;
The thunderbolt makes a single burst, and the earth is cloven
in twain. . . .

poem by Chang Hsien 張憲 (fl. 1341)
in Yü ssŭ chi 玉笥集 3/27b.[1]

Ever since 1790 there have been important contributions, both Chinese and foreign, to the literature on firearms in China. Following the work of Chao I (*Kai-yü ts'ung-k'ao* 1790) and Liang Chang-chü (*Lang-chi ts'ung-t'an ca.* 1848) they have included papers or critical comments by W. F. Mayers (*Jo. of the North China Branch of the Royal Asiatic Society* 1869–1870), E. H. Parker (*China Review* 1900–1901), G. Schlegel (*T'oung Pao* 1902), P. Pelliot (*Bull. de l'école française d'extrême-orient* 1902, *T'oung Pao* 1922), H. Yule (*The Book of Ser Marco Polo*, 1903-1920), and Lu Mou-tê 陸懋德 (*Tsing-hua Jo.* 1928).[2] On several significant points these contributors have been forced to make use of secondary sources, or of fragmentary quotations assembled in later symposia. Our effort has been to ferret out and cite contemporaneous documents, some of them only recently made available through reprinting or photographic reproduction and present an even fuller treatment of the evidence for firearms in China during the earliest period of development, namely from the year 1000 to 1403 A.D.

The Sung hui-yao kao 宋會要稿 185/26/27a states that a lieutenant of the Sung imperial guard, T'ang Fu 唐福, in the year 1000, presented to the throne his newly invented *huo-chien* 火箭, *huo-ch'iu* 火毬, and *huo-chi-li* 火蒺藜. This happened during the war between the empires of Sung and Liao. What these terms cover is not explained in the laconic statement. A military book, entitled Hu-ch'ien ching 虎鈐經 compiled by Hsü Tung 許洞 (*chin-shih* of A.D. 1000), mentions *huo-p'ao* 火炮 in ch. 6/44 under date of the year 1004, but offers no further explanation.

Another well illustrated military work, the *Wu-ching tsung yao* 武經總要,[3] compiled 1044 by

---

[1] We are grateful to Mr. Wang Chung-min of the National Library of Peiping, now at the Library of Congress, for this reference, and for his friendly interest in our study.

[2] These are by no means all. For example, there is a short paper entitled "A study of explosive weapons" 火砲考 by the Korean scholar Yu S'öng-yong 柳成龍 (b. 1542) including in his book Sê ae chip 西厓集 (publ. 1633), reprinted in *Shigakuzasshi* 4: 885–886; but we found this to be without value.

[3] The edition used by us is that found in the photographically reproduced Ssŭ-k'u ch'üan scu chên pên series. The original in the National Library of Peiping, unfortunately incomplete, was printed from the woodblocks prepared in 1232, recut in the years 1403–1424, and repaired in 1439. How old the illustrations are is uncertain; they may date back to 1232, but we suspect that they are not older than the 15th century, possibly later. Chüan 12 is one of the parts missing from the 15th century copy.

TsêNG KUNG-LIANG 曾公亮 and others, goes into more detail. The authors state that the *huo-ch'iu*, made of *huo-yao* 火藥 (explosive powder),[4] weighs 12 catties, and that the *huo-p'ao* is also made of powder. Both *huo-ch'iu* and *huo-p'ao* [5] were thrown from a trebuchet. The authors speak too of the *p'i-li huo-ch'iu* 霹靂火毬 made of a piece of dry bamboo, three nodes long and an inch and a half in diameter. A mixture of gunpowder, pieces of iron, and pottery fragments was attached to the outside of the bamboo, at the middle, and was lighted with a length of red hot iron, which produced a noise like thunder. (Part I, 12/43b; 69b.) The first weapon seems to have been a bomb, the second a trebuchet, and the third a hand grenade.[6]

At the turn of the century the Jurchen began feeling their way against the Liao empire on the southwest and Kao-li (Korea) on the east. It is of interest, therefore, to find the latter — known to have had extensive contacts with the Chinese — ready to use the latest arts of warfare against the foe. In 1104 the Korean officer YUN KWAN 尹瓘 memorialized the throne on the use of *fa-huo têng chün* 發火等軍 "fire despatching and other solders," (*Koryu sa* 高麗史 81/640). What kind of equipment these troops used is not vouchsafed.

In 1126, when the Jurchen attacked the Sung capital, Pien (mod. Kaifeng), a Sung general, YAO YU-CHUNG 姚友仲, made use of *huo-p'ao*, *chi-li-p'ao* 蒺藜炮, and *chin-chih-p'ao* 金汁炮 in defending the city. These were thrown either by trebuchets or by bows.[7] When these were exhausted he ordered soldiers to use *ts'ao-p'ao* (straw fire), which were tied in a bundle about bamboos, set on fire, and cast down upon the attackers. The enemy was repulsed by this means. The Jurchen later acquired this technique. They built elevated platforms, higher than the city wall, from which they flung *huo-p'ao* at the defenders like rain. (Cf. HsÜ MÊNG-HSIN 徐夢莘, 1124–1205, *San-ch'ao pei-mêng hui-pien* 三朝北盟會編 66/12b; 68/4a; 7a; 15a; SHIH MAO-LIANG 石茂良, 12th cent.?, *Pi-jung yeh-hua* 避戎夜話 2b; 4a.)

HsÜ MÊNG-HSIN also reports in the same work (*ibid.* 68/4b, 7a) that YAO YU-CHUNG suggested

the use of *huo-p'ao* at night to destroy the Jurchen catapult frames (*p'ao tso* 礮座), but this proposal was disregarded by his superior officer. To stop the Jurchen from entering the city by means of tunnels laid under the city wall, the defenders burned them to death by means of *huo-p'ao*.

From 1126 on the Jurchen and the Chinese frequently engaged in war. The former even penetrated the valley of the Yangtze and to the south of it. In these struggles the Chinese made extensive use of the *huo-p'ao*. HsÜ MÊNG-HSIN goes farther and insists that both sides employed *huo-p'ao* in 1126 at the investment of Huai-chou (in mod. Honan), when that city was battered to pieces. (*Ibid.* 61/11b.)

In 1127 Ming 洺 prefecture (in mod. Hopei) was besieged by the Jurchen, who were driven away after their machines to attack the city were destroyed by the *huo-p'ao* of the Chinese. (Li Hsin-chuan 李心傳, 1166-1243, *Chien-yen i-lai hsi-nien yao-lu* 建炎以來繫年要錄 7/176.) Three years later (1130), when the Jurchen under LOU-HSIU 婁宿 (Lou-shih 婁室) besieged Shan-chou (in mod. Shensi), and met with strong resistance, they employed *chin-chih-p'ao*, destroying everything within reach of the explosive. (HUNG MAI 洪邁, 1123-1202, *Jung-chai sui-pi* 容齋隨筆 part V, 6/5a.)

In 1161 the Jurchen made an attempt to cross the Yangtze River, but their plan was nullified by the destruction with *huo-p'ao* of their fleet of 600 vessels. (*Chin shih* 65/16b; PAO HUI 包恢, 1182–1268, *Pi-chou kao-lüeh* 敝帚稿略 1/6a). That this did not happen to the Sung fleet was due to a suggestion made in 1127 that the warships be equipped with catapults, *huo-p'ao*,[8] *huo-ch'iang*, and other weapons to resist attack. (*Sung hui-yao kao* 186/29/32a.)

The Chin land forces were also repulsed by the Chinese who made use of a new arm called *p'i-li p'ao* 霹靂砲. This explosive was made of thick paper, filled with lime and sulphur. When these bombs were dropped into the water, they exploded and the lime scattered like a fog. The first use of *p'i-li p'ao* is ascribed to the above mentioned siege of Pien

---

[4] For the composition of this powder, see *ibid.* part I, 12/58a.

[5] The *huo-p'ao* is illustrated *ibid.* part I, 12/56b, reproduced herewith.

[6] We assume a fuse, though nothing is said of one. *Vide infra.*

[7] The Chinese bow, particularly the crossbow, was a very powerful weapon even in ancient times. Cf. DUBS, H. H., *T'oung Pao* 36 (1940), 69 ff. Note in this connection YULE, *The Book of Ser Marco Polo* II, 162, figs. 1–5.

[8] Another use of *huo-p'ao* by the Chinese has already been published by A. C. MOULE in *T'oung Pao* 33 (1937), 115. It seems that a certain official who had been watching the famous Hangchow Bore, and been upset by the appearance of strange monsters disporting themselves in the water, proposed to the emperor in the sixth month of 1239 that powerful crossbows and *huo-p'ao* be used to exterminate them. Mr. MOULE took this passage from the *Hsi hu yu-lan chih* 西湖遊覽志 24/13b by T'IEN JU-CH'ÊNG 田汝成 (*chin-shih* 1526), a native of Ch'ien-t'ang, but it appears much earlier in the *Shan-yu Lin-an chih* 淳祐臨安志 10/9a, compiled in 1252 by SHIH 施諤.

Fig. 1. — *Huo-p'ao*, from *Wu-ching tsung yao*, original compiled in 1044. The illustrations may have been inserted by a later hand between the 15th and 17th centuries.

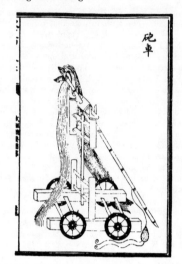

Fig. 2. — *P'ao-ch'ê*, from *Wu-ching tsung yao*, original compiled in 1044. The illustrations may have been inserted by a later hand between the 15th and 17th centuries.

in 1126. The thirteenth century historian YANG CHUNG-LIANG 楊仲良 declares that the defenders, under the direction of LI KANG 李綱, waited until nightfall, then shot it against the besiegers. See *Sung t'ung-chien ch'ang-pien chi-shih pên-mo* 宋通鑑長編紀事本末, publ. 1253, 147/10a.

Another device was the *p'ao-ch'e* 砲車,[9] which ejected fire and cast stones a distance of 200 paces. (YANG WAN-LI 楊萬里, 1124–1206, *Ch'êng-chai chi* 誠齋集 44/8b; *Sung shih* 368/15a.)

The story is told (by the contemporary Hsiang-yang official CHAO WAN-NIEN 趙萬年, in *Hsiang-yang shou-ch'êng lu* 襄陽守城錄 2b, 13b, 14b, 23a) that the Chinese commander CHAO CH'UN 趙淳 delivered the city of Hsiang-yang (in mod. Hupeh) from the Jurchen in 1206 also by using *p'i-li p'ao*. He assumed at the outset of their siege that they must be equipped with *huo-p'ao*; so he took such precautions as to clear the area of huts built of grass and of bamboo, and to distribute pails of water to quench all possible fires. The Jurchen attacked several times, but on each occasion were repulsed. CHAO CH'UN then launched a counter-attack under cover of darkness employing *p'i-li p'ao* and *huo-yao*

[9] For an illustration of the *p'ao-ch'e* see *Wu-ching tsung-yao*, part I, 12/39a and 10/13a, reproduced herewith. This description suggests a proto-cannon.

arrows with telling effect. From two to three thousand of the enemy were slain.

The increased use of arms of this sort, both on water and on land, speaks for their effectiveness in military operation; and the development of the *p'ao-ch'e* indicates a distinct advance over the ancient trebuchet. These arms were employed by the Chinese not alone to fight strong enemies on their borders but as well to pacify lawless elements within. Almost all government forces were apparently so equipped (*Pi-chou kao-lüeh* 1/10b.)

The loss to the Jurchen of the Sung capital and the capture of three thousand members of the court was a shock to the Chinese and provided a stimulus both to reflection and military invention. One contemporary official, who found the investment of so strongly fortified a city as Pien difficult to comprehend, was CH'ÊN KUEI 陳規, co-author of "Record of the Defense of Cities" (*Shou ch'êng lu* 守城錄 4 ch., published by government order in 1172). He is the reputed inventor of the *huo-ch'iang* 火鎗, said to have been a long bamboo tube filled with explosive powder. This was carried by two soldiers. The new weapon was put to the test in 1132 when used to rout bandits who attacked the city of Tê-an (in mod. Hupeh province). Cf. *Shou ch'êng lu* 4/33; *Sung shih* 377/6a.

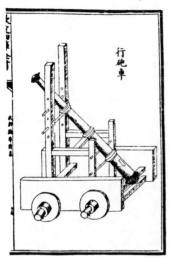

行砲車

Fig. 3. — *Hsing p'ao-ch'ê*, from *Wu-ching tsung yao*, original compiled in 1044. The illustrations may have been inserted by a later hand between the 15th and 17th centuries.

As noticed above, the Jurchen were quick to take over novel military devices. In 1221, when the Jurchen attacked a certain Chinese city they used *t'ieh-huo-p'ao* 鐵火砲 cast in the shape of a gourd of pig iron two inches thick, pierced with a small hole. Its explosion, says an eyewitness, sounded like a thunderbolt, shaking the wall, and the Sung defenders suffered casualties to the number of six or seven out of every ten. (CHAO YÜ-YUNG 趙與褣 [fl. 1221]: *Hsin-ssŭ ch'i-ch'i lu* 辛巳泣蘄錄 21;23.) The late E. H. PARKER pointed out another such instance, without citing his source,[10] many years ago:

"In or about the year 1213 fire-arms were used by the Nüchens 火槍 Their make is thus described: 'Yellow paper to a thickness of sixteen layers is made into a tube over two feet long. In this is placed willow charcoal, iron slag, powdered porcelain, sulphur, arsenic, and such. A string is made fast to the point of the weapon. The soldiers each hang on a small iron can containing fire, and as they approach the enemy's ranks they light up. The flame goes out of the front part of the weapon for over ten feet. The powder is spent but the tube is uninjured. They were used in the attack upon the [Chinese] capital of Pien [K'ai-fung fu], and again on this occasion.'"

(*China Review* XXV: 258.)

[10] It was apparently taken, directly or indirectly, from the *Chin shih* 116/13a. Correct the approximate date to 1231–33.

At this period JENGHIS KHAN (d. 1227) and his successors were invading North China almost at will, but were experiencing difficulty at the many islands of defense: the walled cities. They encountered two weapons which baffled them at first: the *chên-t'ien-lei* 震天雷 made of iron in the shape of a jar filled with powder, which MAYERS and PELLIOT consider to have been bombs or grenades, and the *huo-ch'iang* described above. (*Chin shih* 111/8b; 113/19ab. Cf. MAYERS' translation from the *Wu-pei-chih* of MAO YÜAN-I, fl. 1621–38, *op. cit.*, 91, and PELLIOT, *T'oung Pao* XXI, 1922, 434.)

The *Chin shih* (113/19b–20a) account of the former relates that when ignited there was a flash of fire, the thunder-like sound could be heard over a hundred *li* away, and it would burn an area more than half a *mou* in size. It would even pierce iron armor. Mongol soldiers nearing a city would be torn to pieces by it, and no trace of them be left. The Jurchen would attach one to an iron chain and suspend it from the city wall. LIU CH'I 劉祁 (1203–1250) recounts in his *Kuei ch'ien chih* 歸潛志 11/3b that he personally observed in the fight for the Jurchen capital of Pien (mod. Kaifeng) that the Mongols were repulsed by the *chên-t'ien-lei* of the Jurchen. When hit the Mongols were pulverized. The latter seem indeed to have been a kind of primitive firearm, for they ejected a bullet (or bullets) as well as flame. Possibly the Jurchen had to use paper instead of bamboo because they were without it in the Kaifeng area. A similar instrument must have been used by the Chinese at Shou-ch'un (in mod. Anhui) in the year 1259. Called *t'u-huo-ch'iang* 突火槍 it was made of a long bamboo tube 筒 into which a bullet (or bullets) *tzŭ-k'o* 子窠 was inserted. When the powder was ignited, smoke and fire came forth followed by the bullet. It sounded like a *p'ao* and could be heard more than 150 paces away (*Sung shih* 197/15a).[11]

We also learn from the *Chin shih* 124/16a that in 1236, at the time of the Mongol invasion of the region of (modern) Kansu province, the Jurchen, under the command of KUO HSIA-MA 郭蝦蟆, collected a variety of metals, including gold, silver, bronze, and iron,[12] and cast them to form *p'ao* 鑄 爲砲 in order to resist.

[11] Another reference to bullets is in a lament by WANG YEN 王演, a Yüan poet-official, who mourns the loss of a general who died in battle: "The *p'ao* bullets 礮丸 sound like the clap of a thunderbolt." (*Yüan shih hsüan* 元詩選, *kêng* 上, 14a.)

[12] This reflects the scarcity of appropriate metals in north China. During this dynasty (1115–1234) copper was especially scarce. The rulers needed it for coinage, and went so far as to forbid the wearing of copper buckles.

Another reference to firearms collected to resist the Mongols dates from the time of the raid of URIANGKATAI into Tongking (1257) and his anticipated attack against Sung China on his return from the south. It is made by LI TSÊNG-PO 李曾伯 who was sent to Ching-chiang (mod. Kuei-lin, in Yünnan province) to investigate preparations against the foe. His lengthy memorial of the third month of 1257 contains surprising, indeed almost incredible, information: "With regard to firearms, the iron huo-p'ao of Ching (in mod. Hupeh) and of Huai (in mod. Kiangsu) often number more than a hundred thousand pieces. When I was in Ching-chou, one or two thousand pieces were being turned out every month, and the despatch to Hsiang-yang and Ying (in mod. Shensi) always numbered ten or twenty thousand pieces. But Ching-chiang has at present only 85 huo-p'ao large and small, 95 huo-chien, and 105 huo-ch'iang. This allotment is not enough for one action by a single detachment of a thousand soldiers or less!" (K'o-chai hsü kao hou-chi 可齋續稾後集 5/52a.)

The Mongols must have been as quick as the Jurchen to seize and utilize arms in which they were deficient. In 1233 they used huo-p'ao to set fire to the towers of the prefectural city of Ts'ai 蔡 (in mod. Honan) in their final assault against the Jurchen, and seized a part of the town. (YÜ-WEN MOU-CHAO 宇文懋昭, fl. 1234: Ta Chin kuo chih 大金國志 26/7b.) In 1241, according to M. PRAWDIN (The Mongol Empire, 259), the forces under BATU made use of explosive powder at the battle beside the Sajo against the army of King BELA,[13] and in 1253 MANGU sent to North China for "1000 families of mangonellers, naphtha-shooters, and arblasteers" in preparation for the campaign against Persia. (YULE, The Book of Ser Marco Polo II, 168.) In the years 1268–1273, the Mongols besieged the two Chinese cities of Hsiang-yang and Fan-ch'êng. Among the instruments of defense were huo-ch'iang and huo-pao. (CHOU MI 周密, 1232–1308, Ch'i-tung yeh-yü 齊東野語 18/17b; see also Sung chi san ch'ao chêng-yao 宋季三朝政要 4/7b, by an unknown author.) It is reported, for example, that in 1272 Hsiang-yang was supplied with food, huo-ch'iang, huo-p'ao, and crossbows by CHANG SHUN 張順 and CHANG KUEI 張貴 who made their way by ship via the Han River

through the Mongol blockade until CHANG SHUN exhausted his supply of huo-p'ao and met his end. (Chao chung lu 昭忠錄 16/b, by an unknown author who probably wrote of contemporary events; and CHOU MI, Kuei-hsin tsa-chih pieh chi 癸辛雜識別集 B/43a.)

In the course of time, however, this advantage was offset by the hui-hui p'ao 回回礮 ("man-gonels")[14] constructed to order by the Moslems 'ALĀ-UD-DĪN and ISMA'ĪL. (Cf. PELLIOT, Asia Major 1927, 378 and MOULE, Marco Polo, the Description of the World, I: 318.) These men reached Peking (Khanbaliq) from Western Asia in 1271 or early in 1272, made a demonstration of their new weapons before the five city gates, and then carried them down to the southern frontier to end the sieges of Fan-ch'êng and Hsiang-yang. Each weighed 150 catties, and when touched off made a thundering roar. When the city-tower of Hsiang-yang was struck the whole city shook; and when the ground was hit a cavity seven feet deep opened up. A fact to be noted is that the hui-hui p'ao was not the only new weapon.[15] ISMA'ĪL is credited by the Yüan shih 7/19a with the invention of an improved catapult flinging large rocks (chü shih p'ao 巨石砲) which required little power to shoot a considerable distance. On the day chi-mao 11th moon, 9th year of Chih-yüan (Dec. 7, 1272) the order went out to send a number to the front at Hsiang-yang for use. This is significant as some critics have contended that the hui-hui p'ao was merely a catapult. The authors of the Yüan history indicate otherwise. We have here two weapons of distinct types.

The Yüan history states in another place (Yüan shih 151/18b–19b) that the man fundamentally responsible for the investment of Fan-ch'êng and Hsiang-yang was not a western Asiatic but a Chinese! His name was CHANG CHUN-TSO 張君佐 (d. 1280). He came by his association with explosives naturally, as his father, CHANG JUNG 榮 (d. 1262), a native of Ch'ing-chou (in mod. Hopei), who had joined JENGHIS in the 8th month of 1220 in his campaign in the west, had risen to be

[13] What is the basis of this assertion? H. DESMOND MARTIN (Jo. of the Roy. As. Soc., 1943, 67) says that they used "the fire of catapults and archers," and HENRY H. HOWORTH (History of the Mongols, part I, 149) wrote in 1876 that they "cleared the opposite bank by a battery of seven catapults."

[14] YAO SUI 姚燧, 1238–1314, asserts that the sieges of Fan-ch'êng and Hsiang-yang were ended by the employment of p'ao from the Western Regions (hsi-yü p'ao 西域礮). See his Mu-an chi 牧菴集 13/14b. This assertion occurs in the inscription on the tablet 左承相神道碑 at the tomb of ALI-HAI-YA, the Uigur general in charge of both sieges. (Cf. YULE, The Book of Ser Marco Polo II: 167.)

[15] The Yüan history itself does not make this point. It is an inference drawn from the variant accounts in the basic annals and in the biographical section of the Yüan shih.

Fig. 4. — Bursting cannon ball (?), fired by Mongol led army at a Japanese warrior, campaign of 1281, from *Mōko shūrai ekotoba*, compiled 1292.

commander of both *p'ao* and marine detachments. After the successful fight for these two cities the son, CHANG CHUN-TSO, was charged by the famous Mongol general PO-YEN (MARCO POLO's BAYAN; of. YULE-CORDIER's *Travels* II: 145) with similar siege duties in the attack (in 1274) against Sha-yang (also in mod. Hupeh) and Yang-lo pao (in mod. Hunan). Both places were stormed by means of *huo-p'ao*.

Two other contemporary accounts state that in the same year (1274) PO-YEN ordered a certain *p'ao* artilleryman named CHANG 張 (presumably CHANG CHUN-TSO) to employ *huo-p'ao* 火礮 in the attack against the prefecture of Ying 郢, which stood about where the Yangtze River port of Wu-ch'ang is today. As a result, it is said, smoke and flames filled the sky and the city fell. (*Sung chi san ch'ao chêng-yao* 4/11a; LIU MIN-CHUNG , 1243-1318, *P'ing Sung lu* 平宋錄 1/3a.)

A little later (1277), according to YAO SUI's inscription for the tablet of another tomb (王公神 道碑), ALI-HAI-YA's troops met with such stubborn resistance at Ching-chiang (in mod. Kuangsi) that he destroyed the entire city, burying all but two hundred fifty of the defenders alive. Not only did he spare the latter's lives; he even gave them several oxen and a number of bushels of rice for food in token of their heroic defense. When they had cooked and eaten of the rice and beef, the Chinese, we are told, took a *huo-p'ao* and set fire to it [16] as they sat calmly near by. The resulting explosion sounded like the noise of a hundred claps of thunder. What had been left of the city was shattered. The earth heaved, flames and smoke filled the sky, and many Mongols and Chinese were killed. (YAO SUI, *Mu-an chi* 21/12b. See also *Sung shih* 451/6ab.)

In 1275, at the siege of Ch'ang-chou (in modern Kiangsu), PO-YEN used *huo-p'ao*, bows, and crossbows, day and night (*P'ing Sung lu* 2/3b); and in 1276, when the Mongols arrived at Yang-tzŭ ch'iao (near mod. Yangchow) the Mongols set off *p'ao* 礮 continuously until, we read, "clouds covered the sky, a cold wind blew, and a shower of rain fell." (*Sung chi san ch'ao chêng-yao* 5/14b.) According to CHÊNG SSŬ-HSIAO 鄭思肖 (1239-1316) in his *Hsin shih* 心史 B/47b the Mongols also used *hui-hui p'ao* in this siege, and it did more damage than other *p'ao*. When it struck temples, towers, and public halls, they were broken in pieces. Later *hui-hui p'ao* were employed in the attack on units of the Sung fleet, all of which were sunk. (*Yüan shih* 128/6b; 203/9b-10a.)

[16] Possibly what the Chinese set fire to was the gunpowder stored with the *huo-p'ao*.

What were these *hui-hui-p'ao* or "mangonels"? CHÊNG-SSŬ-HSIAO (*op. cit.* B/71a) offers almost the only description, and that somewhat obscure. He says that they derived from Moslem countries, and were more destructive than ordinary *p'ao*. (Their framework of) huge logs was driven into the ground. The rocks shot from them were several feet in thickness and, in falling, were buried to a depth of three or four feet. To increase their trajectory (the barrels) were pushed back (i.e., elevated) and weight increased. To shoot a shorter distance, they were tilted forward.

The Mongols used the *hui-hui p'ao* in their siege of Khotan, but the Khotanese frustrated the attack by fashioning a net made of the bark of *Trachyarpus excelsa* 椶櫚 to protect the city. We learn from the *Yüan ching-shih ta-tien*,[17] *hsü-lu* 元經世大典 序錄, preserved in the *Kuo ch'ao wên-lei* 國朝文類 41/61b by SU T'IEN-CHÜEH 蘇天爵, 1294-1352, that the finest weapons were the *p'ao* from the western regions and the *chê-tieh* crossbows 摺疊弩, previously unknown. Presumably in the first instance the authors were referring to the *hui-hui p'ao*.

As a protection against the *hui-hui p'ao*, the Chinese made screens of various materials, according to the *Sung shih* 197/15a. Rice stalks were fashioned into ropes approximately four inches thick and 34 feet long. Twenty of these ropes, fastened with a strand of hemp, were suspended in front of certain buildings. Points needing especial protection were guarded by four or five layers. Walls were overlaid with clay.

But these citations tell us little. Were the barrels made of metal? It seems probable. MUS (*Bull. de l'école française d'extrême-orient* 1929, 340 n. 4) quite properly suggests that a comparative study of Moslem and Chinese arms at this time would be of great interest; also of the role played by the people of Turkish and Tartar blood. In any case, the Chinese set about, after the defeat of 1273, to copy the new firearm. It is said that the Sung emperor sent to all frontier cities designs of the *hui-hui p'ao*, and that improvements were even made upon it. (*Sung shih* 197/15a.) What these were we do not know. In 1280 a great loss occurred. A Mongol, who was grinding sulphur,[18] set fire to the arsenal

[17] The *Yüan ching-shih ta-tien* is apparently no longer extant. Besides the *hsü-lu* mentioned here, another part — that on courier stations — was copied into the great manuscript dictionary, the *Yung-lo ta-tien* (compiled 1403-1407). These volumes (nos. 19416 and 19426) are preserved in the Tōyō Bunkō, or Oriental Library, Tokyo. The original work was compiled in 1330-1331 by CHAO SHIH-YEN 趙世延 (1260-1336) and others.

[18] Sulphur itself would not explode. Presumably the arsenal contained a large supply of finished powder.

at Wei-yang (mod. Yangchow) which, with all its *p'ao*, blew up with a resounding bang!

First the *huo ch'iang* caught fire, looking like terrified snakes; then the *p'ao* store house. When the *p'ao* went off there was a noise like mountains falling and a hurricane at sea. Houses even 100 *li* distant collapsed. Martial law was proclaimed as the Mongols supposed that Chinese soldiers were making a surprise attack. The following morning it was observed that the bodies of a hundred guards were torn to bits. The beams and rafters of every structure in the city were fragmented, or blown more than ten *li* away, and the land was a mass of holes. More than two hundred families suffered in this tragic occurrence. (CHOU MI, *Kuei hsin tsa-chih*, part I, 13b–14a.)

The same author records (*ibidem*) another explosion. It seems that the residence of the former Sung prime minister CHAO K'UEI 趙葵 (d. 1266) was on one occasion used to house four live tigers. An arsenal nearby, containing explosive powder, which was spread out to dry, caught fire due to someone's negligence. Again the *p'ao* suddenly went off with a clap like thunder and like an earthquake, houses fell, and the four tigers were killed.

A final reference to *huo-ch'iang* is worthy of a place. When SHIH PI 史弼, 1212–1297, the noted Chinese commander under KUBILAI, was assisting at the siege of Yangchow, in the 6th month of 1276, two Chinese, mounted on horseback and carrying a *huo-ch'iang*, made a lunge at him with the weapon. He warded off the attack with his sword. (*Yüan shih* 162/11b.) This account suggests either a metal or a large bamboo tube.

In the first Mongol expedition against Japan, which set out at the end of 1274 under the command of General HU-TUN 忽敦, [19] many Japanese at Hakata, Kyushu, were slain by *t'ieh-p'ao* 鐵砲, metal or iron [20] firearms, according to *Hsin Yüan shih* 250/6a. This is partly corroborated by a Japanese account by an unknown author of the late 13th century, *Hachiman gudō-kun* 八幡愚童訓, printed for the first time in *Gunsho ruijū* 羣書類從, 13/328, which declares that when the Mongols retreated they shot *t'ieh-p'ao* 鐵砲, to darken the sky. As the noise was loud it astonished and disheartened the attackers, blinded and deafened them,

so confounding them that they did not know which was east and which west.

Another anonymous Japanese account of slightly later date (possibly the first years of the 14th century) — *Taihei-ki* 太平記 39, contained in the *Bugi-bu* 武技部 p. 881 of the *Koji ruien* 古事類苑 vol. 31 — reads in translation as follows: "On the 13th day of the eighth moon of the second year of *Bun-ei* 文永 (1265)[21] the Mongol fleet of more than seventy thousand ships pressed forward simultaneously to the coast of Hakata. Drums were rolled and fighting started, when, by the use of what is known as iron *p'ao* 鐵砲, they flung out iron balls,[22] of the size of a hand ball, two or three thousand at a time, which rolled down the hills with the speed of cart wheels, making a noise like thunder. Thereupon the Japanese troops utilized *huo-t'ung* 火筒 [23] to repulse CHANG at Kao-yu (in mod. Kiangsu). In another place (see below, in account of the fighting in 1362 and 1366) the weapon used is called *huo-ch'ung* 火銃. In Korea, a generation later, as indicated in the same reply, we read of an artillery bureau in which the term employed is *huo-t'ung* 火桶. Does this represent the same firing arm? In the first instance the signific of the first character is bamboo, in the second metal, in the third wood.[23a]

A certain Chinese named YANG 楊, who had served the Mongols as a *p'ao* technician 砲手, deserted in 1356 to CHU YÜAN-CHANG (who finally expelled the Mongols and reigned as first emperor of the Ming, 1368–1398). The latter put him in charge of the detachment of *ch'ung-shou* 銃手, and — in the defeat of CH'ÊN YU-LIANG 陳友諒, d. 1363 — he is mentioned as directing the detachment. (SUNG LIEN, 1310–1381, *Sung hsüeh-shih ch'üan-chi pu-i* 宋學士全集補遺 3/1347.)

From 1356 on there are many references to the employment of explosives and artillery, both on the part of CHU YÜAN-CHANG and of his rivals for power. On March 27 of that year CHU attacked the Mongol fleet off Ts'ai-shih 采石 (in mod. An-

---

[19] YAMADA, NAKABA, *Ghenkō, the Mongol Invasion of Japan*, 106, 141–144, calls him HOL-TON.

[20] It is well to remember in this connection that the Chinese had long been casting iron. Cf. THOMAS T. READ, "The early casting of iron," *The Geographical Review*, Oct. 1934, 544–554. One of his illustrations is of a large iron statue which dates from the middle of the tenth century.

[21] This date is an obvious error for 1275, as the commentator points out.

[22] Cf. illustration, reproduced herewith, in *Môko shūrai ekotoba* 蒙古襲來繪詞, A7, completed 1292, by a man who served in the campaign of 1281.

[23] Another variant is *huo-t'ung* 筒. For example, in the poem by CHOU CHÊN-T'ING 周震霆, composed in the last years of the Yüan; see *Shih ch'u chi* 石初集 5/6b.

[23a] It seems to us not without significance that some of the earliest cannon barrels simulate bamboo stalks in appearance. See especially the one illustrated in *Isis*, 35, 211, lower half of Fig. 3.

hui), on the Yangtze river, with flying *p'ao* 飛礮, and smashed it to pieces. (*Ming shih-lu* 明實錄, T'ai-tsu section, 4.) In the same year his generals routed K'ANG MAO-TS'AI   (1313–1369) with the use of Hsiang-yang *p'ao* 襄陽礮,[24] making him flee towards Chin-ling (mod. Nanking). (Cf. CH'ÊN CHIEN 陳建, *chü-jen* of 1528, *Huang Ming tzŭ chih t'ung chi* 皇明資治通紀 1/26b and KAO JU-SHIH 高汝栻, fl. 1636, *et al., Huang Ming t'ung chi fa chuan ch'üan lu* 法傳全錄 1/22a.) In the 4th moon of 1362 CH'ÊN YU-LIANG attacked (one of?) the city gates of Hung-tu (a city in mod. Kiangsi province, the name of which CHU YÜAN-CHANG changed later to Nan-ch'ang), and made a breach thirty yards in width. TÊNG YÜ 鄧愈 drove off CH'ÊN's troops with *huo ch'ung*. (*Ming shih-lu*, T'ai-tsu sec., 12.)

On July 23, 1362 CHU sent his fleet against that of CH'ÊN YU-LIANG in the neighborhood of K'ang-lang-shan 康郎山 on P'o-yang lake (in mod. Kiangsi). His boats, equipped with explosive weapons 火器, arrows, etc., were divided into eleven flotillas. The day before the fleet went into action CHU said to his commanders: "When approaching the enemy's ships first shoot firearms, next bows and arrows, then — at close quarters — engage them with your short arms." On July 24th HSÜ TA 徐達 (1332–1385), CH'ANG YÜ-CH'UN 常遇春 (1330–1369), and LIAO YUNG-CHUNG 廖永忠 (d. 1375) proceeded to the attack and discomfited the ships in the van. Whereupon YÜ T'UNG-HAI 俞通海 (d. aged 38 *sui* shortly after), seizing the opportunity given by the wind, shot the *huo-p'ao* and burned over twenty enemy craft. (*Ibid.* Cf. *Ming shih* 133/4a and T'AN HSI-SSŬ 譚希思, *chin-shih* 1574, *Ming ta chêng tsuan yao* 明大政纂要, compiled 1559, 1/8b.)

In what appears to be a somewhat lyrical account of this same engagement, though the precise date is not given, SUNG LIEN records in the *P'ing Han lu* 平漢錄 (of T'UNG CH'ÊNG-HSÜ 童承敘, *chin-shih* 1521) 16 that CHU divided his fleet into twenty flotillas under the command of HSÜ TA, CH'ANG YÜ-CH'UN, and LIAO YUNG-CHUNG. As the ships approached CH'ÊN's line, arrows were like rain-drops, the sound of *p'ao* like thunder, the waves became towering, the flare of the flying fire 飛火 reached a distance of a hundred *li*, the color of the water changed to red. . . .

The same historian (in *Sung hsüeh-shih ch'üan-chi* 10/356) reports, in his biography of an augur named CHANG CHUNG 張中, that the latter made

[24] Another name for *hui-hui p'ao*, used so successfully at the siege of Hsiang-yang.

a prediction in 1363 concerning an impending dis-aster, but softened the shock by foretelling that no harm would come to CHU YÜAN-CHANG. True enough, in the sixth moon, the Chung-ch'in storied building 忠勤樓 caught on fire, its contents of *p'ao* and powder were touched off, and the explosion sounded like thunder.

In 1366 CHANG SHIH-CH'ÊNG was besieged at Ku-su (in mod. Kiangsu) by the soldiers under HSÜ TA. The attackers constructed wooden towers from which crossbows, *huo-ch'ung*, and Hsiang-yang *p'ao* were used with terrifying effect. (WU K'UAN 吳寬, 1435–1504, *P'ing Wu lu* 平吳錄 40 and KAO TAI 高岱, *chin-shih* 1550, *Hung yu lu* 鴻猷錄 [preface of 1557] 4/42.)

In 1371 CHU YÜAN-CHANG, now emperor, des-patched an expedition under LIAO YUNG-CHUNG and FU YU-TÊ 傅友德 against MING SHÊNG 明昇 the self-appointed Chinese ruler in Shu (mod. Szechuan). They proceeded by two routes, by river and by land. Both sides seem to have been well equipped. MING SHÊNG, in defense of his own posi-tion, erected three flying bridges on which were wooden platforms holding catapult stones 砲石, wooden pikes 木竿, and iron *ch'ung* 鐵銃. (*Ming shih-lu*, T'ai-tsu sec., 63.) By the 6th moon the imperial fleet was ready to attack. The boats' prows were protected by a covering of iron, and at this spot firearms 火器 were stationed. This fleet won the day, according to the annalist, because of their use of *huo-p'ao* and *huo-t'ung* 火筒. (*Ibid.* 66; *Ming shih* 129/12; HUANG PIAO 黃標, fl. 1544, *P'ing Hsia lu* 平夏錄 19.) By Sept. 20, 1371 FU YU-TÊ had reached Ch'êng-tu and surrounded the city. His forces were met by the enemy deploy-ing soldiers mounted on elephants in the front line. FU broke through this line with the use of arrows and firearms 火器. (*Ming shih-lu*, T'ai-tsu sec., 67.)

On May 28, 1387 the emperor sent out an order to his general MU YING 沐英 (d. 1392), defend-ing the Burma-Yünnan border, to erect higher defenses and dig deeper moats for the towns occupy-ing key positions. Each place was to prepare a mini-mum of one or two thousand *huo-ch'ung*. In Yünnan gunpowder arsenals were to make gun-powder overnight, if necessary, to defend these points. (CHANG TAN 張枕, who served in Yünnan in the years 1382–1398, *Yünnan chi-wu ch'ao-huang* 雲南機務鈔黃 35–36 and *Ming shih-lu*, T'ai-tsu section, 182.)

The following year (May 6, 1388) the Burmese commander SSŬ LUN-FA 思倫發 came with three hundred thousand soldiers to attack Ting-pien 定邊

(in mod. Yünnan). The first line consisted of a hundred elephants, behind which were the men clad in armor, holding shields and carrying bamboo tubes filled with darts. MU YING selected thirty thousand horsemen, formed three lines of men armed with *huo-p'ao* and crossbows, and rushed to meet the Burmese. As the *huo-p'ao* and crossbows were fired, the elephants turned and galloped to the rear. Ssŭ LUN-FA was badly defeated and fled. (*Ming shih-lu*, T'ai-tsu sec., 189 and *Ming shih* 126/19ab.)

Shortly after the death of the first Ming emperor civil war broke out between his successor (a grandson) and his fourth son, the Prince of Yen (who was to reign under the famous title of YUNG-LO, 1403–1424). On May 18, 1400 the imperial army under LI CHING-LUNG 李景隆 (d. *ca.* 1424) met the Prince's army at Pai-kou 白溝 (in southern Hopei), where the former had planted what appear to be a primitive form of land mines. They are called a nest of wasps 一窩蜂 [25] and *ch'uai-ma-tan* 揣馬丹. When men and horses touched them they were at once blown up. (*Ming shih-lu*, T'ai-tsung sec., 6; T'an Hsi-ssŭ, *Ming ta chêng tsuan yao* 11/24b; CH'ÊN CHIEN, *Huang Ming tzŭ chih t'ung chi* 11/31a.)

In another battle between the imperial forces and those under the Prince at Tung-ch'ang (in mod. Shantung), on January 4, 1401, the former won a resounding victory as a result of their superior equipment in firearms 火器. Tens of thousands of YEN's soldiers were slain, and he himself nearly lost his life. (*Ming shih-lu*, T'ai-tsung sec., 7 and *Ming shih* 5/4b.)

Several references have been made above to arsenals constructed either by Mongols or Chinese. Both the annals and the Institutes (or *hui-tien*) throw some light on their contents. Under date of Feb. 21, 1380, for example, the *Ming shih-lu* (T'ai-tsu section, 129) records that CHU YÜAN-CHANG ordered the abolition of the army supply bureau 軍需局 and set up in its place the military weapons bureau 軍器局, which was specially to control the supply of military weapons needed. For every army unit of one hundred men there were to be 10 *ch'ung* 銃, 20 shields, 30 bows, and 40 *ch'iang* 鎗[26]. The *Hsü wên hsien t'ung k'ao, ch'in ting* 續文獻通考, 欽定 134/3994, compiled under imperial auspices

[25] According to CHANG HSÜAN 張萱, *chü-jen* 1582, *Hsi yüan wên chien lu* 西園聞見錄 73/3b, this was a bundle of fire arrows 火箭, 30 to 50 in number.
[26] Judging from the citations below, both *ch'ung* and *ch'iang* were firearms. For some unexplained reason this imperial order is put under date of 1393 in the *Ta Ming hui-tien* 大明會典 192/63a.

in 1747, which has a section on military weapons, tells under date of 1380 of several kinds of *ch'ung* and *ch'iang*; also of Hsiang-yang *p'ao*, hsin 信 *p'ao*, chan k'ou 盞口 *p'ao*, shên 神 *p'ao*, bronze 銅*p'ao*, shên chi 神機 *p'ao*,[27] etc. WANG CH'I 王圻 (*chin-shih* 1565), author of an earlier *Hsü wên hsien t'ung k'ao* (compiled *ca.* 1586), goes so far as to assert that the Ming emperor had several hundred kinds of firearms (*ibid.* 166/18b). Every three years, beginning 1380, the bureau of military weapons and that of cavalry equipment turned out 3,000 bronze *ch'ung* as large as a bowl in diameter, 3,000 bronze *ch'ung* which could be held in the hand, 90,000 bullets 銃箭頭, and 3,000 hsin *p'ao*, etc. In 1393 the regulations for the distribution of weapons were fixed. For a warship they included, in addition to bows and arrows and the like, 16 *ch'ung* which could be held in the hand, 4 *ch'ung* large as a bowl, 20 *huo ch'iang*, 10 *chi-li p'ao*, 1000 *ch'ung* bullets, and 20 *shên chi* arrows. (*Ta Ming hui-tien* 156/125b and 193/69a.) By the third month of 1403 (this was in the first year of YUNG-LO) the emperor ordered, on the suggestion of the ministry of works, that *ch'ung p'ao* 銃砲 be manufactured either from refined copper 熟銅, or from a mixture of unrefined 生 and refined copper. (*Ibid.* 193/68b, *Hsü wên hsien t'ung k'ao, ch'in ting* 134/3994, and *Hsü wên hsien t'ung k'ao* 166/7a.)

One final point requires treatment: evidence for a fuse. It is taken for granted that the early fireworks of the Chinese, well developed by the thirteenth century, included fuses, but no discussion of them has so far been found. The one possible reference, in connection with weapons of war, is in the *Yüan ching-shih ta-tien, hsü-lu*, mentioned above. (See *Kuo ch'ao wên lei* 41/61b.) There

[27] The *shên chi p'ao* seem to be a much later contrivance than the Hsiang-yang *p'ao*. They are the cannon represented in the Peking Historical Museum. See *Isis*, 35, 211, answer to query #105.
(I seize this opportunity to supplement and modify one or two of my comments in that answer, in the light of an article on the Peking Historical Museum published in the *Sixth Annual Report of the Academia Sinica*, 1933–1934, to which Mr. CH'ÊN HUNG-SHUN has drawn my attention. This article [see p. 143] confirms my statement that the cannon preserved in the Shansi Provincial Museum is made of iron and bears a date equivalent to 1377. Another early cannon, made of bronze and 17 English inches in length, is reported from Sui-yüan. The inscription on the metal gives the names of its makers, declares that it was cast in 1379 in Huai-yüan-wei, Fêng-yang [in mod. Anhui] and weighs 3½ catties [approx. 4.7 English pounds]. The article adds that there are over 1400 antique cannon in the collection of the Peking Historical Museum, all of them dating from the reign of HSÜAN-TÊ [1426–1435] and later. L.C.G.)

we read that, in the year 1293, *chih hsin p'ao* 紙 信砲 were acquired from the region of Chiang-Chê, or the delta of the Yangtze river. The term might mean 'paper signal *p'ao*,' or (preferably for our purpose) a '*p'ao* equipped with a paper fuse.' The fact that the *t'ieh-huo-p'ao* of the early part of the thirteenth century is described as having been pierced with a small hole lends weight to the suggestion that this is a fuse. Possibly, of course, there was simply a train of powder leading through the hole to the explosive within.

Conclusion

It is our opinion that there is valid literary evidence for the development, by the thirteenth century, of real firearms in China, although dated examples are lacking until 1356. By the thirteenth century the peoples occupying the Yellow and Yangtze river valleys had the techniques, the materials, and certainly the urge to make them. Warfare in this region was intermittent, indeed almost continuous, throughout the eleventh, twelfth, and thirteenth centuries.

The literature indicates that around the year 1000 the Chinese had flame throwing devices. By 1132 they were using long bamboo tubes filled with explosive powder, and by 1259 bullets were inserted in these tubes and ejected by touching off the powder. In 1236 a Jurchen general, we are told, cast *p'ao* out of such metals as gold, silver, bronze, and iron. To end the long siege at Hsiang-yang in 1272 the Mongols introduced what seem to have been two new weapons. One was an improved catapult capable of flinging large rocks; another was a weapon variously called *hui-hui* (Moslem) *p'ao*, Hsiang-yang *p'ao*, and *hsi-yü* (western regions) *p'ao*. The latter weapon was the immediate charge of a minor Chinese officer, long associated with explosive weapons. It seems to have been particularly effective both against fortified cities and against ships. In the first Mongol expedition to Japan, which set out at the end of 1274, three different accounts speak of the iron *p'ao* used by the Mongols, and one Japanese account relates that these *p'ao* shot out iron balls. (Cf. illustration made by a contemporary author, reproduced herewith.)

At the end of the Yüan dynasty and beginning of the Ming (latter half of the fourteenth century) firearms of many sorts came into increased use. Indeed the founder of the Ming may have driven out the Mongols and crushed his rivals by his superior use and supply of these weapons. He manufactured them by the thousand. Every three years beginning 1380 his arsenals turned out, among other arms, 3,000 small-sized bronze *ch'ung* (a kind of (proto-musket?) and 3,000 large-sized *ch'ung*. They also produced *shên chi p'ao*, a small cannon of which many examples still exist. In 1403 his son, who reigned in 1403–1424, issued orders for the manufacture of *ch'ung p'ao* either from refined copper or a mixture of refined and unrefined copper.

The earliest extant dated cannon (of 1356, 1357, and 1377, illustrated in *Isis* 35, after p. 212) must have had many decades of development behind them, and their makers doubtless drew on both the experience of the Chinese and their neighbors and of the Mongols and their subjects from all over Asia. It seems not too much to hope that other literature besides that drawn upon above may give additional clues as to their prototypes. For example, something might be extracted from the *Yüan ching-shih ta-tien*, if a copy might somewhere be found. The tiny fragment on military weapons copied from this work into the *Kuo ch'ao wên lei* 41/61ab (*vide supra*) is tantalizingly brief.

ON THE INVENTION AND USE OF GUNPOWDER
AND FIREARMS IN CHINA

By Wang Ling 王鈴

1. *Introduction.*
2. *The Invention of Gunpowder.*
3. *The Invention of Fire-crackers and Fire-works.*
   A. *Fire-crackers.*
   B. *Fire-works.*
4. *The Invention of Fire-arms.*
   A. *Fire-arrows, Mines, Smoke-screens and Flame-throwers.*
   B. *Fire-catapults, Incendiary and Poison-gas Projectiles, true Grenades and Fire-barrels (barrel guns).*
5. *How and Where did the Knowledge of Gunpowder Spread?*
6. *Conclusions.*

Mr. Wang Ling's article printed below covers the same ground as the article by L. Carrington Goodrich and Fêng Chia-shêng previously printed (*Isis*, 36, 114–23, 1946), and this implies a certain amount of duplication which was unavoidable, as it would have been unfair to either of the authors to have suppressed parts of his article. Indeed, both articles were composed independently and at the same time, but Goodrich and Fêng worked in New York, while Mr. Wang Ling was working in Lichuang, Szechuan. Naturally enough, the Goodrich MS reached me first, on August 25, 1944, while Wang's was brought to me from China, by Dr. Joseph Needham, on February 23, 1945.

Students are advised to read both articles and not to omit the addenda to the Goodrich one (*Isis*, 36, 250).

The proofs of this article were read by Professor James R. Ware, of Harvard University, Associate Editor of Isis.

GEORGE SARTON

This article was written in the Institute of History and Philosophy of Academia Sinica. For valuable hints and kind help I am deeply indebted to Dr. J. Needham, Prof. Hsiang Ta, Dr. H. Stübel, and Mr. Chang Ch'englang. To all of them I herewith express my heartfelt thanks; Mr. Chang has given me much new information and Prof. Hsiang supplied me with most useful material for tracing the invention of gunpowder to the Five Dynasties.

WANG LING

1. INTRODUCTION

It was generally believed in European countries up to recent times that gunpowder was a Chinese invention. But when it was discovered by certain writers on military subjects that the original Chinese character for cannon 砲 was written with the radical for stone 石 instead of that for fire 火 as it is today, the use of gunpowder with these arms in early China was altogether denied.

An investigation of the question of gunpowder in China, taking into consideration its use with crackers and fireworks as well, seems therefore to be worth while. In the following essay an attempt is made to furnish material from Chinese records necessary for a comparative study of the history of gunpowder in East and West. Certain stages in the development of the use of gunpowder in China are specially dealt with, e.g. the protracted use of gunpowder for its incendiary quality.

2. THE INVENTION OF GUNPOWDER

Gunpowder is composed of sulphur, saltpetre (potassium nitrate), and charcoal. Charcoal has of course been in use for heating purposes since the earliest times of Chinese history. As for sulphur and saltpetre, the two characters *liu* and *hsiao*, 硫 and 硝 were included in the dictionary of Chi Yun, 切韻 in A.D. 605,[1] and also in the 6th century Yuen Pen Yu P'ien 原本玉篇. Furthermore, the names for these substances appear in the Tzu Ching 字鏡 and the Wan Hsiang Ming I 萬象名義 which contain all characters used in the sixth century A.D. Even in the Chin Dynasty 晉 P'i Ts'ang 悼蒼, the character 硝 was to be found. Both materials appear in these early dictionaries as medicines for the curing of diseases. This indicates that sulphur and saltpetre were largely known and used in the Chin 晉 and Sui 隋 dynasties. But the knowledge of sulphur and saltpetre can even be traced back to the 1st century B.C., or earlier, as saltpetre is mentioned in the Shih Chi 史記 and sulphur in the Huai Nan Tzu 淮南子.[2] Saltpetre or nitre apparently became known in the western countries only about the 13th century. (See *Ency. Britannica* Art. gunpowder.) In China, the natural occurrence of salt-

[1] The 切韻 copy discovered at Tunhuang contains the character 硝 for saltpetre. That of 王仁昫, the Tang copied manuscript in 故官, contains the character 硫 for sulphur. The manuscript copy of the latter discovered at Tunhuang, contains both.

[2] 正字通, quoted in 康熙字典.

petre in the soil is surely responsible for its being known at such early times. In this connection the following quotation may be of interest. Henry says:

It has always appeared to me highly probable, that the first discovery of gunpowder might originate from the primaeval method of cooking food by means of wood fires on a soil strongly impregnated with nitre, as it is in many parts of India and China.[3]

Reference to the extraction of saltpetre is already made in the following passage of the Hou Han Shu 後漢書:

From the day of the summer solstice strong fires are forbidden, as well as the smelting of metals with charcoal. The purification of saltpetre has to cease altogether, until the beginning of autumn.[4]

The preparation of saltpetre must thus have been widely practised at the time of the later Han Dynasty since the government prohibited it by order.

In another western book, dealing with the invention of gunpowder (Robert Norton's *The Gunner*), we find the following quotation:

Uffano reported that the invention and use as well of ordnance as of gunpowder took place in the year 85 of our Lord, and was known and practiced in the great and ingenious kingdom of China. In the Maraty province thereof, there yet remain certain pieces of ordnance both of iron and brasse with the memory of the year of founding ingraved on them and the arms of King Vitey, who was said to have been the inventor. And it also appears in ancient and credible histories, that the said King Vitey was an enchanter, and being vexed with cruel wars by the Tartarians, conjured an evil spirit, that he might show him the use and making of guns and powder, which he put into practice in the realm of Pegu, and in the conquest of East-India, and thereby quieted the Tartars. The same being confirmed by certain Portuguese, that have travelled and navigated those quarters, and also affirmed by a letter sent from Captain Artrad written to the King of Spain, wherein are recounted very diligently all the particulars of China, and said that they long since used both ordnance and powder. He affirms further, that he found ancient ill-shaped pieces but that those of later founding are of better fashion and metal than the ancient were.[5]

Based on this report, Norton fixes the date for the invention of gunpowder in China as the year A.D. 85. The invention is thus attributed to an enchanter, and a miraculous story is connected with

it. Iron and brass ordnance is said to have been constructed at the same time. But the whole statement lacks basis, and need not be further dealt with.

According to the book Pao P'u Tzu 抱朴子 of the Chin 晉 Dynasty (265–317 A.D.), saltpetre and sulphur also played a part in Taoist Alchemy.[6] Yet no statement of the inflammability or explosive power of a mixture of the substances is made, though in alchemical processes they certainly were exposed to fire. That at least the inflammability must have been known may be concluded from the following passage from the Pao P'u Tzu

one [Chinese] ounce of saltpetre together with half this amount of sulphur is pulverized, sifted and well mixed. Vinegar is added, and the mixture spread over the inside of a small iron box, 2 inches thick, which is exposed to a strong fire. . . .[7]

It may be significant that the proportion 1 to 2 in which sulphur and saltpetre are mixed, is the same as required for gunpowder (cf. the prescriptions of Sung 宋 Dynasty later on). As it is said that the two substances were well pulverized, mixed and exposed to high temperature after being enclosed in a case, there is a high probability, that the explosive force of the mixture was also discovered. Could acetic acid have taken the place of charcoal here, or was the carbon ingredient purposely omitted from the formulary for the sake of mystification? The T'ai P'ing Kuang Chi 太平廣記 relates that the T'ang emperor Tang Wu-Tsung 唐武宗 who so highly favored Taoism, wanting silver for some alchemical purpose, set many labourers to dig for it in the mountain. When the search proved unsuccessful, an old man appeared and offered his help. During the night a thunderclap was heard, whereupon the mountain broke open, and the silver could be won.[8] Another story relates that a Taoist monk, during his alchemical practices, produced purple flames, which became the cause of a terrible fire.[9] Although such miraculous stories can not be taken as proof of the explosive force of gunpowder being known at that time, they are nevertheless noteworthy, and rather suspicious, especially as such reports of explosions occur quite frequently in early Taoist books.

When we come to the Sung time, official books on military subjects begin to make mention of gunpowder and give detailed prescriptions for its production and use. Thus the process of making gun-

---

[3] 東京帝國大學工科, 大學紀要; Book 7, No. 1; also 有扳鉛藏, 兵器沿革圖說, p. 246.

[4] 後漢書, chap. 15, 禮儀志.

[5] Robert Norton, *The Gunner*. Here the passage was quoted from 兵器沿革圖說, p. 76.

[6] 抱朴子, 內篇, chap. 4, 11, 16.

[7] *Ibid.*, chap. 16, title 小兒作黃金法.

[8] 太平廣記, chap. 74, title 唐武宗朝衛士.

[9] *Ibid.*, chap. 16, title 杜子春.

powder is first mentioned in the Wu Ching Tsung Yao 武經總要. The passage runs as follows:

1 Chin 14 ounces of sulphur, together with 2½ chin of saltpetre, 5 ounces of charcoal, 2½ ounces of pitch and 2½ ounces of dried varnish, are powdered and mixed. Next 2 ounces of dried plant material, 5 ounces of tung oil and 2½ ounces of wax are also mixed to form a paste. Then these ingredients are all mixed together, and slowly stirred. The mixture is then wrapped into a parcel with five layers of paper, which is fastened with hempen thread, and some melted pitch and wax is put on the surface.[10]

Note the precaution against the entry of damp even at this early time. This prescription is copied in the Wu Pei Chih 武備志 [11] as the most reliable method of preparing gunpowder during the Sung Dynasty. The first mentioned book, Wu Ching Tsung Yao, contains also another prescription, in which the amount of sulphur and saltpetre remains the same, while five ounces of charcoal are added.[12] This book was written in the year 1040 A.D. and published five years later under the emperor Rjen Tsung 仁宗, who wrote the preface personally.[13] The publication was meant to serve military purposes, and not only points out a detailed and perfected method for preparing and keeping gunpowder, but also explains its use for different kinds of arms. It is surely necessary to suppose that a considerable period of experimentation and trials had taken place beforehand, and thus the invention of gunpowder and its qualities must be fixed at a somewhat earlier date than the compiling of the above mentioned book, *i.e.*, the eleventh century.

In the Southern Sung dynasty gunpowder seems to have been prepared in large quantities, and stored away in special arsenals. This is proved by the following passage:

The Prime Minister reared four tigers, which were kept in a palisade in the arsenal. On a certain day, while gunpowder was being dried, a fire broke out, and a terrible explosion followed. The ground trembled, and the houses collapsed. The four tigers were instantly killed. This news spread from mouth to mouth and was considered a marvel.[14]

As for the Yuen 元 Dynasty, it can be taken for granted that the Mongols had no knowledge of their own for the use and preparation of gunpowder, but adopted it from China at the time of the wars during the Sung Dynasty. The following sentences may be read in the Kuei Hsin Tsa Shi 癸辛雜識 "The tragedy of the arsenal in Wei Yang 維楊 is much to be regretted. Formerly the positions of gunmakers were all held by southern people [i.e. people of Sung]." Continuing:

But owing to their covetous and deceitful behaviour, they had to be dismissed from their offices and northern people, [Mongols?], had to be employed in their stead. But they understood nothing of the handling of the materials. Suddenly, one day while sulphur was being triturated, it caught fire, and the flames flashed like frightened snakes. The workmen thought it amusing, laughed and joked, but after a short time, there was a noise like the fall of a mountain, and the rumbling of the waves of the sea. The whole city was frightened, and the people feared the approach of an army. Even at a distance of a hundred li, tiles and houses trembled with a clacking sound. Fires were raging here and there, martial law was proclaimed, and the disturbance lasted the whole day. After order had been restored, an inspection was held, and it was found that one hundred men of the guards had been blown to bits, beams and pillars had been split, or carried away by the gale, to a distance of over ten miles. The smooth ground was scooped into pits and trenches, more than ten feet deep. Over two hundred families living in the neighbourhood were victims of the unexpected disaster. This was indeed an unusual occurrence.[14]

The cause of this terrible explosion was thus apparently the inexperience of the Mongol workers, who did not know how to handle the inflammable materials, in the absence of the people of Sung. This passage serves indeed as good evidence that the Mongols learned the method of preparing gunpowder from the Chinese!

In sum, therefore, one may say that all the constituents of gunpowder were known in China at least as far back as the first century B.C. There are suggestions that the Taoist alchemists of the third century A.D. knew the explosive properties of the mixture, but this cannot yet be proved. In the eleventh century knowledge of the mixture is well established, and practical application of it is being made; the actual discovery must thus be placed somewhere in or before the tenth century. As we shall see below, the term "fire-drug" first occurs in the early years of the seventh century, but only in connection with fireworks which were not necessarily explosive.

3. THE INVENTION OF FIRECRACKERS AND FIREWORKS.

(A) *Firecrackers.*

It is the general belief that in China the custom of firing off crackers led to the earliest use of gun-

---

[10] 武經總要, part I, chap. 12.
[11] 武備志, chap. 130.
[12] 武經總要, part I, chap. 11.
[13] 龜公武郡齋讀書志, chap 14.
[14] 周密, 癸辛雜識, 前集, p. 14 (稗海).

powder, preceding its application to weapons.

The name for cracker, Pao Chu 爆竹 literally means "bursting" or "decrepitating" bamboo. It is a well known fact that a piece of fresh bamboo, put into the fire, bursts with a cracking noise (decrepitation). For the modern cracker, though consisting of gunpowder, wrapped round with several layers of paper, the ancient name has survived.

There are several passages in the old records, referring to decrepitating bamboo. The Shen I Ching 神異經 of the Han 漢 Dynasty, states: "Bamboo was put into the fire and the noise Bi-Boo" (indicating the sound) "frightened the spirit of the mountain." [15] The Ching Ch'u Sui Shih Chi 荊楚歲時記 of the Chin Dynasty contains the following sentences: "On New Years Day, with the first cock crow, as soon as the people get up, there is the bamboo bursting in the courtyard, to frighten the spirit of the mountain." [15] The Feng Su T'ung 風俗通 remarks: "The cracking of bamboo is like the roar of the wild animals." [16] In the poetry of the T'ang Dynasty, further quotations referring to the custom can be found. "The New Year has just begun, the small courtyard still shows the embers of the cracking bamboo." [17] In this sentence the expression Pao Kan 爆竿 is used, of which the character Kan means the stem of the bamboo. So it seems clear that the cracker in ancient times contained no gunpowder.

It is in the records of the Northern Sung Dynasty (960–1126 A.D.) that a new term for cracker appears, Pao Chang, 爆仗 "All of a sudden a clap was heard, it was called Pao Chang." [18] Thus runs the earliest mention of a new type of cracker, which, according to more detailed descriptions in later records, certainly contained gunpowder. The expression "clap," (in Chinese Pi-li 霹靂), also points to some violent explosion. The term Pao Chang is also used by Meng Liang Lu 夢粱錄. "The sound of Pao Chang shuddered like thunder. Pao Chang and fireworks were sold on the streets." [19] From this remark it can be seen that the new invention had already become popular in the capital. Furthermore the Wu Lin Chiu Shih 武林舊事 gives the following description of New Year celebrations in the capital: "At the West Lake the young people competed in firing Pao Chang." [20] This passage as well as the following, deals with events at the time of the Southern Sung

dynasty and refers to Hangchow, which was then the imperial capital.

Towards the end of the year, the Pao Chang fire crackers were made in a new shape, that looked like fruits, men, and other things. They were connected by a continuous fuse. When this was lit, one cracker after another exploded, without interruption. [20]

This description calls to mind the popular present-day cracker, called Pien Pao 鞭爆. The fuze (in Chinese 引線 "leading" or "lighting" string, Yin Hsien) contains gunpowder. Though in the former quotations we are not informed how the Pao Chang differed from the Pao Chu, the latter passages unmistakably indicate the use of gunpowder for the Pao Chang. It was therefore probably made of paper. As no date for the new invention is fixed, it can only be estimated that the Pao Chang made its first appearance about the time of the Northern Sung dynasty 北宋 when it first occurs in the records.

In the Wu Yuan, 物原 Ma Chün 馬鈞 a great inventor at the time of the Three Kingdoms (三國; third century A.D.) is named as the inventor of Pao Chang [21] but Pei Sung-Chih 裴松之 in his commentary on the history of the Three kingdoms, though recording five of Ma Chün's inventions, does not mention the Pao Chang among them. [22] Neither does the Tien Lüeh 典畧 containing a biography of Ma Chün. The author of Wu Yuan may therefore be wrong in his statement. During the Sung dynasty the term Pao Chu was not only retained, but also the use of the decrepitating bamboo was continued. Sometimes the old term "cracking bamboo" may have been applied to the new model, instead of its new name: the case in the works of the Sung poets [23] in which we may take Pao Chu in either sense. The following line of a poem "A bamboo five feet was cut and burnt," [24] indicates that the use of bamboo persisted after the invention of Pao Chang; the exact date for the invention is there-

[15] 神異經, p. 9. 荊楚歲時記, p. 1 (漢魏叢書本).
[16] 應劭, 風俗通.
[17] 全唐詩, part 10, 冊 2, 來鵠詩, 早春.
[18] 孟元老, 東京夢華錄, chap. 7.
[19] 吳自牧, 夢粱錄, chap. 6.
[20] 周密, 武林舊事, chap. 3, title 歲除.

[21] 羅頎, 物原, chap. 14, title 兵原; according to 清嘉錄 this was quoted from 事物記原 but actually nothing is mentioned in the latter.
[22] 三國志, chap. 29, 杜虁傳, note.
[23] 蘇軾詩 "爆竹驚鄰鬼"; 陳去非詩 "城中爆竹巳殘更"; 王安石詩 "爆竹聲中一歲除." The above mentioned term "pao chu" may be explained either as cracking bamboo or as fire cracker.
[24] 范成大, 村田樂府, "截筒五尺煨以薪" 范成大詩, "幸無爆竹驚寒夢". 爆竹有寒灰 范成大, 吳郡志 "是夕爆竹及催田問" also 元好問詩: "輸林漲薪爆竹聲", also 周弼詩, "庭前爆紙青竹爆" 方岳, 立春詞, "引出千花萬草擁光椒盤竹爆".
In all these quotations, the reference is to the cracking bamboo and has nothing to do with the present day firecracker.

fore difficult to fix, but it can be said that with the time of the Northern Sung Dynasty (11th century A.D.), gunpowder began to be put in the cracker, which in its original form had consisted only of bamboo.

(B) *Fireworks.*

Fireworks are represented by the Chinese characters Yen Huo 烟火 which literally mean "smoke fire." This expression is used as early as the fourth century B.C. Referring to the New Year celebrations it is said that Yen Huo and Tao Shen 桃神 were arranged at the door.[25] Here however, Yen Huo can hardly stand for fireworks, but probably for the smoke produced by a fire of straw and wood. The Wu Yuan 物原 ascribes to the time of the emperor Yang Ti 煬帝 (A.D. 605–616) of the Sui 隋 Dynasty the invention of the Huo Yao Tsa Hsi, 火藥雜戲 or "fire drug play," apparently a kind of theatrical performance.[26] It is well known that this emperor spent a large amount of money on all kind of amusements. At the New Year festival "Fire Mountains" were burnt,[27] and it is quite plausible that fireworks in the modern sense were invented to satisfy the extravagance of this emperor. A poem, written on the occasion of the 15th of the first month by the emperor himself, has been handed down, containing the following passages: "Trees of lamps shine with a thousand lights, flames and sparks shoot forth from the seven branches." [28] While the first sentence is probably merely a description of lamps hanging on trees, the second, in which the character hua 花 appears, as well as kai 開 meaning "spark" and "to break out" respectively, indicates some kind of sparkling fire. The expression "seven branches" 七枝 may be interpreted as the name of a particular kind of firework. Furthermore, there is a reply to the emperor's poem, written by a high official, referring to the same event: "Flames of fire move round the wheel, peach blossoms spring forth from the falling branches. Clouds of smoke move around the house, and the fairy lake reflects the floating lights." [29] Here there can be no doubt that the poet alludes to a firework, apparently something like the present-day Catherine wheel. Fireworks in China thus made their first appearance during the seventh century A.D.

As for fireworks during the T'ang 唐 Dynasty, the poem of Su Wei-Tao 蘇味道,[30] which the expressions Huo Shu, 火樹 (fire tree) and Yin Hua 銀花 (silver sparkle) occur, is often quoted by scholars. The former, may well mean a tree decorated with lamps, and need not necessarily mean fireworks but "silver sparkle" surely does. The inventions of the Sui 隋 Dynasty were no doubt taken over and perfected during this following period. The use of gunpowder with these fireworks is not impossible, since we have seen that its ingredients were known even in the Han dynasty, but there is no definite evidence in favour of it although the term "huo yao" or fire-drug is in fact also the later name for gunpowder.

At the time of the Northern Sung 北宋 fireworks indeed attained a high perfection. The Tung Ching Meng Hua Lu 東京夢華錄 dealing with the events of the year 1103 A.D. has handed down a detailed report of a theatrical performance including one of the above-mentioned "Fire-Drug Plays."

After a cracker clap, the fireworks began. There appeared ghost-like figures with white masks, scattered hair, and sharp-pointed protruding teeth, dressed in trousers of gold and black, while the upper garments, short on the back, were of blue colour, dotted with golden flowers. Their feet were bare, and they carried a huge gong, dancing and marching forward. This was called 'Carrying the Gong.' . . . And again the sound of crackers announced some figures with blue and green faces, the masks showing golden pupils. Their garments were decorated with leopard-skins and embroidery. Some carried a sword or an axe, others a pestle or a stick. Some were running and pursuing, while others were looking and listening. At another sound of crackers a strange apparition was introduced, with a mask and a long beard. His gown was green and he wore court shoes. He held an official tablet, and looked like the king of the ghosts. Another figure, standing next to him, beat upon a small gong, and the two danced together. This was called 'The Dancing Judge of the Ghosts.' Again two or three lean figures appeared, their bodies being covered with white powder, and their heads looking like skulls. They ran and jumped like actors on the stage, miming with each other. This was called 'dumb show,' or pantomime. After another clap of crackers, seven shapes appeared amidst fire and smoke. Their hair was uncombed, their bodies stained with many colours, and their blue gauze dresses showed embroidery. One of them, wearing a small hat with a golden flower, carried a white flag. The others wearing turbans, attacked and seemed to pierce each other with swords.[31]

---

[25] 荊楚歲時記, p. 3 (same edition).
[26] 物原, chap. 14.
[27] 太平廣記, sect. 奢侈.
[28] 全漢三國晉南北朝詩, vol. 20, 全隋詩, chap. 1. 隋 煬帝, 正月十五日放通衢建燈夜升南樓 "燈樹千光照, 花燄七枝開"
[29] Chap. 3. 諸葛穎奉和通衢建燈

"逐輪時徙燄, 桃花生落枝, 飛煙繞定室, 浮光映瑤池".
[30] 全唐詩, pt. 2, 册 2.
[31] 東京夢華錄, chap. 7.

The above detailed description attests the highly developed technical skill of fireworks at the time of the Northern Sung. Was such a perfection ever attained in western countries? The prominence of colours (like those of flares, "Bengal matches," etc.) in the account may throw a light on the original meaning of the expression "fire-drug," for just as medicinal substances may affect the colour of a patient, so certain substances such as strontium, barium, copper, etc. must have been found to produce flames of colour different from the usual orange-yellow. Then later, the term may have been transferred to a substance with even more astonishing properties. As there are contemporary statements of the use of gunpowder on the battlefield, its use with the Sung fireworks may well be assumed.

That fireworks served as an amusement, not only at the imperial court or in the streets of the capital [32] but were also popular in places remote from the capital, is proved by the following remark by a scholar of the Sung Dynasty. "A native of Wu Chou 婺州 called Chou Shih, 周四 knew how to manage fireworks. The local official, Tang Chung-Yu, 唐仲友 asked him to come, and several thousands of taels out of the public exchequer were spent on the performance of his plays." [33]

In conclusion it may be said that firecrackers and fireworks including gunpowder were both well known at the time of the Sung Dynasty, i.e. the eleventh century A.D. Fireworks, probably without gunpowder originated during the Sui Dynasty in the beginning of the seventh century A.D. Decrepitating bamboo, oldest of all, goes back to the Han Dynasty (first century A.D.)

4. THE INVENTION OF FIREARMS.

(A) *Fire-arrows, Mines, Smoke-screens and Flame-Throwers.*

Gunpowder seems to have been used with the so-called "fire-arrow." Though a weapon of this name occurs very early in history, there is no evidence that gunpowder was used with fire-arrows of the Chou 周 Dynasty or the period of the Three Kingdoms 三國 (220–260 A.D.). At the time of the Division between North and South 南北朝 (420–560 A.D.) as well as during the T'ang Dynasty, fire-arrows were also in use [34] but the description of

them is so vague that they may well have been simply daubed with some kind of combustible material, such as tar or resin. The so-called "powder-arrow" of the epoch of the Three Kingdoms is not said to have had gunpowder either; it may have been similar to a poison-arrow, used already during the Han Dynasty in the fight against the Huns, 匈奴. [35]

What must have been regarded as an improved fire-arrow is mentioned in the Military Memoirs of the Sung Dynasty 宋史兵記. "In the third year of the K'ai Pao 開寶 period of the reign of Sung Tai-Tsu 宋太祖, the general Fêng Chi-Sheng 馮繼昇 together with some other officers, suggested a new model of fire-arrow. The Emperor had it tested, and (as the test proved successful) presents of gowns and silk were bestowed upon the inventors." [36] In the Wu Li Hsiao Chih 物理小識 it is further stated, that in the second year of K'ai Pao (A.D. 969) Yo I-Fang 岳義方 presented a certain fire-arrow to the emperor. He was rewarded with a gift of silk. [37] Probably this refers to the same event as the preceding passage from the Sung History. Assuming that this newly invented fire-arrow was of a different type, it may have been the kind to which the Wu Ching Tsung Yao refers: "There are fire-arrows, to the point of which gunpowder is applied." [38] Here the term Huo Yao 火葯 for gunpowder is clearly stated, allowing no different interpretation. This book, being an official publication, enumerates all kinds of weapons used by the army. It appeared in the year 1040 A.D., 70 years after the new type of fire-arrow had been presented.

Another type of fire-arrow, however, is also mentioned, the Huo Yao Pien Chien 火葯鞭箭, the description and name of which clearly indicate the use of gunpowder. It is possible that it was the force of the exploding gunpowder which projected it, as it is stated, that five ounces of gunpowder were applied to the end of the arrow. [39] If so, it was really a rocket weapon (see Fig. 1). The word *pien* means a whip.

In the Military Memoirs of the Sung Dynasty, 宋史兵記 a third type of arrow, the so-called San

[32] 吳自牧, 夢粱錄, chap. 6.
[33] 晦庵先生朱文公文集, chap. 18. 按唐仲友第三狀.
[34] 周官, title 司弓矢: "凡矢枉矢絜矢利火射" 三國志 魏志 chap. 28, 諸葛誕傳; chap. 3, 明年紀 note, contains the following sentence: "昭 (郜昭) 以火箭逆射其雲梯" 宋書, chap. 87, 殷孝傳; 周書, chap. 18 王思政傳; 北史, chap. 62, 王思政傳; 通典, chap. 152, 守拒法.
In these chapters the ho chien (fire arrow) is mentioned.

[35] 虞世南, 北堂書鈔, chap. 125, title 箭, contains the quotations from 英雄記 and 東觀記.
[36] 宋書, chap. 197, 兵志, 150.
[37] 方以智, 物理小識, chap. 8, title 火爆; and 通雅 chap. 35.
[38] 曾公亮, 武經總要, 前集, chap. 13.
[39] *Ibid.*, chap. 12.

Kung Ch'uang Tzu Nu 三弓牀子弩 is mentioned, which is projected by a crossbow. In the Sung History no mention is made of the use of gunpowder. But the same arrow is fully described in the Wu Ching Tsung Yao which states that this arrow *may* also be projected by the force of gunpowder, if the elasticity of the crossbow permits,[40] a rather enigmatic remark.

The Sung History 宋史 relates how these arrows were used on the battlefield. "In the 5th year of Ch'un Hua 淳化 [A.D. 994] an army of 100,000 men besieged the city of Tzu T'ung 梓橦. A fierce attack was made, and the people in the city were greatly alarmed. Chang Yung 張雍 ordered the hurling of stones by machines, and succeeded in pushing back the invaders. At the same time, fire-arrows were shot off, whereupon the enemy retreated."[41] These fire-arrows must certainly have been those described in the Wu Ching Tsung Yao, as this book deals specifically with the weapons of the Sung Dynasty. The quotations from the various books, as given above, permit the conclusion that gunpowder was used for military purposes at least as early as the middle of the tenth century.

During the period of continuous fighting which now set in, fire-arrows were largely put into use, by the Sung. 宋 Yuen 元, and Chin 金 armies. Records run as follows: (1), "In the fourth year of Chien Yen, 建炎 [A.D. 1130] the Chin Tartars shot with fire-arrows at the covers of large boats. The army of Han Shih-Chung 韓世忠 [a general of the Sung] was upset and numerous soldiers were burnt or drowned."[42] (2), "During the period Shao Hsing 紹興 [A.D. 1161] Li Pao 李寶 ordered his army to shoot with fire-arrows at the ships of the Chin Tartars. Flames and smoke broke out, and as a result several hundreds of ships were burnt."[43] (3), "During the second year of K'ai Hsi, 開禧 [A.D. 1206] Chao Chun, 趙淳 a general of the Sung, secretly told his officers and soldiers that the Huo Yao Chien 火藥箭 [arrows with gunpowder] would be prepared. Later, he ordered the firing of these arrows in order to burn down the wood, straw, and catapults of the enemy." Or again, (4), "In the year 1206 A.D. the Tartars burnt the ships near the northern entrance of the city by means of fire-arrows."[44] And (5), "In the ninth year of Cheng Ta 正大 [A.D. 1232], new defensive measures were taken by the Chin against

the effect of fire-arrows and fire-catapults."[45] (6), "In the year 1274 A.D. in a battle between the Sung and the Yuan, A Chu, 阿珠 a General of the Sung, used fire-arrows. The sails and masts of the ships caught fire. The sky was covered with smoke and flames."[46] (7), In the first year the Teh Yu [A.D. 1275] "fire arrows were used to attack the private quarters of the emperor."[47] (8), During the 26th year of Hung Wu 洪武 [A.D. 1393] a law was proclaimed, by the imperial court, that each ship should be armed with twenty fire-arrows 火攻箭.[48] This means, that at any rate from the year 1393 on, fire-arrows were in general use all over Chinese territory.

In the western countries, gunpowder, and fire-arms making use of its explosive and projective force, appear almost simultaneously. In China, however, gunpowder was used during a long period for its incendiary properties only, as the fire-arrow shows.

In the time of the Sung Dynasty, a new weapon, which may be called the forerunner of both the modern mine and the modern smoke screen also makes its appearance. It is true that a Ming Dynasty book Wu Pei Chih 武備志 mentions the use of mines in much earlier times repeating the general belief of the people, that Chu-ko Liang 諸葛亮 invented a kind of steel wheel, used in connection with a mine.[49] Whatever this may have been, it was certainly not connected with gunpowder. But the mine of the earlier Sung Dynasty, called Pi Li Huo Ch'iu 霹靂火毬 the "bursting fire-ball" was dug into the ground, and contained gunpowder. Since it also contained pieces of broken porcelain it might be termed a stationary grenade. On exploding, the gunpowder ignited a mixture producing flame and smoke with which the mine was filled. The smoke was fanned in the direction of the enemy, especially when he tried to make a breach in a city wall or engaged in tunnelling operations.[50] This kind of composition was quickly applied to projectiles shot from catapults.

Apart from the mine, the army of the Sung used another strange weapon which was called Meng Huo Yu 猛火油 the "fierce fire oil machine." (See Fig. 2.) It seems to have been the ancestor of modern flame-throwers. The tank D in the diagram (Fig. 2) was filled with inflammable oil, while part A, called the "fire-tower," was provided with

[40] *Ibid.*, chap. 13.
[41] 宋史, chap. 307, 張雍傳.
[42] 續宋中興編年賚治通鑑, chap. 2.
[43] 文獻通攷, chap. 158.
[44] 趙萬年, 襄陽守城錄, p. 7, p. 9 (指南).

[45] 金史, chap. 113, 赤盞合喜傳.
[46] 續通攷, chap. 131.
[47] 宋季三朝政要, chap. 5, p. 3.
[48] 續通攷, chap. 134.
[49] 武備志, chap. 134.
[50] 武經總要, chap. 12.

gunpowder. When the machine was used, the gunpowder was kindled, and the handle of the pump B was pushed, so that the oil was conducted from D out towards A. At this end, the oil, was ignited, and terrible flames, according to the descriptions, shot out.[51] Of this instrument, mention is made in the Ch'ing Hsiang Tsa Chi, 青箱雜記 a book of

high flash-point for this use, it is extremely probable that petroleum was obtained from natural seepages. These are not uncommon all over Kansu and Shensi, and some of them produce a relatively light and volatile oil; they were known to the Chinese in the first century A.D. and are mentioned by many authors, including Marco Polo. Natural petroleum was

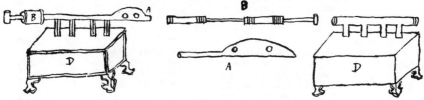

FIG. 2

the year 1004 A.D. It tells of two persons Chang Tsun 張存 and Rjen Ping 任丼 who obtained official positions by private connections, though their education was of a very low standard. An unknown person wrote a pamphlet to ridicule the two candidates as follows: "Chang Tsun knows how to shoot the whirlwind catapult, Rjen Ping is merely able to kindle the fierce fire oil machine the Meng Huo Yu." [52] Since the name of the machine could be used in such a connection, it is obvious that the instrument was well known before the year 1004. Besides, a weapon, very similar to that described and illustrated above, already appears in the Wu Yueh Pei Shih, 吳越備史 for the year 919 A.D. Though the use of gunpowder as a conductor is not specially mentioned, it seems very likely that the tank referred to worked on the same principle. The passage in question runs as follows: "Wu Su-Wang 武肅王 led his fleet of over 500 ships to fight against Huai Tien 淮甸 near Lan Shan 狼山. The enemy was attacked by [burning] oil which had been brought from Arabia to the south of the sea. This was projected from an iron tank. Coming in touch with water, the flames became more vivid. . . . Wu Su-Wang decorated the mouth of the tank with silver [as a ruse]. If the tank fell into the hand of the enemy, they would scrape off the silver but leave the tank intact. So the oil would be saved." [53] If gunpowder was in fact used with this flame-thrower, the year 919 A.D. would mark its first use in Chinese armies.

The use of inflammable oil is itself of much interest. Since no vegetable oil would have a sufficiently

known as "weak water." That it was extremely valuable appears from the ruse of Wu Su-Wang.

(B) *Fire-catapults, Incendiary and Poison-gas Projectiles, true Grenades, and Fire-barrels (barrel-guns).*

The other firearms which have to be considered are the P'ao 砲 or cannon, and the fire-catapult. A character P'ao 炮 with the radical fire 火 meaning to singe, to roast or to bake, appears as early as the Han Dynasty, in the Shuo Wên dictionary 說文 and has retained its form and meaning until today. That character is identical to the present-day P'ao 炮 for cannon, while in ancient days the characters for cannon 礮砲礟 all had the stone radical, instead of that for fire. The most ancient weapon of B.C. 707 or before, represented by the character 旝 has also the meaning of hurling stones, according to the opinion of Hsü Shen 許慎 and Chia K'uei 賈逵. The radical "stone" was sometimes substituted by the hand radical, such as P'ao 抛 or P'ai 拍 meaning to throw and to strike with the hand respectively. The character P'ao 砲 before the Ming time designates therefore the catapult or balista, for hurling stones. We meet with a variety of names for this weapon in the Chinese texts. It is called 霹靂車, 發石車, 石車, 抛車, 拍車, 石高車, 擂石車, 礧石, 廣雲旝, 將軍砲, 發石機, 砲,[54] but all the names point the same procedure of throwing stones.

[51] *Ibid.*, chap. 12.
[52] 吳處厚, 青箱雜記, chap. 8.
[53] 林禹, 吳越備史, chap. 2.

[54] 說文左傳正義, chap. 6, "旝動而鼓" 說文 character 旝(詩大雅)殷商之旅, 其旝如林 後漢書, chap. 104, 袁紹傳 "霹靂車", "石車" 三國志, 魏志, chap. 28, 諸葛誕傳, "發石車" 宋書, chap 87, 殷琰傳, "石車" "礌車" 梁書, chap. 33, 王僧辯傳 "礧石" 陳書, chap. 11, 南史, chap. 66, 黃法㧑傳 "拍車" 新唐書, chap. 136, 李光弼傳 "擂石車", chap. 220, 高麗傳 "抛車", chap. 84, 李密博 "廣雲旝三百具, 以機發石爲攻城械號將軍砲", 竇

During the fighting of the Sung, Chin and Yuan time the Huo P'ao 火砲 was constantly put into action, and with this weapon gunpowder was used, though it had no relation with the present-day gun or cannon with its metal barrel. The Huo P'ao resembled in its construction a balance or scale, one end being attached to extensible thongs, and the other carrying the projectile. But the projectile of the Huo P'ao or fire-catapult was something containing gunpowder, in fact a grenade or an incendiary shell, while that of the older P'ao or catapult was simply a stone.

There were different kinds of grenades or incendiaries projected by means of the fire-catapult. One of these was named the "barbed fireball 蒺藜火毬."[55] This contained gunpowder, and the active part of it consisted in a number of barbed iron hooks. The gunpowder was apparently wrapped into paper around the surface of the grenade.[56] In this case the gunpowder may have been used for its incendiary quality rather than for its projective force. The hooks may have been intended to attach the projectile to targets such as thatched roofs rather than to be dispersed explosively as missiles. Besides this fire-ball other projectiles were in use with the fire-catapult. There was the "iron-beaked bird" 鐵嘴火鷂 the frame of which was constructed of wood with an iron bill and a tail of patty stalks. The latter filled with gunpowder. Or there was the "bamboo fire-bird" 竹火鷂 which had an oval meshed basket-like body, containing one pound of gunpowder, wrapped in several layers of paper. It also had a tail of stalks three to five pounds in weight.[51]

Then there was the "smoke ball" 煙毬 which weighed 5 pounds, the inner layer consisting of a mixture of sulphur, saltpetre, charcoal powder and dried mint plant, shaped into a ball, while its surface was made of paper, hemp fibre and charcoal. Another kind, the "poison-drug smoke ball," 毒藥煙毬 when set on fire produced a gas which caused bleeding from the mouth and nose.[56] The composition of this was as follows:

	Chin	Liang
Sulphur	0	15
Saltpetre	1	14
Aconite tubers, dried	0	5

Croton oil beans, powdered	0	5
Langtu, a poisonous plant	0	5
Tung oil	0	$2\frac{1}{2}$
Other vegetable oil	0	$2\frac{1}{2}$
Charcoal	0	5
Pitch	0	$2\frac{1}{2}$
Arsenic oxide	0	2
Beeswax	0	1
Bamboo fibre	0	1

Still another chemical warfare device was a smoke ball of very peculiar composition, thus:

	Chin	Liang
Human faeces, dried, powdered and sieved	1	0
Langtu, as before	0	8
Aconite tubers	0	8
Croton oil beans	0	8
Pods of *Gleditschia sinensis*	0	8
Arsenic sulphide	0	8
Arsenic oxide	0	8
Pan Mao (?)	0	4
Lime	1	0
Perilla oil	0	8

This was boiled before storing and presumably fired off mixed with gunpowder. On burning, the fumes would, it was said, penetrate through the cracks of armour and cause blistering. The man firing the balls was recommended to protect himself by sucking black feathers and liquorice.

It would be very interesting to know just how effective these forms of chemical warfare were.

Lastly, there were "signal balls" or "beginning fire balls" 引火毬 which were shot off at the commencement of a bombardment in order to determine the exact range.

All the projectiles just mentioned had little or no iron parts, but consisted mainly of bamboo and paper, resembling in this respect the firecrackers and fireworks from which they probably originated. Form and construction were still quite primitive. As to the date, when at least some of the above-mentioned projectiles, namely the fire-ball, fire-arrow, and barbed fire-ball made their first appearance, we have an exact statement in the Sung History which runs as follows: "In the 3rd year of Hsien P'ing 咸平 [A.D. 1000] T'ang Fu 唐福 a certain captain of the navy, presented models of the fire-arrow 火藥箭, fire-ball 火毬 and barbed fire-ball 火蒺藜 to the emperor. A reward of money was given to him." [57] Thus the year A.D. 1000 must be looked upon as marking the beginning of the use of the catapult for shooting projectiles containing gunpowder.

The fire-catapult was put into use on the battle-

治通鑑, chap. 188, 唐紀四 "大炮飛石重五十斤擲二百步" 囊治通鑑, chap. 293, 後周紀四 "世宗自取一石, 馬上持之以供砲".
The above quotations all show the hurling of stones.
[55] 武經總要, chap. 12. Diagram (2) is also found in Yule's *Travels of Marco Polo*, edited by H. Cordier.
[56] 武經總要, chap. 12.

[57] 宋史, chap. 197兵志.

field at the time of the Southern Sung. In the Lao Hsueh An Pi Chi 老學庵筆記 we find the following description: "For attacking the enemy a Huei P'ao 灰砲 was invented [lime P'ao] which sent off as a grenade a case of thin and breakable tile, which contained poison, lime, and barbed hooks. With this the ships of the enemy were attacked, the lime causing smoke and mist so that the enemy could not open his eyes." [58] This mixture was a kind of "wet fire" used without gunpowder. The following account of the battle of Ts'ai Shih 采石 against the Chin 金 Tartars, A.D. 1161 in which the P'i Li P'ao 霹靂砲 was used by Yü Yun Wen 虞允文 has been erroneously looked upon by Ch'ing 清 Dynasty scholars as the first record of the use of the fire-catapult. [59] The catapult was discharged on board ship and the grenades contained lime and sulphur, which caught fire as soon as it came into contact with the water. An explosion followed, causing a huge spray of water and smoke, involving the surroundings in a dense fog. A contemporary writer described this battle as follows:

"A weight of ten thousand piculs dropped into the water, like a meteor from a blue sky. With a noise like thunder it sprang up again, straight into the air, spreading a drizzling mist, so that nothing could be distinguished, even at the distance of a foot. The robbers [i.e. the Chin Tartars] were terribly frightened, and did not know where to go. [60]

The explosion here described, was probably not caused by gunpowder. The inflammable mixture used in this case, closely resembles that of the so-called "Greek Fire" applied in warfare during the early European middle ages, especially in the two sieges of Constantinople (668–675 A.D. and 716–718 A.D. respectively) the deliverance of which city was chiefly described to the efficacy of Greek Fire. [61] It was also known under the name of "Wet Fire," or "Sea Fire," and consisted of a mixture of sulphur, naphtha, and quick-lime. The heat of quenching of the quick-lime presumably raised the temperature of the mixture sufficient to ignite the naphtha. Inflammable compositions of destructive power and liquid fires have been known in the Near East since the time of the Assyrians and even are

said to be represented on an Assyrian bas-relief. [62] Nevertheless, the invention of Greek fire, in which saltpetre or nitre is lacking, did not, like that of gunpowder, effect a total revolution in the art of war and the history of mankind. [63] The fire-catapult, first recorded during the Sung Dynasty, was imitated later by the Chin Tartars and taken over by the Yuan. In what follows, a chronological arrangement of the mention of the fire-catapult will be given, according to the historical records. As we have already seen, the fire-catapult was used by the soldiers of the Sung Dynasty, before A.D. 1040.

(1) A.D. 1126. A battle between the Chin and Sung: "The Chin Tartars burnt the watch towers by using the fire-catapult." [64]

(2) A.D. 1161. Battles between Sung, Chin and Yuan: "An army of the Chin Tartars attacked Hai Chow 海州; a movable catapult was invented by Wei Hsing 魏勝 a general of Sung. . . . Fire and stones were hurled." [65]

(3) "The enemy [refers to Yuan] was approaching. Finding us [refers to Chin] quite unprepared, they hurled fire grenades on our ships. Seeing his ship in flames, leaving no means of escape, Cheng Chia, 鄭家 a general of Chin, jumped into the water. He died at the age of 41." [66]

(4) A.D. 1176. Fire-catapult used as a metaphor in literature: "There was a tremendous noise, like the shooting of a fire-catapult." [67]

(5) A.D. 1206. A battle between Sung and Yuan: ". . . The approach of the Yuen Tartars, armed with fire-catapults being imminent, huts and wooden structures near the city, as well as the granary store houses were removed. . . ." [68] The thunderclap-gun (p'i li p'ao) was used whereupon the Yuen Tartars destroyed their fire-catapults, over ten in number, and fled during the night. Besides the P'i Li P'ao the Sung soldiers used fire-arrows, which they shot into the camp of the enemy, while his fortifications were attacked by the P'i Li P'ao." [69]

(6) "At the foot of the city, the so called 'Ten Stick fire-catapults' were put up." [70]

(7) A.D. 1231. A battle between Chin and

[62] *Encyclopaedia Britannica*, 7th edition, Article "Greek Fire."

[63] Gibbon, *Decline and the Fall of the Roman Empire*, vol. 6, chap. I, III, p. 101.

[64] 徐夢莘, 三朝北盟會編, chap. 66, 光緒四年越東集印本 "又用火炮燔樓櫓"; 光緒三十四年許涵度校本 "又楳大砲燔樓"
顧祖禹舊抄精本 "又飛火砲燔樓櫓".

[65] 宋史, chap. 368, 魏勝傳.

[66] 金史, chap. 65, 鄭家傳.

[67] 癸辛雜識　別集上.

[68-70] 趙萬年, 襄陽城守記, 附錄措置事目.

[58] 陸游, 老學庵筆記, chap. 1.

[59] 梁章鉅. 浪跡叢談, chap. 5.

[60] 趙翼, 陔餘叢攷 chap. 30.

[60] 楊萬里, 誠齋集, chap. 44, 海䲡賦.

[61] Gibbon. *Decline and the Fall of the Roman Empire*, vol. 6, chap. I, II. In a miniature in a manuscript of the 11th century, the fleet of Emperor Michael II is seen destroying the ships of the usurper by means of the so-called "Greek Fire" which was shot out like the explosion of gunpowder, — from *Propylaën Weltgeschichte*, p. 186.

Yuen: "During the fighting a new type of artillery appeared, called Chen T'ien Lei, 震天雷 which means 'heaven-shaking thunder.' Their [refers to Chin] weapon for defending the city, a certain fire-catapult, called the heaven-shaking thunder, consisted of an iron vessel filled with gunpowder. The powder was kindled, and the vessel hurled away, producing a thundering sound which could be heard at a distance of a hundred li. The ground, to the extent of half an acre, was affected, and even iron and metal penetrated [by the splinters of the iron vessel]. The Mongols had used ox-hides in defence, at the foot of the city wall, and drove a narrow tunnel, through which only one man could pass. The Chin Tartars were unable to fight the Mongols from the city wall, until some one suggested, that iron chains should be attached to the heaven-shaking thunder bomb and lowered down close to the tunnel. When the bomb was brought to explosion, the ox-hides were pierced, and together with the men were blown into pieces, so that nothing was left." [71] Thus it seems that with the heaven-shaking thunder bomb gunpowder was used for the first time for its explosive power. This occurred in the first few years of the thirteenth century.

(8) A.D. 1231. A battle between Chin and Yuen: "E K'o 訛可 [a general of Chin] fled away from the ship, together with his defeated army of three thousand soldiers. The Mongols pursued them with clamour and uproar. The shower of arrows and stones could be seen on the north bank of the river. Several miles away, the fleet of the Mongols cut off the retreat of their enemy, who was unable to get through. The heaven-shaking thunder bombs were fired on board the ships. The catapults and fire could distinctly be seen." [72]

(9) A.D. 1232. A battle between Yuan and Chin: "The Chin Tartars used ox-hide for their protection, and could not be approached. But the Yuen soldiers succeeded in defeating them by means of the fire-catapult. A fire broke out, which could not be stopped." [73]

(10) A.D. 1236. The battle between Yuen and Chin: "A local official of the Chin Tartars held the isolated city against the Yuen, who attacked it with all their might. Though believing resistance almost impossible, he collected all the gold, silver, bronze and iron, and had it cast into P'ao, 砲 in order to resist the attack." (Here P'ao probably stands for projectiles.[74]

[71] 金史, chap. 113, 赤盞合喜傳.
[72] *Ibid.*, chap. 111, 完顏訛可傳.
[73] *Ibid.*, chap. 113, 赤盞合喜傳.
[74] *Ibid.*, 郭哈瑪爾傳.

(11) A.D. 1266. A battle between Yuen and Chin: "In the second year of the period Tien Hsing, 天興 the Yuen Tartars attacked the western cities, with all their force, and set fire to the towers on the city wall, by means of the fire-catapults." [75]

(12) A.D. 1272. A battle between Yuen and Sung: "Two captains of the Sung, Chang Kuei, 張貴 and Chang Hsün 張順 loaded their ships with ammunition, sailed towards the city and met the enemy midway. Chang Hsün using a red lamp as signal, gave the order to fight, and kill the enemy. With fire-catapult and gunpowder arrow, they fought against the northern soldiers [soldiers of Yuen] who fell into the water." [76]

(13) "When the Yuen Tartars besieged Hsiang Yang, 襄陽 each was armed with fire gun 火鑢 and fire-catapult 火砲.[77]

(14) A.D. 1274. A battle between Yuen and Sung: "In the evening the wind blew strongly. Pai Yen 伯顏 ordered his men to construct a Chin Sha P'ao 金沙砲 and used it to set the houses on fire. The smoke and flames covered the sky. The city was captured." [78]

(15) "The general Chang Yung 張榮 made an attack with his artillery, and burnt the city by using the fire-catapult. The houses were completely demolished. The defender of the city was overcome. The fire-catapult was again put into action during the attack on the fortress Yang Lo 陽邏 which was taken." [79]

(16) A.D. 1275. A battle between Yuen and Sung: "The Yuan general Pai Yen 伯顏 ordered his men to construct more fire-catapults, in order to attack Lin An 臨安." [80]

[75] *Ibid.*, 四朝別史, 大金國志, chap. 26.
[76] *Ibid.*, 癸辛雜識, 別集下.
 The Sung Chi San Chao Cheng Yao, 宗季三朝政要, contains the following statement: "In the 5th month of the 8th year of Hsien Chun 葳淳 (A.D. 1271), the general Chang Shun 張順 and Chang Kuei came to the rescue of Hsiang Yang 襄陽. . . . Every ship was equipped with fire guns 火倉 and fire catapults 火砲." (宋季三朝政要, chap. 4.)
[77] 宋史, chap. 450, 張順傳.
[78] 元史, chap. 127, 伯顏傳.
 The Sung Chi San Chao Cheng Yao contains the following statement: "In the 11th month of the 10th year of Hsien Chun (A.D. 1274) there was, in the army of Pai Yen, a physiognomist by the name of Li Kuo-Yung 李國用 who was able to direct the wind [into a favourable direction]. Through his practices — the wind blew violently. Pai Yen ordered the gunner Chang Yuan-Shuai to attack the enemy (the Sung army) with fire catapults 火砲. The smoke and the flames covered the sky. The city was captured."
[79] 元史, chap. 151, 張榮傳.
[80] 元史, chap. 127, 伯顏傳.

(17) A.D. 1277. A battle between Yuen and Sung: "Lou Chien, 婁鈐 a general of Sung, ordered his soldiers to use fire-catapults; when these were shot off, a tremendous noise was heard like roaring thunder. The cities were shaken. The sky was obscured by smoke and gas. Many of the foreign soldiers [refers to Yuen] were frightened to death. When the fires had burned down, not even the ashes were left, to the astonishment of everybody who came to see it." [81]

(18) A.D. 1287. A battle between Yuen and Sung: "In the year 1287 Li T'ing 李庭 led ten strong men armed with a fire bomb into the camp

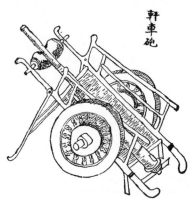

軒
車
砲

FIG. 3

of the enemy by night. When the bullet exploded, the enemy was very frightened. They killed each other, and ran away in confusion." [82]

According to the records it may thus be said that the fire-catapult was first used by the Sung army. Then it was taken over by the Chin Tartars, and finally adopted by the Yuen people. Gunpowder was used during the earlier stages for its incendiary and inflammable qualities; later for its true explosive force. Incendiary projectile consisted mainly of paper or bamboo, while iron was used by the Chin Tartars. The first true bomb or grenade seems to have been the so-called "Heaven-shaking thunder" bomb (see quotations (5) and (7) above); hurled also by the catapult, since its shape, described as like a pitcher, would certainly not have allowed its passing through a barrel. This evidently appeared in the early years of the thirteenth century.

We come now to the question of the invention of

[81] 宋史, chap. 451, 馬墍傳.
[82] 元史, chap. 162, 李庭傳.

a tube or barrel, through which the projectile should move. This invention is of cardinal importance since, as has often been pointed out, it led ultimately to the development of the external and internal combustion engines, the piston being merely a tethered projectile in the barrel-cylinder. Pistons have, of course, a much older history than this, as in the square-sectioned box-like air-bellows for metalworking among the Chinese, which go back to far

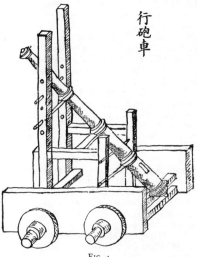

行
砲
車

FIG. 4

antiquity, or the water-pumps of the Romans (see Fig. 47 in Ch. Singer, *Short History of Science*, Oxford, 1941, p. 110). But in all these cases, pistons were the movers and not the moved, dominating liquids or gases and not dominated by them.

The book Wu Yuan 物原 states that the invention of the barrel was made by Lü Wang 呂望 as early as the Chou 周 dynasty, but all proof is lacking, and the statement [83] is not worth discussing. It is however most remarkable to find the following diagrams of Hsuan Chê P'ao (chariot gun) 軒車砲 and Hsing P'ao Chê (movable gun carriage) 行砲車 in the Wu Ching Tsung Yao 武經總要 (chap. 10) of 1040 A.D. (See Figs. 3 & 4.)

These figures illustrate an almost modern cannon, and are entirely different from the fire-catapult. There is a barrel, and the hole looks like a touchhole. But no explanation accompanies these diagrams. Probably they were inserted by 16th or 17th century

[83] 物原, chap. 14.

editors or publishers, after barrel guns had been adopted from western countries, for the present edition is derived from the Ming Dynasty copy.

It is recorded in the Sung History, 宋史 that on the 16th of the 10th month in the 4th year of Chien Tao 乾道 (A.D. 1168) Chê Pao 車砲 (P'ao on wheels), Huo P'ao 火砲 (Fire P'ao), and Yen Chiang 煙槍 (Smoke Chiang) were displayed before the emperor,[84] The Hsien Chun Yu Chiao Chi 咸淳御教記 contains a similar statement. The Chê P'ao (P'ao on wheels) may be accepted as just a catapult on wheels and has nothing to do with the barrelled Hsuan Chê P'ao or Hsing P'ao Chê of the Wu Ching Tsung Yao just mentioned. But in the records of Southern Sung 南宋 a new form of barrel does appear. The passage runs as follows: "About the year A.D. 1231, the Flying fire Gun came into use [Fei Huo Chiang 飛火槍] by the Chin Tartars. Powder was applied, and when kindled, the gun was pushed forward [surely a mis-reading for backward (recoil)?] about ten feet [by the explosive force]. Nobody dared to approach, and the Yuen Tartars were greatly frightened." [85]

"In the year 1233 the Chin Tartars also used a 'fire gun,' 火槍. During the fight the soldiers rushed with the fire gun into the army of the Northerners, who were unable to defend themselves and dispersed. In order to kindle their gun the Chin soldiers preserved some fire in a small iron receptacle. The flames that shot out of the gun had a length of ten feet." [86] It is quite probable that the "fire gun" and the "flying fire gun" were similar weapons, if not closely alike. Both could be used within a distance of twenty feet and produced flames in order to frighten the enemy. It is clearly stated that the latter had a barrel, two feet in length. This barrel was made of sixteen layers of yellow paper and was filled with willow charcoal, and a mixture of pulverized iron, porcelain powder, sulphur and arsenic.[86] Because of the absence of saltpetre, this mixture, though highly inflammable, can not be regarded as gunpowder. These fire barrels seem not to have been so frequently used as the fire-catapult. This kind of tube had of course little in common with the present-day gun, projecting a shell or bullet through an iron barrel. It was essentially a flame-thrower, and therefore more akin to the "fierce fire oil machine" of the tenth century already described. But another type of gun, called T'u Huo Ch'iang 突火槍 (rushing-out fire gun) mentioned in the Sung History, can be looked upon as the forerunner

of the modern gun. In the records we find the following statement: "In the first year of the K'ai Ch'ing period, 開慶 [A.D. 1259] the T'u Huo Ch'iang was constructed." This weapon, as its name indicates, had a tube or pipe, which however seems to have been made of bamboo. The very loud and far-reaching sound of this kind of gun is especially mentioned.[87] The projectiles must have been very different from those used with the fire-catapult, since gunpowder was used for the propulsive power. The use of bamboo for the first gun barrels recalls another aspect of ancient Chinese technology, namely the use of bamboo piping for the conveyance of natural brine, as at the famous area of Tzeliuching 自流井 in Szechuan. Bound with fibre, such bamboo tubes will stand a pressure of 80 lbs. per sq. in.

According to the book "Diagrams and Explanations of Historical Weapons," 兵器沿革圖説 guns with a metal barrel were used by the Mongols. Unfortunately no details about the nature and material of the barrel are given. In the Chinese Histories, the extensive use of barrel cannon or barrel guns by the armies of the Sung, Yuen and Chin, has hitherto been little mentioned and known, but according to the "History of the Humiliation of Japan" 日本國辱史, quoted later on, barrels were largely brought into use during the invasions of the Mongols in the years 1274 and 1281 A.D. We have the following two Yuan poems describing the use of iron projectiles and fire barrels:

The black dragon brought forth an egg as big as a tou [unit of measurement]. The egg burst, a dragon flew out, with peals of thunder rolling. It rose like a blazing sun with fire flashing red. A clap was heard, and heaven and earth were shaken, as if by the fall of mountains and rivers.

Or, "The iron gate of the entrance was still closed, the fire pipe 火筒 was already red hot [from constant use]. Hua Yao 花猺 [a barbarian] was overcome and fled away. Puddles were filled with blood running through the city, in all directions. Fire like a cloud, spread over ten li with fire-birds flying." [88] The black dragon and the fire pipe probably refer to the same thing, and the description of the bursting egg shows that some kind of grenade or shell was used. This serves to refute the theory that the origin of Huo P'ao and Huo Chiang (both understood as involving an iron barrel) can be traced back only to the Ming Dynasty.[89] It is unfortunate that we know nothing of the metallurgical difficulties overcome by the makers of the first iron gun barrels.

---

[84] 宋史禮志, chap. 121.
[85] 金史, chap. 113, 赤盞合喜傳.
[86] 金史, chap. 116 蒲察官奴傳.

[87] 宋史, chap. 197, 兵志.
[88] 張憲, 王筒集, chap. 3.
[89] 吳承志, 遜齋文集, chap. 5.

Summarizing the contents of this section, we may say that while stone-throwing catapults go back beyond the Christian era and while incendiary arrows date from at least the third century A.D., gunpowder is not found until the tenth. Gunpowder seems first to have been used for its incendiary properties, on arrows (970–1040 A.D.), some of which may have used the rocket principle, and in flamethrowers (919–1000 A.D.) Between 994 and 1393 A.D., when fire-arrows became regular battle equipment in the Chinese forces, there are many descriptions of their use. Mines and smoke-screens appear at the beginning of the eleventh century.

At this time, too, the projectiles hurled by catapults develop from stones to various kinds of incendiary objects, some containing gunpowder; there are many accounts of this between 1000 and 1287 A.D. Compositions resembling Greek fire are noticed around 1161 A.D. Not until 1206 or 1231 A.D. can the true explosive grenade be identified. In 1231, too, developing from the flamethrower, we find a tubular weapon, which by 1259 has become a true barrel gun, though of bamboo, and by 1275 a gun with a metal barrel.

In 1240 the Mongol invasion of Europe was in full swing.

5. HOW AND WHERE DID THE KNOWLEDGE OF GUNPOWDER SPREAD?

H. G. Wells, in his *Outline of World History*, supposes that it was through the Mongols, at the time of their invasion, that gunpowder became known to Europe. This is still a question open to discussion. The Mongol invasion began in the year 1235 A.D. while the first mention of gunpowder in Europe appears in Roger Bacon's treatise *De mirabile potestate Artis et Naturae* in 1242. But Roger Bacon did not know the propulsive force of gunpowder. Berthold Schwarz constructed the first firearms in 1354, and probably for this reason the invention of gunpowder has also been attributed to him.

Apart from the Mongols, other countries forming the link between Asia and Europe must also be taken into consideration regarding the early use of gunpowder and the spread of knowledge concerning it. For the period of the Crusades no historian, either Christian or Moslem, alludes to gunpowder. Nevertheless there can be little doubt that the so-called Hui Hui P'ao 回回砲 which played a part in the fighting between China and the Mongols, had been adopted by the latter from Arabia. What kind of a P'ao this weapon actually was, has not yet been decided. The scholars of the Ming Dynasty supposed that it was a kind of P'ao in which gunpowder was used.[90] In the books Lang Chi Ts'ung T'an 浪跡叢談 and K'ai Yu Ts'ung K'ao 陔餘叢考 it is even said that the Hui Hui P'ao, which was used in the attack on Hsiang Yang, 襄陽 marked the beginning of the use of Huo P'ao 火砲 (fire cannon).[91] The historian Visdelou shares this opinion.[92] According to the Japanese writer 有坂鉛藏 the Hsiang Yang P'ao 襄陽砲 or Hui Hui P'ao 回回砲 consisted of an iron barrel, through which the projectile was sent off by the explosive force of gunpowder.[93] Various descriptions of the weapon in question follow:

(1) "The P'ao was put up against the south-east of the city. The projectile weighed one hundred and fifty pounds and its sound shook heaven and earth. It broke whatever it struck and penetrated seven feet deep into the ground. Li Wen-Huan 李文煥 the general of Sung was alarmed, and surrendered the city." [94]

(2) "One P'ao [P'ao here stands for the projectile] reached the watch-tower, and made a noise like thunder. Great confusion reigned within the city. Many jumped over the walls and surrendered." [95]

(3) "The large P'ao was used to attack them. The sound could be heard over a hundred miles away. The army of Sung was forced to move." [96]

(4) "The insurgents [referring to the Mongols] shot the Hui Hui P'ao at the city of Hsiang Yang and destroyed the fortifications, towers and pillars." [97] In all these descriptions the tremendous sound and the effect of the projectile in penetrating into the ground or demolishing fortifications are emphasized. Because of this effect, some scholars have assumed that gunpowder was used with the P'ao in question, but there is absolutely nothing said

[90] 續文献通攷, chap. 134.
[91] 浪跡叢談, chap. 5, 陔餘叢書, chap. 30.
[92] Visdelou, *Supplément de l'Histoire de la Tartarie*, p. 188. [93] 兵器沿革圖說, chap. 3, sec. 2.
    Mr. Wu Chêng-Chih held the same opinion that the Huei Huei P'ao was a kind of P'ao hurling stones. But he identified this P'ao with the Huo P'ao which the Mongols used in attacking the city of Pien Ching. To the latter view, I can not agree (see note 89, chap. 5). The Huei Huei P'ao is recorded in the Yuan History as a "new" P'ao. From this, Mr. Chiang concluded that the Huei Huei P'ao differed by the use of fire from the former P'ao. (See 姜先生全集, vol. 12, 湛園札記, chap. 22.) But it is evident that the expression, "new P'ao," refers to an innovation in the technical construction of the P'ao; and no way points to the use of fire with it.
[94] 元史, chap. 203, 亦思馬因傳.
[95] *Ibid.*, chap. 128, 阿里海牙傳.
[96] *Ibid.*, chap. 127, 伯顏傳.
[97] 鄭所南, 心史, 卷下雜文大義略叙.

in the descriptions about fire and explosions, which would form the strongest argument in favour of gunpowder. The smashing of towers and walls accompanied by a loud noise, could as well be accomplished by the throwing of huge stones. In the Yuan History there is the following passage: "A big stone p'ao, which is identical with the Hui Hui P'ao, could be shot at a distant target with little strength." [98] In this case there is no doubt that a catapult is meant, a "huge stone p'ao" used for throwing stones. In the *Travels of Marco Polo*, "mangonels" are mentioned, capable of throwing stones of 300 lbs. of weight, causing the crash of buildings and showing considerable penetrating power. [99] With such a description the interpretation of a "barrel" "gun" would be absurd. Sung Ying-Hsing 宋應星 the author of the 天工開物, a scholar of the Ming Dynasty, however, supposed the Hsiang Yang P'ao to have had a kind of barrel, but Ch'iu Chun, 邱寶 who had probably seen some ancient diagram, says definitely in his Tai Hsueh Yen I Pu 大學衍義補 that it was simply a stone-throwing machine. [100] Probably at the time of publication of the former book, the construction of the Hsiang Yang P'ao had entirely changed and its original form fallen into oblivion. There is in fact sufficient evidence to believe that the Hui Hui P'ao or Hsiang Yang P'ao, adopted from the Arabs, was a stone hurling machine, of some particular construction, and there is no reason to believe that with it gunpowder was introduced into China.

Another important fact may be considered here. The Arabian word "Thelg as Sin" for saltpetre, means "Snow of China," and indicates that saltpetre was originally introduced from China. It is known that saltpetre and sulphur were largely exported from China in ancient times. We know of an edict, dating of the fourth year of Chih P'ing 治平 (A.D. 1067) prohibiting the export of the two materials to foreign countries. [101]

Liang Chang-Chu 梁章鉅 a scholar of the Ch'ing Dynasty put forward another theory, namely that gunpowder was introduced from the west together with fireworks. [102] Fireworks of some sort appear as early as the third century A.D. in Europe. Claudian, writing in the fourth century, gives a

poetical description of whirling wheels and dropping fountains. [103] Poems dealing with the same subject in China appear in the 7th century only. However, until the time of the Crusaders, who carried back with them the knowledge of incendiary compositions, no mention of fireworks can be traced in Europe after the year 476. Moreover saltpetre became known in Europe only in the thirteenth century. [104] The above mentioned Roman firework had therefore nothing in common with the fireworks of a later time in China, known as the Fire Drug Plays, 火藥雜戲. The book of Marcus Graecus, *Liber ignium*, in which saltpetre, sulphur and charcoal are mentioned (together for the first time) was formerly said to have been published in the year A.D. 846, but according to George Jacob, it is a publication of the year 1250. [105] H. W. L. Hime also insists that gunpowder was not known in Europe till the thirteenth century.

China, with its naturally occurring saltpetre, had of course the greatest advantage over all other countries for an early knowledge of the mixture of saltpetre, sulphur and charcoal. This fact has always been acknowledged in the west. E. Littré in his *Dictionnaire de la Langue Française* writes: "I am not at all surprised that the Chinese invented gunpowder fifteen hundred years ahead of us; their soil is full of an excellent saltpetre. . . ." [106] In Europe the knowledge of gunpowder was immediately followed by the construction of firearms making use of the explosive and propulsive force of the powder, while in China progress was much slower [107] and for centuries gunpowder was used

[98] 元史, chap. 7, 世祖本紀.

[99] Yule, *Travels of Marco Polo*, edited by H. Cordier, Book II, Part III, chap. 70, p. 159. Cf. 宋應星, 天工開物, chap. 8, title 砲.

[100] 續文獻通攷, chap. 134. Cf. Fadl Albak Rashid-eddin, *Histoire des Mongoles*, edited by Blochet, tome II, p. 65.

[101] 宋史, chap. 186, 食貨志.

[102] 浪跡叢談, chap. 5, 陔餘叢攷, chap. 30.

[103] *Encyclopaedia Britannica*, art. "Fireworks."

[104] *Encyclopaedia Britannica*, art. "Gunpowder."

[105] George Jacob, *Oriental Elements of Culture in the Occident*.

[106] E. Littré, *Dictionnaire de la Langue Française*, word "Saltpètre."

[107] Though it has been proved that the fire catapult was known and often used since the Northern Sung, it was apparently not yet generally adopted as a necessary part of the army by the time of Southern Sung. The Ch'un Yu Lin An Chih 淳祐臨安志 (chap. 6), and Hsien Ch'un Lin An Chih, 咸淳臨安志 (chap. 14), give the various classes of the army as infantry, cavalry etc., but no "gunners" are mentioned. The K'ai Ch'ing Szu Ming Hsu Chih 開慶四明續志 in the section 出戍 which enumerates the arms forming the equipment of the soldiers sent to guard the frontiers, makes no reference to P'ao. The same book in the section 作院 gives a record of the production of arms in the manufactory at the capital from the 10th month of the 6th year of Pao Yu up to the 5th month of the first year of K'ai Ch'ing. This statement is very detailed but no fire weapon or part of it is recorded. It seems that firearms were not used as a chief arm until they became a standing institution with the army of the Yuan Dynasty. This shows again the slow development of firearms in China.

for its inflammable properties only. It is undisputed that the Fo Lang Chi 佛郎機 Hong I P'ao 紅夷砲 etc. from the time of Wan Li 萬曆 were made after the European pattern. Nevertheless, the early use of Huo P'ao in the beginning of the Northern Sung Dynasty and that of the fire-barrel in the Yuen Dynasty allow no denial.

With regard to the possible introduction of gunpowder from India, it is said that Alexander the Great suffered a defeat because the Indians used firearms.[108] But that does not necessarily mean that they used gunpowder; moreover, Alexander according to other reliable records, never suffered any defeat in India. It is true that saltpetre was known at an early date in India. As mentioned above, the Arabian word for saltpetre means "snow from China." The Persian language calls it "Salt from China or India" yet the Sanskrit word "pakya," in the sense of "Saltpetre," does not occur in any literary text. This suggests that the word was probably coined by lexicons of a late date for the sake of completion, i.e. to enumerate all known substances. The word seems to have been invented in a rather late period.[109] In China, as we have seen, the character for saltpetre is used before the 1st century B.C.

An iron or copper barrel is recorded in the Ming History, under the name of Shen Chi Ch'iang 神機鎗 which means "Magical-Machine Gun." It is said to have been adopted from the natives of Chiao Chih, 交趾 (Annam).[110] But before the expedition to Chiao Chi the barrel had already been used by the army of Ming. This is proved by the following passages: "In the 3rd month of the 21st year of Hong Wu [A.D. 1388] Mu Ying 沐英 used Huo Tung Shen Chi Chien 火銃神機箭 to defend himself against the elephant army of Ssu Lun-Fa 思倫發." [111] In the year A.D. 1393 the number of the so called "Gunners" 砲手 was fixed by military regulation. In the Tien Lueh 典略 it is further said that the arsenal produced every three years "round-hole bronze barrel" (guns) 碗口鎗銃 and "bronze barrels with handles" 手把銅銃 to the number of 3000 each. In the same book the names of other types are given as follows: "Magical Gun" 神鎗 "Magical Barrel" 神銃 "Barrel Shooting Horse," 斬馬銃 "Iron Barrel with Handle" 手把鉄銃.[112] All these refer to a time before the

expedition to Chiao Chi. The foreign barrel gun may have been superior to the Chinese, but the early existence of the latter can not be denied.

Another source dating from the Ming Dynasty remarks that the so-called Niao Ch'iang, 鳥鎗 the barrel of which had the shape of a bird's bill, originated in Japan. But Ch'i Chi-Kuang, 戚繼光 a Chinese general, reports having seen this type of weapon in a Chinese arsenal before the time of the Japanese arrival.[113] At a time when China made a wide use of gunpowder, it was still unknown in Japan.

It is generally acknowledged by Japanese scholars that gunpowder was introduced into Japan by the Mongols. The following quotations are all from Japanese sources.

(1) From Pa Fan Yu Tung Hsun 八幡愚童訓: "In the 11th year of Yung P'ing 永平 [A.D. 1274] the Mongols cast the iron P'ao which caused light and noise when fired. The people were bewildered at the sight, the ears were deafened by the noise. The soldiers lost their nerve, and could no more distinguish right from left." [114]

(2) The T'ai P'ing Chi 太平記 says that, "the iron P'ao shaped like a bell, made a noise like a car rolling down boards, or like a clap of thunder. Two or three thousand bullets were shot out. Many of the soldiers were burnt to death. The towers of the city wall caught fire, and could not be extinguished.[114]

(3) In the "Illustrations of Takesaki Goro" 竹崎五郎 a rough diagram is shown, to demonstrate the action of the shells. Yano Jinichi and Tsubai have proved that the diagram illustrates a kind of explosive shell in which gunpowder was used.[115] (See fig. 5.)

FIG. 5

(4) Another Japanese book, 日本國辱史 "History of Japan's Humiliation," contains the following detailed description: "During the 11th year of Chih Yuan [A.D. 1274], in the battle of Tsu Shima, 對馬島 fire-barrels and fire-arrows were used by the Mongol fleet. When attacking Iki Shima 壹岐島 the Mongol army was equipped with fire-barrels and poisoned arrows, while the Japanese only defended themselves with hoes and spades. . . .

---

[108] 工兵沿圖說, p. 75.

[109] Monier Williams, *Sanskrit-English Dictionary*, character "pakya."

[110] 明史, chap. 92, 兵志; 丘濬, 大學衍義補, chap. 122.

[111] 皇明實錄. 洪武二十一年三月沐英以火銃神機箭防禦思倫發之象陣 (中央研究院歷史語言研究所校本)

[112, 113] 續通攷, chap. 134.

[114, 115] 兵器沿革圖說, chap. 3, sec. 2.

Mongol soldiers, over ten thousand in number, shot off fire-barrels and arrows, with a noise like the roaring of the sea. . . . The Mongols surrounded the city using fire-barrels and arrows to frighten the inhabitants. . . . In the attack with 火箭 fire-arrows and poisoned arrows were shot at the defenders. While disembarking at 博多灣 poisoned arrows and fire-arrows were used simultaneously, while fire-barrels and fire-catapults protected the landing soldiers. . . . The Mongol army fired the powerful fire-catapults, whereupon the horse of General 青屋 was so frightened that it jumped and ran away, and could not be controlled. . . . Bullets from catapults together with poisoned arrows came down like rain; the Mongol battalions concentrated their force, and made a mighty attack under the protection of barrels and catapults. In the 4th year of Hung An 弘安 [A.D. 1281] during their second invasion of Japan, the Mongols, twenty thousand in number, equipped with barrels, catapults, poisoned arrows and fire-arrows, made their landing. . . . While landing at Ike Shima, the Yuen army made great use of the barrel-gun, terrifying the Japanese. . . . The Yuen soldiers bombarded the Japanese army with huge cannons, on board of their ships, so that the sky was dark with smoke, and one could not see and distinguish, even within a distance of one foot. . . . In the defence against the Japanese attack, the Yuen soldiers shot off the fire-barrels. . . . In the battle of 能古 and 志賀, the Yuen general Fan Wen-Hu started his attack by using the big P'ao, small barrel poisoned arrow, and fire-arrow. . . ." [116]

(5) In the new Yuan History 新元史 it is mentioned that in the twelfth year of Chih Yuan 至元, (A.D. 1275) Yuan Tartars shot iron bullets, Tieh P'ao 鉄砲 and killed numerous Japanese soldiers.

(6) In A.D. 1281 the Yuen army used fire-catapults while bombarding the Japanese, who suffered a defeat.[117] It must have been in these fights that the Japanese got acquainted with gunpowder, and learned its use.

Another passage is quoted by the same authors from the "History of Korea": "During the third year of Hsin Yu" 辛褟 [A.D. 1377], the office of a Huo T'ung Tu Chien, 火㷱都監 [meaning an officer in charge of barrel-guns] was first established. This was suggested by a certain Ts'ui Mao-Hsuan, 崔茂宣 who had lived in the same city with Li Yuan, 李元 who was known for getting saltpetre for the Yuan army. It is from him that Ts'ui Mao-

Yuan learned the procedure of preparing gunpowder, trained his own workmen, and proposed the above-mentioned establishment of a "fire-barrel Office," [118] *i.e.*, Huo T'ung Tu Chien. This same statement is contained in the book Kao Shih Ts'o Yao.[119] This is reliable proof that the Koreans learned the use of gunpowder from the Yuen.

There is still some more to be said about the Mongols. Gustav Schlegel, in his article, "The Invention and Use of Gunpowder in China," mentions the Mongol expedition to Java in which, according to Hageman's *History of Java*, the sound of the P'ao is recorded. Schlegel further quotes Pauthier's *Marco Polo* to the effect that in the year 1237 A.D. ten Mongols, armed with guns, so frightened their enemies, that they all fell down. Schlegel draws the conclusion that the Mongols at the time were actually in possession of firearms. Java probably learned their use from them.[120] As for the third Mongol expedition to Europe, Souboutai, 速不台 the general who was commander-in-chief and achieved such remarkable success, was the same who had previously played an important part in China. He used the fire-catapult in attacking the Chin Tartars. It was he too, who was attacked by the "heaven-shaking thunder bomb" and forced to retreat. In short, he was well accustomed to the use of firearms, before the invasion of Europe. Paying attention to the dates of the various campaigns, it appears that in the year 1232 A.D., three years prior to his inauguration as deputy commander-in-chief of the army which was to invade Europe, this general lost a great number of his men during a vain attempt to take the city of Pien Ching 汴京 which was defended by fire-catapults producing tremendous explosions. In the same year the shooting of the fire-catapult in order to burn the ox-hides occurred under his command (as mentioned above). In the following year the general occupied the city of Pien Ching which was apparently a center of ammunition works. By imperial order, he was prohibited from killing the workmen, and specifically ordered to spare their lives. Thus the skilled workers were saved. In the year 1234 he destroyed the Chin Tartars at the city of T'sai Chou, 蔡州 by making use of the fire-catapult,[121] and it was one year after the conquest of this city that he was called back in order to lead the great army and overrun Europe. It is surely obvious

---

[116] 石榮暲, 元化征倭記, chap. 5, 7. This is a free translation of a part of the 日本國歷史.

[117] 新元史, chap. 250.

[118] 高麗史, chap. 133, quoted in 兵器沿革圖說.

[119] 李晬光, 芝峰類說.

[120] G. Schlegel, On the Invention and Use of Firearms and Gunpowder in China Prior to the Arrival of Europeans, *T'oung Pao*, series II, vol. III.

[121] 明史, chap. 92, 兵志.

that a general who had gained such wide experience in the use of the fire-catapult on the battlefields of China would make use of it in the same way in Europe.

Though it is recorded in western history that the catapult was used during the Mongol wars, unfortunately no details are given as to the kind of projectile or its effect. Howorth describes the battle of Waradin as follows:

They [referring to the Mongols] then bombarded the citadel with seven ballistas until they breached its walls and finally stormed it.[122]

D'Ohsson gives the following statement:

In the year 1241, at the bridge of the Sayo river, the Hungarians were chased by a battery of seven catapults. . . . They [referring to the Mongols] surrounded this place [the fort of Waradin] and bombarded day and night with seven ballistas a part of the wall which had just been repaired. When it was overthrown, they took the fortress by assault. . . . The Mongols first pitched their camp at a certain distance from the city [Strigonie], until thirty catapults had been constructed. After that they surrounded the place, and their prisoners erected a rampart of fascines, under the protection of which they put up their machines which were in action without interruption.[123]

Though these accounts are rather vague, it is clear that the catapult or ballista played a prominent part also on the battlefields of Europe. With the same catapult stones or grenades could be hurled. The "catapult" of D'Ohsson's book was translated in the Yuan Shih I Wen Cheng Pu 元史譯文証補 [124] and the Meng Wu Er Shih Chi 蒙兀爾史記 [125] as "P'ao" 砲, while in the biography of Cadan 合丹 in the New Yuan History 新元史 it is rendered as "Huo P'ao," 火砲 meaning fire-catapult.[126] The Meng Wu Er Shih Chi is based on western sources, and contains the following important passage: "In the year 1237 the Mongols attacked with the P'ao during five days. On the sixth day the city was taken. The powerful soldiers threw the Huo Kuan 火罐 (fire-pot) and rushed into the city, crying and shouting, in spite of the smoke." [127] These fire-pots may have been similar to the above mentioned "heaven-shaking thunder bomb" which contained gunpowder. Considering the general skill of the

Mongols in war, their use of the stirrup and their famous bows and arrows, it is quite possible that they also used fire-arrows. The Mongol invasion brought about a more intensive study of arms in Europe, and at least prepared the way for the use of gunpowder, if it was not directly responsible for its introduction to western countries.

6. Conclusions

1. Saltpetre and sulphur appear in Chinese texts before the first century B.C., although the former became known in Europe only about the 13th century A.D.

2. In the third century A.D. saltpetre and sulphur were mixed by the alchemists in the same proportion as in gunpowder. As these mixtures were exposed to high temperatures, preliminary conditions for a discovery of gunpowder were created.

3. At the beginning of the 7th century, some kinds of fireworks, bearing the name of "fire-drug" later used for gunpowder itself, are described in the records, but the actual use of gunpowder in them cannot be established.

4. In the year 919 A.D. (in the time of the Five Dynasties) and the beginning of Sung Dynasty, Chinese troops used on the battlefield a flamethrower worked with natural petroleum, in which gunpowder may have been involved as an igniter. The invention of gunpowder may thus be traced to the beginning of the tenth century.

5. In the year A.D. 969 fire-arrows containing gunpowder were invented and widely used in the fighting between the Sung, Chin and Yuan armies.

6. In the year 1000 A.D. the catapult hurling an incendiary projectile made its appearance. This new weapon played an important part in the battles of that time, and as gunpowder was contained in the projectile, it became necessary to prepare the substance on a large scale.

7. In the year 1040 A.D. the Wu Ching Tsung Yao was written, stabilising the name of Huo Yao 火藥 for gunpowder. Detailed formularies and methods of preparing gunpowder are contained in this book.

8. In the year 1103 A.D. fire-works containing gunpowder were used in the capital, while its explosive quality is first mentioned in connection with crackers at the same time.

9. The year 1231 A.D. marks a stage in the use of gunpowder with the "heaven-shaking thunder bomb," the first true explosive grenade. It was used with catapults.

10. The true barrel gun appears with bamboo

---

[122] Howorth, *History of the Mongols from the 9th to the 19th century*, chap. 2, Batu Khan, p. 49.

[123] D'Ohsson, *Histoire des Mongols*, book 2, chap. 3, p. 142, p. 143, p. 149, p. 154, p. 619.

[124] 洪鈞, 元史譯文証補, chap. 5, 拔都補傳.

[125] 屠寄. 蒙兀尔史記, chap. 35.

[126] 新元史, chap. 111, 金丹傳.

[127] 蒙兀尔史記, chap. 35, 巴禿傳.

tube in 1259 A.D. and with metal tube in 1275 A.D.

11. The Mongols acquired their knowledge of gunpowder from China, and introduced it to Japan, Korea, Java and Europe. Evidence of the origin of gunpowder in India is not convincing.

12. In Europe acquaintance with gunpowder was immediately followed by the construction of relatively efficient weapons, taking advantage of its propulsive and explosive power, its quality as an incendiary being considered of secondary importance. But this was perhaps because the knowledge of gunpowder had already undergone a long evolution in China, passing through different stages. Known since ancient times because of the natural occurrence of saltpetre in the soil, it was first used for fireworks, crackers, and a number of rather fantastic weapons serving incendiary and smoke-producing purposes. Reliable sources reach back to the tenth and eleventh centuries. With the 12th century the last stage was reached, when gunpowder was used for its projective force. The western countries, learning of its properties in the 13th and 14th centuries, soon surpassed China in the construction of firearms, e.g., in the 15th, 16th and 17th centuries, but her priority in the invention of gunpowder and its use in war can not be denied.

An Account of the Salt Industry
at Tzu-liu-ching
Tzu-liu-ching chi

BY LI JUNG

(Introductory Note and Translation By LIEN-CHE TU FANG *)

INTRODUCTORY NOTE ‡

I. SALT IN CHINA

In most agrarian civilizations, salt has been a matter of major concern. Its universal use and the conspicuous manner of its distribution make it a decisive source of government revenue and an ideal article for fiscal management.

China's experience offers no exception to this general rule. The origins of certain North Chinese centers of salt production are associated with the country's legendary

* This translation was made for the Chinese History Project in connection with its work on the Ch'ing dynasty. The Project, under the sponsorship of the University of Washington and in cooperation with Columbia University, is engaged in writing a documentary History of Chinese Society.

‡ The Chinese characters corresponding to the transliterations will be found at the end of the article.

rulers, and the early discussions of Chinese statesmen show great interest in the public management of "salt and iron." Chinese historiography, being essentially practical and administrative, has provided us with an unbroken stream of information on salt production and salt revenue.

Salt in China has been and still is produced mainly in three forms. Along the seacoast from Manchuria down to Kwangtung, salt is evaporated from sea water. In the Northwest, in Kansu, Shensi, Kokonor, and Mongolia, it is evaporated from the water of salt ponds or salt lakes. In the southwest, it is derived from rock salt reached by wells. In Szechwan, which is particularly well-endowed with salt deposits, the rock salt occurs in some places together with natural gas. Among the areas so favored, Tzu-liu-ching is outstanding. Here the salt producers have developed ingenious techniques for accelerating the process of evaporation by burning the gas.

The Tzu-liu-ching site is more than 300 *li* or over 100 miles west of Chungking [1] and covers some 60 square miles.[2] The salt industry in this region involves operations of both production and circulation, some of which require a considerable amount of labor and capital.[3] The number of wells has varied over time; it has also varied with the sources of information. Some claim there are between 1000 and 2000 wells, others set their estimate as high as 4000.[4]

Many Chinese and Western authors have discussed the production of salt at Tzu-liu-ching. Li Jung's account, however, is remarkable, and perhaps unique because of its extraordinary attention to the geological and technical aspects of the industry. It is for this reason that a complete translation of his account of Tzu-liu-ching is offered here.

2. THE AUTHOR OF THE ACCOUNT [5]

Li Jung was born around 1820 in Chien-chou in Northern Szechwan. In 1843 he passed the provincial examination and obtained the degree of *chü-jen*. Four years later he passed the metropolitan examination and became a *chin-shih*. During the Taiping Rebellion, Li Jung assisted the leader of the imperial forces, Tseng Kuo-fan, and in 1864, in recognition of his meritorious behavior, he was appointed Lieutenant-Governor of Hunan. His incumbency was short; he was dismissed the same year, having aroused the displeasure of certain rich individuals through his eagerness to raise funds in order to maintain troops. In his later years, Li Jung devoted himself to teaching in his native province, first in the Academy of the Chiang-yu district, then in K'uang-shan Academy in Mien-chou. He died in 1889, at about seventy years of age.

Li Jung's collected works, entitled *Shih-san-feng shu-wu ch'üan-chi*, have appeared in two editions, one printed from wooden blocks and published in 1890, the other, a lithographic edition, published in 1899. The present translation is made from the earlier edition, *Wen-kao* section, pages 1a–5a.

AN ACCOUNT OF TZU-LIU-CHING

Ninety *li* northwest of the district seat of Fu-shun, there are five clusters of salt wells: the T'ai-yüan wells, the Chan wells, the Wang wells, the Hsü wells, and the Fu-i wells. The last is also known as the Tzu-liu wells [or the Tzu-liu-ching]. A story which has been handed down tells us that on the west bank of Jung-ch'i, salt water once bubbled up in a manner

---

[1] V. K. Ting, *New Atlas of the Chinese Republic*, 1934, map 34.

[2] *Richard's Comprehensive Geography of the Chinese Empire*, 1908, p. 114.

[3] For an analysis of the economic structure of Tzu-liu-ching, cf. K. A. Wittfogel, *Wirtschaft und Gesellschaft Chinas*, Leipzig 1931, pp. 530–535.

[4] The Chinese Government Bureau of Economic Information, *The Chinese Economic Monthly*, Dec. 1926, p. 521.

[5] For the Chinese characters see list at the end of this article. The transliterations are alphabetically arranged.

similar to that of the white and black salt wells in Yunnan. Later, stones and rubble from a landslide covered this area, and the salt spring dried up. From that time on, the natives drilled wells in this region.

These wells are under five registry offices: the T'ung-tzu Tang, the Lung Tang, the Hsin Tang, the Ch'ang-fa Tang, and the Ch'iu Tang. Taxation on salt is administered and recorded by official registrars. In modern times the places where the archives are kept in the different government boards are called *tang*. This is the origin of the salt *tang* of Tzu-liu-ching.

Before drilling a well, an examination of the area should be made in order to set aside space for erecting the necessary buildings. These structures serve six purposes: [the building] where money is handled is called the cashier's quarters; that immediately above the well is called the pestle quarters; that where the buffaloes turn the wheel to bring up the brine is called the wheel and buffalo quarters; that where the salt is stored in buckets is called the bucket quarters; that where the brine is evaporated is called the oven quarters. Only when the six quarters are all complete can a well be operated.

In a salt yard there are four departments with managerial functions. To survey the land and to supervise the drilling of the well is the task of the department of the well. To take charge of the brine and the ovens and to calculate the amount of salt after evaporation is the task of the department of ovens. To install bamboo pipes for transporting the brine from various distances is the task of the department of pipes. To supervise the storage and transportation by water or land and the sale of salt is the task of the department of sales. These four departments of the well, the ovens, the pipes, and the sales shoulder the greatest responsibilities of the salt business.

Men who make decisions on policies of either government or private concern in the salt yards are known as registrars. Men who determine the time of the sales and the prices are known as brokers. And men who are in charge of employees for miscellaneous errands, and of interviewing and receiving customers, are known as contact men.

Men who do the drilling are called mountain-workers, those who evaporate the salt are called salt-workers, and those who install the gas pipes and gas rings are called gas-workers. The gas-workers hand down their skill from father to son and learn no other trade.

To drill a well, an opening is dug at the surface three feet in diameter and nine feet in circumference. First, stone rings shaped like mortars but open at the bottom are prepared. For the first hundred feet or so, these rings are fitted in the well one above the other. Below this level and for some 300 more feet, wooden rings with hollow centers and shaped like cylinders are fitted one above the other. This is done to keep the fresh water out. Then a board for the lever is installed above the well and to this a bamboo rope is attached with an iron drill tied to the other end. Some ten people take turns in treading it. In shape this device resembles a pestle, and that is the reason why the shelter above the well is called the "pestle quarters."

Below the stone rings large drills known as "fish-tail drills" are used. They are 12 feet long and weigh 200 catties (267 lbs.) each. Below the wooden rings, smaller drills are used. These are called "silver-ingot drills." They are 9 feet long and weigh 100 catties (133⅓ lbs.) each. Below 1000 feet the pull is so great that it is no longer safe to use a large drill for fear that the bamboo rope may break.

One day's labor may drill more than a foot, 7 to 8 inches, or 4 to 5 inches. Sometimes it may take several days to drill one inch. When brine is reached it is called "success." Usually it takes 4 or 5 to some 10 years to reach the stage of "success." There have been cases where it has taken tens of years with several changes of ownership. If, at a depth of 3000 feet, the amount of brine is still insufficient, the well is abandoned and labeled "useless well."

In drilling wells the different strata of the earth should be noted. The red-stone stratum comes first; then the tile-gray; then the yellow-ginger, at which oil appears; then the grass-white; then the yellow-sand, at which a weak gas called "grass-surface fire" appears; then the black-sand, then the white-sand, at which level yellowish water appears; then the coal stratum, the hemp-hoop and the dark-smoke; and then the green-bean, at which level dark water appears. The red-stone stratum is formed by red stone and sand. All the designations of the strata, such as tile-gray, yellow-ginger, hemp-hoop, and green-bean, are so called because of their resemblance to these items. The coal can be burned as fuel and the black-smoke looks like fine powder. Not all the strata are found in drilling any one well, but the yellow-ginger and the green-bean strata should always be present. Sometimes a fibrous stratum [asbestos?] appears, and it is the most difficult to drill. Ten feet of this fibrous stratum may take a whole year's time to drill through.

The brine which appears at a depth of some 700 to 800 feet is known as "grass-surface water." One bowl of this brine can be evaporated to 4 or 5 *ch'ien*[6] of salt. Two hundred and eighty bowls of this brine make one *tan*,[7] which is worth some 5 or 6 *fen*[8] of silver. The brine

---

[6] A unit of weight, less than half an ounce.
[7] A load carried on a shoulder pole with two buckets attached to each end of the pole.
[8] One hundredth of a tael of silver.

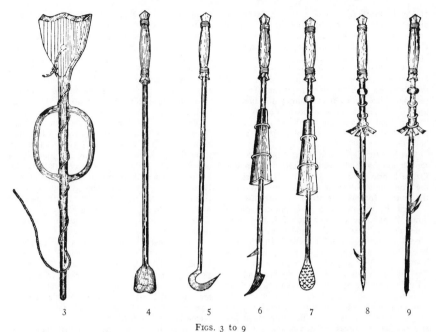

3   4   5   6   7   8   9

FIGS. 3 to 9

The figures above, like those on the plates, are taken from the *Ssŭ-ch'uan Yen-fa Shih, or History of the Salt Administration of Szechwan*, 1882. Reading from left to right: FIG. 3, "Fish-tail drill"; FIG. 4, "Silver-ingot drill"; FIG. 5, "Sweeping scythe"; FIG. 6, "Side-blades"; FIG. 7, "Wooden-dragon"; FIG. 8, "Fish-hook sword"; FIG. 9, "One-legged rod."

which appears at a depth of 1200 to 1300 feet is known as "yellow water;" one bowl of it can be evaporated to something over one *liang*[9] of salt, and one *tan* of this brine is worth more than one *ch'ien*.[10] The brine which appears at a depth of 2600 to 2700 feet is known as "dark water." One bowl of this brine can be evaporated to more than 2 *liang* of salt, and one *tan* of it is worth more than 3 *ch'ien* of silver. The bowls and *tan* vary in capacity, and the different kinds of brine vary in quality. Nevertheless, they are judged by the three above-mentioned criteria. The price of brine is determined by its weight. The so-called "grass-surface water" is brine that is lightest in weight. A well of 2600 to 2700 feet in depth reaches the heaviest brine.

As for the so-called "grass-surface fire," it is a weak type of gas. In a gas well, when a depth of 2600 to 2700 feet is reached, abundant gas is obtained. In drilling a gas well, when "success" is reached, the gas rises from the bottom of the well with a rumbling noise. It

[9] One-sixteenth of a catty, about one-third of an ounce.

[10] A monetary unit, one-tenth of a tael of silver.

looks like a steaming cauldron, has a disagreeable odor similar to that of natural salt, and is icy cold. It is not really fire, but gas. When ignited by real fire it burns. With enough gas, the flame is not extinguished even in storms. When the gas is weak, it can be extinguished either by immersing in water or smothering with earth. Extinguished, it again assumes its gaseous form. Therefore it can be carried either in bamboo or wooden pipes and the pipes do not burn. It can also be stored in a pig or cow bladder. If the bladder is moistened daily with alcohol, it can be kept for months. If the bladder is pricked with a needle, then the gas emerges from the hole and, when ignited, can burn for a few hours. It ceases to burn only when the gas is exhausted. It resembles a kind of toy made from saltpeter.

As soon as gas emerges from a well, a wood basin is placed upside down on top to hold it in. The basin is 10 feet high, 10 feet in diameter, and 30 feet in circumference. The lower part [of the basin] is wide, but it becomes narrower towards the top so that the gas is controlled. Bamboo pipes are installed along the sides of the basin to carry the gas to the evaporating ovens. In the center of the bottom of the basin there is also a hole 3 inches in diameter. This hole, the opening for the well, is lined with a stone ring and banked with earth. If there is also brine in the well, it can be bucketed out as usual.

Gas is measured by the number of boilers it can serve. For one boiler the gas can be rented for about 40 taels of silver per year. The exceedingly rich gas wells can supply the fuel for some five, six, or seven hundred boilers. The less productive ones can supply a few tens of boilers, and the poorest can only support a few boilers.

The longest pipes reach more than 1,000 feet, shorter ones several hundred feet, or even less than a hundred feet. In installing the pipes for the brine it is better to work downward from the source. For gas it is better to work upward.

Some of the gas wells also produce oil in liquid form. The oil may be of four colors: "rice-soup oil" is white, "green-bean oil" is green, "gardenia-seed oil" is yellow, and "black-lacquer oil" is black. Oil of green, yellow, or black kind has an odor similar to sulphur. The odor of white oil is much weaker, and the oil itself more transparent. Cakes of dried dung of horse or cow, after being soaked in this oil, will burn even when floating on water. The oil can also cure skin troubles such as itches and scabies. In times of high prices, one catty of this oil costs about 80 cash.

The Hai-shun well is the richest in gas. It supplies more than 700 boilers. However, the well which produced all three — brine, gas, and oil — was the Mo-shui well, but both the brine and the oil gave out within two to three years. The gas produced from this well still supplies more than 400 boilers. It has been in use for 20 years and the gas is still abundant. The Te-ch'eng well once produced a strong-smelling gas and brine, which could cause death. It supplied some 500 boilers. At one time the brine from this well spurted to a height of 30 to 40 feet. In a day and a night it could yield more than 1000 *tan*. Nevertheless, in one year both the brine and gas were exhausted.

The Shuang-fu well also used to spurt as much as some 1000 *tan* of brine in a day and a night, but for a year now the flow has ceased. By using buffaloes to turn the wheel, it still yields some 100 *tan* in a day and a night. The brine from the Ju-hai well also emits a gas which is strong enough to cause death. A man, being suffocated by the gas either from the brine or gas wells, can be revived by cool air from a big fan. When a bird flies into the gas, it falls to the ground. Should a person pick it up, set it on his palm, and blow on it, the bird would be able to fly away. These are the mysteries of the wells.

Generally both the pipes for brine and for gas are made of bamboo, but occasionally gas pipes are made of wood. Around the pipes, bamboo splints are first wound, then hemp, and on top of this putty is applied. After this treatment no rain or water can seep in, nor can the brine or gas leak out. The pipes go high up into the hills and mountains or far down into the ground. They are joined together, hooked up, and supported by posts. They are more spectacular than the sight of the rainbow dipping into mountain torrents or autumn waterfalls rushing down the gully.

The wells at Hsiao-ch'i produce no gas. Therefore the brine is carried by pipes for more than 10 *li* to the west bank of Jung-ch'i. From there the pipes, covered by grooved stones, pass under water to the east bank and then to the gas wells under the registries of Lung Tang and Hsin Tang. Both banks rise as high as some 200 feet and the pipes traverse all that distance without a break, at times invisible and then again open to view. It constitutes a wonderful sight.

Trouble may develop either in the completed or uncompleted wells. When there is trouble work should be stopped. Such wells are known as "suspended wells," and must be treated according to their defects. For example, when the big drill breaks loose, the "sweeping-scythe" is used; when the small drill falls off, the "side-blades" are used; when the bamboo tube bucket breaks loose, the "wooden-dragon" is used; when the chain falls off, the "fish-hook-sword" is used; and when the bamboo rope is broken, the "one-legged-rod" is used. The various devices are indescribably clever, and the remedies are improvised without recourse to any set pattern. There are such troubles known as "falling stones," "flowing sand," or "oozing water." In case of falling stones, putty is employed for patching. In case of flowing sand or oozing water, work should be suspended until the sand ceases [to appear] or the water dries up. Otherwise there is no cure.

As to the decline and prosperity of a well, in regard to nature, it is a question of time, and in regard to men, it is fate. There are instances in which one owner drilled a well to the depth of some 2000 feet without seeing "success." As soon as it changed hands "success" was immediately reached. There have also been extremely prosperous wells of brine or gas which, as soon as their owners changed, began to decline. Concerning the cost of drilling wells, shallow ones come to thousands of taels, and the deep ones come to tens of thousands of taels. Sometimes, the cost runs to 30 or 40 thousand taels without reaching "success." Such is the uncertainty of this business.

During the K'ang-hsi and Yung-cheng reign periods [1662–1735] a tax was collected on a set number of wells. When a well was on the decline or when the brine was exhausted, the owner was still obliged to pay taxes on his nonproductive well. It was at the end of the

Ch'ien-lung period [1736–1795] that the Provincial Salt Controller, Lin Chün, conducted a personal investigation in this area and requested that taxes from newly opened wells should replace taxes from exhausted wells. The number of salt boilers were set at 508½ as before. Therefore the owners of the wells, even to the present time, worship him at a shrine.

Invisible gas fire was mentioned as early as the Han dynasty. However, there is no clear statement about it in the Wei, Chin, T'ang, and Sung dynasties. In the present dynasty, early in the Tao-kuang period [1821–1850] only weak gas was known. At that time wood or charcoal was chiefly used for evaporating salt. Only about one-tenth of the evaporating plants utilized gas from the wells. Down to the 7th and 8th years of Hsien-feng [1857–1858] the use of gas became more general. In the early T'ung-chih period [1862–1874] the use of gas increased tremendously. During the T'ai-p'ing Rebellion, Szechwan salt found markets in the districts of An-lu, Ching-chou, Hsiang-yang, Yün-yang, I-ch'ang and Ching-men in Hupeh and also Li-chou in Hunan. The salt business became very prosperous, and it was a great aid in meeting the various demands for military supplies.

Among all the salt-producing areas in Szechwan, evaporation by gas fire is regarded

---

Transliterations and Chinese Characters

An-lu 安陸

Chan 詹

Ch'ang-fa tang 長發檔

Chiang-yu 江油

ch'ien 錢

ch'ien 錢

Chien-chou 劍州

Ching-chou 荊州

Ching-mên 荊門

Ch'iu tang 邱檔

fên 分

Fu-i 富義

Fu-shun 富順

Hai-shun 海順

Hsiang-yang 襄陽

Hsiao-ch'i 小溪

Hsin tang 新檔

Hsü 徐

hua-yen 花鹽

I-ch'ang 宜昌

Ju-hai 如海

Jung-ch'i 榮溪

K'uang-shan Academy 匡山書院

li 里

Li-chou 澧州

Li Jung 李榕

liang 兩

Lin Chün 林儁

Lung tang 龍檔

Mien-chou 綿州

Mo-shui 磨水

pa-yen 巴鹽

Shih-san-fêng shu-wu ch'üan-chi 十三峯書屋全集

Shuang-fu 雙福

Ssŭ-ch'uan yen-fa chih 四川鹽法志

T'ai-yüan 太原

tan 擔

Tê-ch'êng 德成

Tsêng Kuo-fan 曾國藩

T'ung-tzu tang 桐梓檔

Tzu-liu-ching chi 自流井記

Wang 王

Wen-kao 文稿

Yün-yang 鄖陽

as the best method. A soap-bean [*gleditschia sinensis*] soup is used to purify the brine. Then the extraneous matter is skimmed off and the impure carbonate of soda removed. The type of salt which is known as *hua-yen* [crystal salt] needs only a day and a night to evaporate. It is snow-white, clear, and has a delicious flavor. Another type of salt known as *pa-yen* [cake salt] takes two days and two nights to evaporate. It cakes up like a coating of slightly burned rice that adheres to the bottom of the pot. The color of this latter type is not very white but its flavor is even better than that of *hua-yen*. On the borders of Kweichow where salt is in demand the natives tie a string around a cake of *pa-yen*, dip it into soup for a little while, and then pull it out and put it away. They use it sparingly just for a slight salty flavor. Salt is rare and much treasured there.

About 10,000 brine carriers are engaged. These men have great strength and can carry loads of some 300 catties. Going back and forth in this manner, they can make 1000 cash in a day. The number of men working on the salt boats is double this number. And the men engaged in carrying salt are again double the number of the men on the boats. Their wages are slightly lower than those of the brine carriers. There are some 10,000 men who are engaged as salt-workers, mountain-workers, and gas-workers. Their wages are much higher. Several hundred families there have invested a huge capital in the salt business. Several hundred more families are engaged there in metal works as carpenters or stone masons and in other miscellaneous works. Another several hundred families serve as merchants and traders of textiles, grain, domesticated animals, bamboo and wood, and paints and hemp. Altogether there must be a population of three to four hundred thousand people. Workers who drill the wells have only two days off in the year, the last day of the old year and the first day of the new year. Workers who evaporate the salt have no holidays at all the year round. It is the hardest labor in the world.

Fig. 1, An Early Phase of the Drilling

Fig. 2, Evaporation by gas

FIRE-ARMS AMONG THE CHINESE. A supplementary note. — Since publication of the paper on "The early development of fire-arms in China" (*Isis*, vol. 36, pt. 2, no. 104, Jan. 1946) the writer has visited China again and been able to add somewhat to his information on the subject. He had the opportunity of seeing several dated cannon not reported in Answer to Query no. 105 (*Isis*, vol. 35, pt. 3, no. 101, Summer 1944) or in the above.

1. An iron piece, bearing a date equivalent to July 1372 (Hung-wu 5: 6), about 18 inches long, 1 inch in diameter, with hole for fuse. Museum of Harvard-Yenching Institute, Peiping.
2. A bronze piece, bearing the date Hung-wu 5: 12 (26 Dec. 1372 – 23 Jan. 1373); length 44.6 cm., muzzle diameter 3.9 cm. National Central Museum, Nanking.
3. A bronze piece, dated 1379 (Hung-wu 12), about 16 inches long. Historical Museum, Peiping.

These three pieces all resemble in shape the one pictured in *Isis*, vol. 35, p. 211, fig. 3, lower half. The second bears the longest inscription which may be translated as follows:

"The Chung-shan (i.e. Purple Mountain, Nanking) garrison, number *chung* 130. The long *ch'ung* (or fire-arm) barrel weighs 3 catties 6 (approx. 4 lbs. 8 oz.). Manufactured by the Pao-yüan bureau [1] on a lucky day in the 12th moon, 5th year of Hung-wu."

Through the kindness of Dr Walter Fuchs, then research member of the Harvard-Yenching Institute, Peiping, he acquired an offprint of the article by Kuroda Genji on early cannon, cited in *Isis*, vol. 35, pt. 3, 211, but not previously seen. This is a valuable discussion with fifteen illustrations. The earliest Chinese fire-arms pictured are dated 1409, 1414, 1421, 1423, and 1426. To this should be added a useful paper inadvertently omitted: John L. Boots, "Korean weapons and armor," *Trans. of the Korea Branch, Royal Asiatic Soc.*, 24, Dec. 1934, pt. 2, 37 pp., 41 plates. Dr Boots discusses the introduction of explosive powder and cannon from China (sometime prior to 1392), the establishment of a "department of cannon and manufactured cannon and weapons" in the 3rd moon of 1377, and the use of grenades and fire arrows, and supplies illustrations of early fire-arms taken from a Korean work published in 1474 and of

[1] This bureau controlled the casting of drum, minting of money, etc. The capital was then located in Nanking.

undated cannon, mortars, and matchlocks housed in the Government museum in Seoul.

While at Nanking the writer saw a cannon half buried in the grounds of the Academia Sinica, which resembled the large piece pictured in *Isis*, vol. 35, p. 211, figure 2. No one on the spot knew its history. If it is one of the cannon unearthed in Nanking in 1851-1861, it is difficult to believe that it dates from the mid-fourteenth century, so dissimilar is it from any which rest on more abundant evidence. For the time being this observer is inclined to reject the date of 1356/7 proposed. Still the puzzling remark of Athanasius Kircher, S.J. (1602-1680) in his *China . . . Illustrata* (1667) will not down: "Our missionaries report having seen in various provinces, above all at Nanking, great cannons cast *ab immemorabili tempore*." Could the good fathers (such men as Boym, Grueber, and d'Orville) have been deceived into thinking that cannon, introduced by the Portuguese, Dutch, and others, or made in like fashion, were older than they actually were?

Japanese archeologists who in 1937 explored the area of Shang-tu, where Kubilai was proclaimed emperor in 1260, and which the Polos visited in 1275, reported in 1941 the find of two stone balls which, they surmise, may have served as cannon balls. (*Shang-tu, The summer capital of the Yüan dynasty in Dolon Nor, Mongolia*, Tokyo, p. 24 and fig. 21.) One of the balls measured 4½ inches in diameter, and weighed 55.65 ounces; the other measured 3 inches in diam., and weighed 21.2 oz. As most of the datable finds reported are of early manufacture [2] — the city was destroyed by fire in 1358 — this seems a likely guess. It will be remembered that the Chinese in the twelfth century were flinging stone balls from trebuchets, and that Chêng Ssŭ-hsiao (1239-1316) wrote of the stones shot by the "mangonels" of the Mongols.

A Chinese friend, Mr Chu Shih-chia, has drawn attention to a file of interesting correspondence preserved in the U. S. National Archives (Record Group no. 59, General Records of the Department of State. Diplomatic despatches, China, vol. 64, 1 Feb. 1883-31 March 1883). Captain J. B. Campbell of the U. S. Army, 4th Artillery, having been appointed instructor of ordinance (*sic*) and gunnery at the Artillery School, Fort Monroe, Va., sought information through our Legation in China on the origin, development, and use of explosive powder in China. The reply to his specific questions was made by Dr Joseph Edkins (1823-1905), member of the London Missionary Society stationed in Peking, and lifetime student of Chinese. He acknowledges making extensive use of the well-known paper of W. F. Mayers, pub-

[2] A Sung coin, a stele put up in 1322 or shortly after, cobalt blue tiles similar to Persian tiles of the same period; also tiles decorated with leaves and stems. The authors recall that in Marco Polo's account is a description of the palace "painted with figures of men and beasts and birds, and with a variety of trees and flowers." (Yule — Cordier, *The Travels of Marco Polo* 1: 298.)

lished in 1869, but parts company with it at several points. In his own words:

The differences are the following — Gunpowder first known to the Chinese. — Gunpowder was used in Artillery with propulsive effect in China A.D. 1234. — The military use of gunpowder was a Mahommedan invention of the 12th century. It was introduced to China by the armies of Genghis Khan. The discovery of the composition of gunpowder is probably due to the Chinese Alchymists when trying the powers of various mineral substances under the action of fire in order to make gold.

For those interested in the development of fire-arms in China after 1403 two or three recent publications may be recommended. In 1940 Mr Chêng Chên-to of Shanghai reproduced by photographic process a collection of rare works dating from the end of the Ming dynasty, both manuscript and previously printed matter, entitled *Hsüan-lan-t'ang ts'ung-shu*, which includes two short books on fire-arms. One, originally published from woodblocks in 1598, and submitted to the emperor, entitled *Shên-ch'i p'u* (Treatise on fire-arms), is by Chao Shih-chên, a secretary in the Wên-hua-tien (one of the four throne-halls in Peking). It gives detailed information on weapons, mostly matchlocks, available at that time. Nine of the illustrations, showing a man in foreign (Arab?) garb preparing to shoot, are due, says Chao, to To-ssŭ-ma, obviously a non-Chinese name. Another shows a European (Portuguese?) in the act of firing, and three others depict Chinese soldiers shooting from a kneeling position. The second work, *Shên-ch'i p'u huo-wên*, by the

and that beginning with 1529 they started to manufacture ones like them, which they called *ta-chiang-chün*, great general. The above-mentioned works fill in gaps in our knowledge as to the types then favored. Shortly thereafter the tottering house of Ming enlisted the services of a few Portuguese soldiers and commissioned certain Jesuit priests to make cannon based on the latest European models. A useful study dealing with this effort appeared recently in *Le Bulletin Catholique de Pékin*, no. 389, mai-juin 1946, pp. 160–174, by P. Bornet, S.J., entitled: "Au service de la Chine, Schall et Verbiest maîtres fondeurs. I. Les canons." The earliest seventeenth century cannon still in Peiping, which Father Bornet reports, is made of iron, and dates apparently from the period of T'ien-ch'i (1621–27). It was cast in south China (Liang Kuang), bears an escutcheon, and may have been brought to the north in 1625 by the small Portuguese and Chinese contingent which sought to aid the Chinese in their struggle against the Manchus. There are 13 other cannon in Peiping which were made in the years between 1621 and 1644, when the Ming dynasty fell.

L. Carrington Goodrich
(Columbia University)

KEY TO CHINESE CHARACTERS

Chung-shan	鍾 山
chung	鍾
ch'ung	銃
Pao-yüan	寶源
Chêng Chên-to	鄭振鐸
Hsüan-lan-t'ang ts'ung-shu	玄覽堂叢書
Shên-ch'i p'u	神器譜
Chao Shih-chên	趙士楨
To-ssŭ-ma	朶思麻
huo-wên	或問

same author, is a reproduction of a manuscript, dated the 7th moon of 1599, but including memorials to the throne of slightly later date. It presents a series of questions and answers on the use of fire-arms and has illustrations of both matchlocks and field pieces. The author was exercised over the invasion of Korea by Japanese troops under Hideyoshi and sought to arouse the imperial house to a proper defense of its frontier. For this reason he appends a map of the Chinese coast, Liaotung, Korea, and adjacent islands.

We know from the official history of the Ming (*Ming shih* 92/11a) that the Chinese were struck by the superior quality of the arms brought by the Portuguese to the China coast

A Second-Century Chinese Illustration of Salt Mining

BY RICHARD C. RUDOLPH*

O NE of the most ingenious methods whereby the Chinese produce large quantities of salt is that of drilling wells in order to obtain salt brine which is
evaporated. These salt wells are to be found in southwestern China, especially
in the province of Szechwan. Chinese literature abounds with references to these salt
wells and entire works have been written on them and their administration.

Perhaps the most useful of the early references are those in the *Hua yang kuo chih*
(Records of the Kingdoms South of Mt Hua). This, the earliest extant Chinese
work on local history and geography, was compiled in A.D. 347 or shortly thereafter
and covers Szechwan and adjacent regions. The section on Shu, the old name for the

FIG. A

region now known as Szechwan, contains numerous references to the location of salt
wells but the most interesting are those that can be dated. The earliest of these in
the *Hua yang kuo chih* is as follows: "When the Chou house had been extinguished,
Hsiao Wen Wang (King) (250–249 B.C.) of the Ch'in dynasty made Li Ping prefect
of Shu. He was conversant with astronomy and geography . . . and knew how to
control the course of streams. He bored salt wells in Kuang-tu. . . ."[1] Li Ping is

* University of California at Los Angeles.
[1] *Hua yang kuo chih* (Ssu pu pei yao ed.), 3

3b. Characters for Chinese expressions appear in
the glossary at the end of the article.

one of China's most famous hydraulic engineers and it was he, and his son, who designed the vast irrigation system that makes the Chengtu plain one of the most productive areas of China. The compiler of the *Hua yang kuo chih* was a native of Szechwan and thus indebted to Li Ping, so it is possible that he ascribes to him the drilling of salt wells in the third century B.C. in order to shed glory upon him. Be that as it may, the same work makes numerous other references to Szechwan salt wells beginning with 127 B.C. In 67 B.C., for example, 20 wells were drilled in Lin-ch'iung along the P'u River.[2]

It is even possible that there were salt wells in Shu before the time of Li Ping because there is a tradition that the government salt and iron monopoly initiated by Kuan Chung in the seventh century eventually led to the use of salt wells in Szechwan in the fourth century B.C.[3]

The official history of the former Han dynasty (206 B.C.–A.D. 24), compiled in the first century, has a section devoted to the biographies of financial successes, and there a case is cited where a man made great profits from the salt wells of Shu.[4] Wang Ch'ung, an eclectic who wrote in the first century also mentions the salt wells of Shu in one of his arguments.[5] These few sources will suffice to show that obtaining salt from wells was a well established industry in Szechwan by the first century. This industry has continued to grow during the centuries and is now one of the largest sources of income for this region.[6]

Although there is frequent mention of "salt well" (*yen ching*) and "salt pond" (*yen ch'ih*) in the early sources, they contain no detailed information on the actual recovery of salt or illustrations of this process. Some excellent illustrations of salt manufacture appeared as early as the fourteenth century but salt wells are not included.[7] In 1637 the *T'ien kung k'ai wu*, an illustrated work on various industries and labor-saving machinery, was published and this apparently was the first work to illustrate the Szechwan salt wells and their method of operation.[8]

------

[2] *Op. cit.*, 3–5a.

[3] Rudolf P. Hommel, *China at Work* (New York, 1917), 117. It should be noted that there were also "salt ponds" (*yen ch'ih*) where salt brine came to the surface without mechanical aid. Some of the present wells are extremely shallow and these depths probably could have been reached with primitive equipment.

[4] *Ch'ien han shu* (Ssu pu pei yao ed.), 91.7b.

[5] *Lun heng* (Ssu pu ts'ung k'an ed.), 13.8b.

[6] For T'ang (618–906) and later Chinese sources, see Chang Hung-chao, *Shih ya*, v. 2 (*Lapidarium Sinicum* (in Chinese), Memoirs Geological Survey of China, Series B, no. 2, 2nd ed., Peking, 1927). For a detailed and illustrated description of the industry today, see, *inter alia*, Wallace Crawford, "The Salt Industry of Tzeliutsing," *China Journal of Science and Arts*, 4 (1926), 169–75, 225–29, 281–90; 5 (1926), 20–26.

[7] The *Ao po t'u* as the title suggests, is concerned with recovering salt from sea water. Completed in 1334, it contains forty-seven beautifully drawn illustrations depicting this process. The *Yung lo ta tien* manuscript copy has been reprinted as volume three of *Shih an ts'ung shu* (1914–17). Another manuscript copy with forty-two illustrations is in Harvard Library (Sarton, *Introduction*, 3, 761).

This was reprinted in 1927 and a revised edition appeared in 1929. The wood block illus-

trations give an accurate idea of the various processes in boring the wells, raising the brine and evaporating it. The illustrations used in Lienche Tu Fang, "An account of the Salt Industry at Tzu-liu-ching," *Isis* 39 (1948), 228–234, are from the *Ssu ch'uan yen fa chih* published in 1882.

GLOSSARY

Ao po t'u	熬波圖	Lun heng	論衡
Chang Hung-chao	章鴻釗	Shih ya	石雅
Chi shih an ts'ung shu	古石叢書	Shu	蜀
Ch'ien han shu	前漢書	Ssu ch'uan yen fa chih	四川鹽法志
Ch'iung-lai	邛崍	Ssu pu pei yao	四部備要
Hsiao Wen Wang	孝文王	Ssu pu ts'ung k'an	四部叢刊
Hua yang kuo chih	華陽國志	T'ien Kung k'ai wu	天工開物
huo ching	火井	Wang Ch'ung	王充
Kuan Chung	管仲	yen ch'ih	鹽池
Kuang-tu	廣都	yen ching	鹽井
Li Ping	李冰	Yung lo ta tien	永樂大典
Lin-ch'iung	臨邛		

FIG. B

Recently. however. two bricks have been unearthed in Szechwan which bear reliefs depicting the salt industry as it was practiced there in the first or second century. These bricks. or tiles. were used in the construction of two tombs of the Later Han dynasty (A.D. 24–220). The fact that these scenes accompanied the dead indicates the importance attached to the industry at that time. It is even possible that the occupants of the tombs had been owners of salt wells.

The accompanying illustration is a photograph of a rubbing made from one of the tiles. It measures 18 by 14 inches and was found at Ch'iung-lai. a small town about 40 air-miles southwest by west of Chengtu. the provincial capital of Szechwan. Ch'iung-lai is the modern name for Lin-ch'iung. mentioned above. where 20 wells were operating in A.D. 67. On the left can be seen a derrick erected over a salt brine well with four men standing on scaffolding hauling up brine in buckets. It appears as if they are using counterbalancing buckets with the rope passing over a pulley mounted at the top of the derrick. To the right of the second stage of the derrick is a hopper supported by props. The brine is poured into this hopper and then conducted by pipes to the evaporating pans in the damaged lower right corner of the tile. The pipes were undoubtedly made of bamboo as they still are today. One man. standing. is watching the pans while a second. kneeling. is tending the fire of the long oven. In the upper right are hunters with crossbows and below them people. probably carrying loads of salt. are walking along the mountain trails.

The second brick measures 18 by 16 inches and was found near Chengtu.[9] It is essentially the same as the one illustrated here but is of particular interest because the evaporating oven is shown very clearly. There are four straight lines leading into the mouth of the oven and it is possible that these represent gas conduits. Natural gas is frequently encountered when boring these wells and is utilized as fuel in the evaporating process. The *Hua yang kuo chih* mentions "fire wells" (*huo ching*) of Han date and says that the people used this natural gas fuel for various purposes including the boiling of salt brine.[10] The elevation of each pan as they recede from the mouth of the oven also suggests the use of gas as a fuel. This same arrangement is found necessary today in order to force the gas. lighter than air. to the far end of the long oven.[11]

These reliefs thus give contemporary pictorial substantiation to the terse references in Han literature about Szechwan salt wells and they also give added technical information that otherwise would have been lost.

[9] This is the better of the two but the rubbing of it is not available at this time. It will appear in *Han Tomb Art from West China* by Professor Wen Yu and myself. now being published by the University of California Press.

[10] *Op. cit.. 3. 7b–8a.*
[11] The modern illustration of an evaporating oven in *Isis. loc. cit..* shows bamboo gas pipes tipped with clay leading to each evaporating pan.

EARLY CANNON IN CHINA

By L. Carrington Goodrich *

Several years ago I published an illustrated " Note on a Few Early Chinese Bombards " (*Isis*, 1944, *35*: 211). A little later Mr. Fêng Chia-shêng and I followed this up with an illustrated article entitled " The Early Development of Firearms in China " (*Isis*, 1946, *36*: 114-123; see also, *Isis*, 1946, *36*: 250). On the heels of this came the contribution of Wang Ling, " On the Invention and Use of Gunpowder and Firearms in China " (*Isis*, 1947, *37*: 160-178). The following year I published a supplementary note on several dated cannon preserved in China (*Isis*, 1948, *39*: 63-64). The earliest dated pieces reported were those of July 1372, 26 December 1372–23 January 1373, and 1379.

Because of the interest in this subject it seems appropriate to draw attention to the most recent finds in China, pictured opposite. The illustration is taken from an article appearing in *Wên Wu*, 1962, 3: 41-44, by Wang Jung.

The earliest is a bronze cannon, *t'ung p'ao* (Figs. 1 & 2), which bears a date equivalent to 11 March 1332. The inscription indicates that it was destined for a " Pacifying the border anti-bandit military unit. Ma-shan.[1] No. 300." Its weight is 6.94 kg., length 35.3 cm., diameter of the muzzle 10.5 cm. The piece is housed in the Historical Museum, Peking. In reply to a letter which I sent to Mr. Fêng Chia-shêng (now a member of the Institute of Nationalities, Academy of Science), asking about its provenance, he wrote that, according to a friend, it was " discovered in a temple, Peking West."

The second, also made of bronze, called *huo t'ung* (Figs. 3, 4, 5), bears a date equivalent to 1351. It is inscribed " to pierce one hundred thicknesses and produce a sound shaking the nine heavens. Flies like a divine being. " T'ien shan." [1] The weight is 4.75 kg., length 43.5 cm., and the muzzle diameter 3 cm. It is preserved in the People's Revolutionary Military Museum, Peking.

The third (metal not indicated) (Figs. 6 & 7), has an inscription which reads: " Han. Left squadron of the Navy. Large bowl-mouth cannon, Number *chin* series 42. Weight 26 catties. Made by the Pao-yüan bureau on a lucky day in the 12th month, fifth year of Hung-wu." This date is equivalent to some time between 26 December 1372 and 23 January 1373. Its weight is 15.75 kg., length 36.5 cm., and diameter of muzzle 11 cm. The piece is housed in the same museum as the second. (The date and place of manu-

---

* Columbia University.
[1] Ma-shan and T'ien-shan may refer to two different armies stationed on the frontier. A T'ien-shan army, created in 714 under the T'ang dynasty, was located in the region of Kara-khojo, near Turfan, in central Asia.

元明的火銃（文見11期）

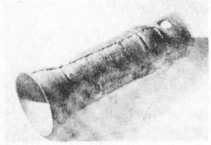

1. 元至順三年造"銅炮"

2. 元至順三年造"銅炮"銘文拓片

3. 元至正十一年造銅火銃

4. 元至正十一年造銅火銃銘文拓片

6. 明洪武五年造"大碗口筒"

7. 明洪武五年造
"大碗口筒"銘
文拓片

5. 元至正十一年造銅火銃銘文拓片

facture are identical with one reported in 1948, but the measurements differ.)

It may be of some interest to compare these with the one dated 1377, preserved in the library-museum, Tai-yüan, Shansi province, which I photographed in 1931, and with the one preserved in Korea; see *Isis*, 1944, *35*: Figures 1 & 3 facing pages 212-213, and Plate 21 in an article by J. L. Boots in *Transactions of the Korea Branch of the Royal Asiatic Society*, 1934: *24*. See also Figures 34, 35, and 39 in the article by T. L. Davis and J. R. Ware in the *Journal of Chemical Education*, 1947, *24*: 522. As to cannon used by the navy in early times, see the illustrations (planches 17 and 23) in L. Audemard, *Les Jonques chinoises*. I: *Histoire de la jonque* (Rotterdam, 1957), and my brief review in *Journal of the American Oriental Society*, 1959, *79*: 207. These illustrations may be traced ultimately to the *Wu-pei chih* (On Military Preparedness), Chapter 117, by Mao Yüan-i, preface of 1621. Wang Jung, the author of the article in *Wên Wu* cited above, quotes the book *Ping lu* (On Warfare) by Ho Ju-pin (preface of 1606) as saying that naval guns were mounted on a swivel.

Cultural
Interaction

The Astronomical Instruments of Cha-ma-lu-ting, their Identification, and their Relations to the Instruments of the Observatory of Marāgha[‡]

BY WILLY HARTNER *

IN the 48th chapter of the [†] *Yüan-shih* [a], fol. 10 r° ff., under the heading: *Hsi-yü I-hsiang* [b], "pictures of instruments from the western region," reference is made to drawings (or models) of seven astronomical instruments which, in 1267 A.D., were transmitted from one of the western countries to the court of the Mongol Emperor Kublai (Yüan Shih-tsu[c]) by a certain Cha-ma-lu-ting.[d] The name of this man seems to be mentioned nowhere else, nor has any other information about him thus far been discovered. George Sarton (*Introduction*, Vol. II, Washington 1931, p. 1021) interprets Cha-ma-lu-ting, of which he also gives the variant spelling Cha-ma-li-ting, as a Chinese transliteration of the Arabic name Jamāl al-Dīn. There can be no doubt that this is correct. Because the scarce information obtainable on this Jamāl is drawn from Chinese, rather than from Arabic or Persian sources, Sarton prefers to list him under the Chinese form of his name. The note which he devotes to him reads:

Persian astronomer who devised for Kublai Khān in 1267 a new calendar — Yeh-lü's calendar had never been adopted — called the Ten thousand years calendar, Wan-nien li (12486, 8301, 6923). It was probably so named because it was a development of an ancient Persian system based on a period of ten thousand years; or the words ten thousand may simply mean perpetual.[1] This calendar is lost. He introduced seven Persian astronomical instruments, among them an armillary sphere constructed for the latitude of 36°, probably to be used in the college of P'ing-yang (9310, 12883), Shansi, in latitude 36°6'. See my note on Yeh-lü Ch'u-ts'ai (first half of thirteenth century).

Owing to the circumstances under which this article was written, I have been able to consult only one of the three literary references cited by Sarton: Y. Mikami's *Development of Mathematics in China and Japan* (Leipzig 1913, p. 100f.) which reads as follows:

Cha-ma-li-ting constructed in 1267 some Arabian astronomical instruments in China. Of these we may mention a celestial globe, a sun dial, gnomons for the equinoxes and solstices and an instrument for the observation of the starry heavens. In the "Historical Records of the Yuen Dynasty," section on astronomical subjects, the Arabian names of these instruments are recorded.

I have had no direct access to the two other references: A. Wylie, *Chinese Researches* (part 3, 16, 1897), and Yule's *Marco Polo* (3rd ed., vol. 1, 455, 1903). According to a communication from the late Professor Otto Franke, Wylie gives solely a description of those instruments which were manufactured by order of

---

[‡] Max Dehn Septuagenario, 13 November 1948.
* Institut für Geschichte der Naturwissenschaften, University of Frankfurt.
[†] All transliterations of Chinese characters are followed by a superscript letter or letters. This enables transliterated characters to be identified in the *Key to Chinese Characters* to be found at the end of this article.
[1] The latter alternative is the more probable because no such Persian system seems ever to have existed, whereas *wan* [see Chinese characters sub (aj)] was always used to express very high numbers, or infinity.

later Chinese rulers. The Persian items he mentions only *en passant*, though he was conversant with the Yüan-shih passage. His statement that seven *instruments* were submitted to the emperor is erroneous because it is clearly said that only drawings or models (*i-hsiang*[e]) were brought to the court of Khānbalig. Yule, moreover, in his *Marco Polo* deals only with the later instruments, especially the ones devised by the Jesuits. Finally, I believe I remember that Gaubil in the 2nd vol. of Souciet's *Observations mathématiques, astronomiques, etc.* (Paris 1729–32), which is also unavailable to me now, alludes to the instruments of Cha-ma-lu-ting.

To the best of my knowledge no serious attempt has thus far been made to identify these astronomical instruments, of which the Yüan-shih not only lists the foreign names in Chinese transliteration, but also gives the Chinese equivalents, or short descriptions of their use, or both.[2]

The seven terms and their Chinese equivalents are: [3]

1. *tsa-t'u-ha-tzŭ* (or *lao*) *chi;* [f] in Chinese *hun-t'ien i.*[g]

2. *tsa-t'u-shuo-pa-t'ai;* [h] in Chinese *ts'ê-yen chou-t'ien hsing-yao chih ch'i.*[f]

3. *lu-ha-ma-i-miao-wa* (or *yao*)-*chih* [j]; in Chinese *ch'un-ch'iu-fên kuei.*[k]

4. *lu-ha-ma-i-mu-ssŭ-t'a-yü;* [l] in [4] Chinese *tung-hsia-chih kuei.*[m]

5. *k'u-lai-i-sa-ma;* [n] in Chinese *hun-t'ien-t'u.*[o] In an additional note the instrument is called a *wan-hsieh,*[p] to be made from bronze.

6. *k'u-lai-i-a-êrh-tzŭ;* [q] in Chinese *ti-li chih.*[r] Additional note: "A globe to be made from wood, upon which seven parts of water are represented in green color, three parts of land in white color, with rivers, lakes, etc."

7. *wu-su-tu-êrh-tzŭ* [s] (or *la*; see footnote 3, and the name of the first instrument). "The Chinese name has not been worked out. An instrument to be made from bronze, on which the times (hours) of the day and the night are engraved."

The following analysis will mainly proceed in the same sequence as above. In some cases a comparison between two or more of the terms will prove useful.

1. The Chinese term *hun-t'ien i* [t] is one of the common names given to the *armillary sphere*; it is also sometimes applied to the *celestial globe* (in modern Japanese, for example, this compound, read *kontengi* may mean both). But this duality occurs rarely in the older literature, wherein the solid (celestial) globe is usually designated by *ch'iu,*[u] or else by the general term *t'u,*[v] "picture" (see below, § 5). The armillary sphere proper, which is a Greek invention, may have been known in post-Han China. I am at a loss to discover unambiguous information on it before

---

[2] My attention was called to the subject by a letter from Professor H. H. Schaeder, whom Professor Otto Franke had consulted because he wished to publish the solution of the problem in the 5th volume of his *Geschichte des Chinesischen Reiches* (the manuscripts of Vols. 4 and 5 were ready for print by the beginning of 1944; they are still unpublished. Otto Franke died in 1946). Unfortunately, not even a copy of the *Yüan-shih* was available to me when I entered upon the subject in 1944, and until this day I have not succeeded in obtaining this first-hand source of information. Thus I had to rely upon the data which were kindly transmitted to me by Franke, and his additional statement that the *Yüan-shih* apparently does not yield any other information pertinent to the subject. It proved equally impossible to obtain photostatic copies of Arabic or Persian MSS, or even to borrow the most common European reference books from public libraries. What I did not have in my own notes and records, or in my own library or that of my Institute, was practically beyond my reach. The following analysis will therefore not be exhaustive and

must remain a truncated torso. If I publish it as it is, it is only because I believe that its main result will remain valid even if other sources are taken into consideration, and that it will perhaps throw some light on the scientific relations between the Near and the Far East during the Mongol period.

[3] As concerns the Chinese characters used for rendering the sounds of the seven foreign technical terms, ambiguities arise in two cases:
Case 1. *tzŭ* [see Chinese Characters, *sub* (f), 4th ideogram] is often confused with the similar *la* [ibidem, 5th ideogram], even in carefully printed texts. Both readings must therefore be considered possible.
Case 2. The 6th ideogram *sub* (j) is pronounced either *wa* or *yao*. The intended reading will become evident only from the context.

[4] From the fact that Franke writes (*t'a*) [see Chinese Characters, *sub* (ak)] but transliterates *ta*, it could be inferred that the original text has a similar character which is pronounced *ta* [ibidem, *sub* (al)] (without aspiration). The difference is not essential.

the Yüan period. Zinner's statement [5] that an armillary sphere was devised by *Lo Hsia-hung* [w] (2nd half of 2nd century B.C.) is certainly erroneous. His instrument was a model of the *kai-t'ien* [x] system, consisting of a hemisphere or cupola resting on the horizon. Nor do all the later improvements cited by Zinner refer to the armillary sphere, but to solid hemispheres or — starting with *Wang Fan* [y] (fl. about 250 A.D.) — to complete celestial spheres, i.e. *globes*. Under the Yüan, however, the armillary sphere as an astronomical instrument for observation and for measuring becomes common. One constructed by *Kuo Shou-ching*,[z] dated 1279, is still extant in Pei-ping.[6] I consider it practically beyond doubt that Kuo's inventive genius was stimulated by the study of Cha-ma-lu-ting's model, for this first of his seven instruments was nothing but an armillary sphere. As a matter of fact, we easily recognize in the first two syllables of *tsa-t'u-ha-tzŭ(la)-chi* the Arabic *dhātu* (in Persian pronounced *zāt[u]*), female of *dhū*, "the owner," which very frequently occurs in technical terms designating instruments (*ālāt*). If we remember that, ever since the time of al-Battānī, the armillary sphere used to be called *dhāt al-ḥalaq*, "the owner of the rings," the last doubt as to the true reading and interpretation of our term in question will be removed. The fourth character is to be read *la*, not *tzŭ*, and thus the Chinese transliteration becomes correct almost to the letter: *tsa-t'u-ha-la-chi* = *dhāt al-ḥalaq*, or, in classical reading, *dhātu 'l-ḥalac̣(i)*. If we wonder why the *l* of the Arabic article was omitted, we have the choice between two explanations: an additional character *êrh* [aa] may have appeared too clumsy to the author of the Chinese text, or, more probably, it was not the correct Arabic form that he tried to render according to Cha's pronounciation, but a more vulgar, colloquial Arabo-Persian *iḍāfa* construction: *dhāt-i-ḥalaq* or *dhāt-u-ḥalaq*.[7] It is true, an *iḍāfa* construction after *dhāt* will never be found in good Persian, but this does not exclude the possibility that a man who is not, strictly speaking, erudite, would nevertheless use it. And, as will be seen, there are reasons to believe that our Cha-ma-lu-ting was not erudite, but a man of rather limited abilities who had obtained a sufficient training in the elements of astronomy to be sent to the court of Khānbalig in order to demonstrate those models.

2. In the transliterated term *tsa-t'u-shuo-pa-t'ai*, we easily recognize the same *tsa-t'u* = *dhātu* as in our first case (instrument No. 1). The compound *shuo-pa-t'ai* sounds like the dual of a female noun ending in *-bat*: *-batai(n)*. Contrary to the first instance, however, the Chinese equivalent: *ts'e-yen chou-t'ien hsing-yao chih ch'i* [i] is not a technical term, but only a rather vague description of the instrument's use: "Instrument for observing (and examining, i.e. measuring, *yen* [ab]) the rays of the stars of the celestial vault." Because this does not seem to give us any direct clue, our sole choice will be to try to find the Arabic terms that can be hidden behind *shuo-pa-t'ai*, and then to compare them with the Chinese explanation. There are two possibilities that I can hit upon:

a. *Shuo-pa-t'ai* = *thuqbatai(n)* (Persian pronounciation: *suqbatai(n)*, "the two holes," or "the two diopters." Remembering that *shuo* [ac] is pronounced in most of the southern dialects with a final *-k* (*shok, sok*), and admitting the possibility that the

[5] Ernst Zinner, *Geschichte der Sternkunde*, Berlin, 1931, pp. 233–34.

[6] It is one of the two of Kuo's items which, after having been removed from the Chinese capital by the German troops during the Boxer Rebellion and taken to Potsdam as "war spoils" (a euphemistic term for theft), were restored to China according to the Versailles Treaty. See Sarton, *Introduction* II, p. 1022.

[7] The *iḍāfa* vowel, it will be remembered, is only theoretically an *i*. Most commonly it is pronounced like a short *ĕ*, but any other nuance may occur, according to the merits of the case and the nationality of the speaker. In our instance, just the influence of the classical form *dhātu* may have caused the pronounciation with *u*. Even in the spoken modern Persian, the classical Arabic *-u* ending, in *dhātu*, is still preserved in all the manifold technical terms where the word occurs, such as *dhātu 'l-ḥalaq*, *dhātu 'r-riyat* (pneumonia), etc.

scribe was a southerner, we note that the correspondence would seem almost perfect. *Thuqba* is the common term for the diopter-hole of an alidad. And *dhātu 'th-thuqbatai*(*ni*) "the instrument with the two holes" is a technical term designating Hipparchus' *dioptra* as mentioned by Ptolemy in *Alm.* 5, 14, and described in detail by Proclus in his *Hypotyposis*.[8] More commonly, indeed, it is called by Moslem astronomers "the instrument with the movable diopter"; [9] but, to quote one example, Mu'ayyad ad-Dīn al-'Urḍī, in his description of the instruments of the observatory of Marāgha,[10] expressly lists both terms as synonyms.[11]

This dioptra, as is well known, serves the purpose of determining the apparent diameters of the sun and the moon, and of observing (especially measuring the magnitudes of) solar and lunar eclipses (with the aid of two pierced disks, whose apertures correspond to the apparent diameters of the sun and the moon). This special use, however, does not seem to be alluded to in the Chinese explanation as mentioned above: *chou-t'ien hsing-yao* [ad] means all the stars of the sphere including the planets and the sun and the moon, not only the latter two. Therefore this first attempt of ours to identify the instrument probably misses the target.

b. The second alternative is: *shuo-pa-t'ai* = *shu'batai*(*n*), "the two legs." The "instrument with the two legs," *dhātu 'sh-shu'batai*(*ni*), is another Arabic name for Ptolemy's parallactic ruler (ὄργανον παραλλακτικόν, *triquetrum*), which al-Battānī usually calls *al-'idāda aṭ-ṭawīla*, "the long ruler." The term is found in al-'Urḍī's manuscript,[12] where he claims that the improvements of the instrument are due to his own skill. The name itself, however, was not invented by him; it occurs already in a MS by al-Kindī,[13] in which it applies to the original model as described by Ptolemy.

Apart from minor improvements upon the stability of the instrument, its safer dressing in the plane of the meridian, etc., the chief advantages of al-'Urḍī's *triquetrum* were that the division into partes, minutae, etc., was no longer on the vertical, but on the lower of the two movable rulers, and that this latter was made long enough to measure all zenith-distances between 0° and 90°, whereas the Ptolemaic ruler did not allow to measure angles exceeding 60°.

If, again, we check with the Chinese description, it becomes evident that it must be this instrument which is intended. The function of the *triquetrum* is to observe and measure the stars; the term *yen* [ab] refers obviously to the measuring of the position, or, speaking more exactly, of the zenith-distance of a star at its culmination. The phonetic congruence is as perfect as it can be. The hypothesis that a southern

---

[8] Ed. Manitius, Leipzig. 1909, p. 127ff.

[9] If my memory does not deceive me, the expressions *dhāt al-libna al-mutaḥarrika*, and *dhāt al-hadafa al-mutaḥarrika* are both found in MSS.

[10] There exists only one MS: Paris, *de Slane's catalogue*, No. 2544, 10, fol. 60b–79a. Despite its extraordinary importance, no edition of it was ever made, nor do any photostatic copies of it seem to exist. I had to rely on the two translations extant: Jourdain, "Mém. sur les instrumens employés à l'observatoire de Méragah" (*Mag. encyclopédique ou Journal des Sciences*, Tome VI, Paris 1809, p. 43–101), which is not complete; and H. J. Seemann, "Die Instrumente der Sternwarte zu Marāgha nach den Mitteilungen von al 'Urdī" (*Sitzungsberichte d. Physik.-medizin. Sozietät zu Erlangen*, Bd. 60, 1928, p. 15–126). Jourdain gives no Arabic terms at all, and Seemann in exceptional cases only. This lack is a very serious drawback because, as every translator of oriental MSS should know, he would make a valuable contribution to the establishment of a dictionary of technical terms by adding them in brackets or footnotes as abun-

dantly as possible. He would be sure of the gratitude of his readers who thus no longer would have to grope about in the dark.

[11] This *dioptra* is the 5th of the ten instruments of Marāgha described by al-'Urḍī; Seemann, *loc. cit.*, pp. 61–71.

[12] al-'Urḍī lists this instrument as No. 7; Seemann, *loc. cit.*, pp. 81–87 and pp. 104–108. Again, Seemann does not give the Arabic term, but there can be no doubt that it is *dhātu 'sh-shu'batain*, and nothing else. On p. 83, Seemann's translation reads: "Zu den Instrumenten, die ich selbst hergestellt habe, gehört das Instrument mit den beiden Schenkeln (Fatûn)." What *fatûn* is supposed to mean is a riddle to me. I never came across this term in manuscripts or in dictionaries. I suppose it is a misprint.

[13] Codex Leyd. 199, fol. 29 b–36 [a] (Catalogue, Vol. III, p. 82); translated by E. Wiedemann: "Über eine astronomische Schrift von al Kindī," *Beiträge z. Gesch. d. Natw.*, XXI, in *Sitzungsberichte der Physik.-mediz. Sozietät Erlangen*, Bd. 42, p. 294ff. (1910).

scribe chose the character *shuo* [ac] on account of its final *-k* as spoken in his home dialect is no longer needed here: The Pekinese pronounciation *shuo*[4] renders with the very greatest accuracy the first syllable of *shuʻbatai*, where even the sound of the " *ʻain quiescens*" is expressed by the fourth (sinking) tone.

3 and 4. The two terms *lu-ha-ma-i-miao-wa* (*yao*)-*chih* and *lu-ha-ma-i-mu-ssŭ-t'a-yü* will have to be analyzed jointly because, as is seen from the word *kuei*'s [ae] occurring in both of the Chinese equivalents, they obviously refer to two varieties of gnomons or sundials.

Lu-ha-ma is of course the correctly rendered Arabic *rukhāma*, a very frequently used word for the plane sundial (*basīṭa*). It appears in the writings of al-Farghānī, Muh. b. Mūsā al-Khwārizmī, and al-Battānī. One of the most elaborate treatises dealing with the *rukhāma* was composed by Abū ʻAlī al-Ḥasan b. ʻAlī b. ʻUmar al-Marrākushī, the well-known Moroccan astronomer who died *c.* 1262.[14] His work is the first in several centuries that marks a definite progress of gnomonics; it contains descriptions of new types of dials to be traced on horizontal as well as vertical and inclined surfaces placed in various positions with regard to the meridian and the prime vertical; furthermore it treats of projections of the hour-lines on conical, cylindrical, and other surfaces. His *opus*, which probably was composed about the middle of the 13th century, was undoubtedly known to the astronomers of Marāgha who were anxious to equip the new observatory of the Īl-Khān with the most modern instruments. Previously I made reference to the probability of a scientific connection between Marāgha and the court of the Yüan at Khānbaliq. My assumption will be supported and supplemented by further details at the end of this study.

The meanings of *miao-wa* (*yao*)-*chih* and *mu-ssŭ-t'a-yü* then become evident. The former (where the reading with *wa* has to be chosen) corresponds to the Arabic *muʻawwaj*, the latter, to *mustawī*. These two terms refer, as the student of Moslem astronomy will know, to the two kinds of hours used in antiquity and during the Middle Ages. The *sāʻa mustawiya* is the *equal hour* (synonymon of *s. muʻtadila* or *sāʻat al-iʻtidāl*, Gr. ἰσημερινὴ ὥρα, Latin *hora aequinoctialis*), and *sāʻu muʻwajja*, the *unequal hour* (syn. *s. zamāniya*, Gr. καιρικὴ ὥρα, Lat. *hora temporalis*). *Muʻawwaj* (of which *miʻawwaj*, as resulting from the Chinese transliteration, is only a corruption frequently resulting among people who do not speak Arabic as their native tongue) is a less correct form substituted for the learned *muʻwajj*. Both have the same meaning, and both are derived from the same root *ʻwj*, *muʻawwaj* being the passive participle of the second, *muʻwajj*, the active participle of the ninth stem.

Thus we have the equations:

(3) *lu-ha-ma-i-miao-wa-chih* = *rukhāma-i-muʻawwaj* (short for the correct *rukhāma-i-sāʻāt-i-muʻwajja*), "sundial for unequal hours," and

(4) *lu-ha-ma-i-mu-ssŭ-t'a-yü* = *rukhāma-i-mustawī* (short for the correct *rukhāma-i-sāʻāt-i-mustawiya*), "sundial for equal hours."

In comparing these interpretations with the corresponding Chinese terms, we see at once that the two must be interchanged in order to make sense: *Ch'un-ch'iu-fên kuei* [k], meaning word by word "the sundial for the vernal and autumnal equinoxes," obviously belongs to No. 4, and *tung-hsia-chih kuei* [m], "the sundial for the winter and summer solstices," to No. 3. No doubt the confusion was made by the scribe who did not understand all of Cha-ma-lu-ting's long explanations and, adding a little fiction

---

[14] Sarton, *Introduction* II, 621. French translation of his treatise by J. J. Sédillot: *Traité des instruments astronomiques des arabes composé au treizième siècle par Aboul Hhassan Ali de Maroc* (Vols. I–II, Paris 1834–35; suppl. 1844).

It is the main source of Karl Schoy's "Gnomonik der Araber" (in *Gesch. d. Zeitmessung u. d. Uhren*, Bd. I, ed. E.v. Bassermann-Jordan, Berlin and Leipzig 1923.)

of his own, wrote down the terms as he believed them to make sense. No exaggerated imagination is needed to reconstruct the course of the conversation of two men who probably did not know each other's languages and therefore had to use an interpreter, who again did not know much of astronomy. It ran as follows:

Cha-ma-lu-ting demonstrates two models of sundials which, except for the lines and curves traced on their surfaces, are practically alike. The scribe wants to know the difference between them. Cha pointing first at the model of an ordinary *rukhāma* as used by the Moslems since olden times, says that it is the one designed for the *unequal* hours: then, pointing at the other model, he says that it is a *rukhāma* for the equal hours. The scribe notes down the two names. Because the Chinese ever since the remotest antiquity were accustomed to dividing the day and the night ($\nu\nu\chi\theta\acute{\eta}\mu\epsilon\rho o\nu$) into 12 equal (double) hours — which are counted from the sun's culmination (midnight and noon), not, as with the Moslems, from sunset and sunrise — the scribe does not understand why the Moslem astronomers distinguish between two kinds of hours. He therefore requests an explanation. Cha then, starting for pedagogical reasons from the Chinese hours, with which both he and the scribe are familiar, explains that these are the ones called in Arabic "equal" (*sā'āt mustawiya*) or "equatorial" (*sā'āt al-i'tidāl*) because they yield themselves automatically from the unequal hours when the sun stands in the equator (*dā'irat al-i'tidāl*), in other words at the times of the vernal and autumnal equinoxes (*al-i'tidāl ar-rabī'ī wa 'l-i'tidāl al-kharīfī*). The scribe, assuming that there must be something more behind Cha's words than just an explanation of the well-known Chinese equal hours, writes down what he believes he caught from Cha's words: "Gnomon (sundial) for the vernal and autumnal equinoxes." By a pure mistake, or because Cha had shown him first the sundial with the lines of the unequal hours, he writes these words under the transliterated name of the latter instrument (No. 3), where they are still found today.

Cha then continues explaining:

On the surface of the other dial you find the lines or curves of the *unequal* hours, which are the ones commonly used with us. Their length varies from day to day. During the summer, the day hours are longer than the night hours, during the winter, the night hours are longer than the day hours. The difference is greatest at the times of the winter and summer solstices.

The scribe, who again has not succeeded in understanding all of Cha's explanation as mutilated by the interpreter, seizes the last words, which seem to make sense to him because they complete the antitheses of equinoxes and solstices — belonging to the elementary stock of astronomical knowledge and found in any Chinese astronomical text (*ch'un-ch'iu fên*[af]*-tung-hsia chih*[ag]). Hence he writes down — again in the wrong place, of course — "Gnomon for the winter and summer solstices."

The fact that a model of its own was designed for the sundial indicating equal hours deserves special attention. As I pointed out, the first progress of gnomonics worth mentioning was made by Abū 'Alī of Morocco about the year 1250. He gave directions for tracing the lines of the equal hours,[15] of which little seems to have been known before his time. As is seen from the Yüan-shih passage, the theory worked out by him in the Far West of the Moslem world, had become known in the Far East, at the court of the Yüan hardly more than twenty years later.

⁓§

5 and 6. After the above analyses, it is a rather trivial problem to identify the terms *k'u-lai-i-sa-ma* and *k'u-lai-i-a-êrh-tzŭ*, because there are no longer any unknown quantities in our two equations.

5. *K'u-lai-i-sa-ma* = *kura-i-samā'*, "a celestial globe." This is confirmed by the Chinese *hun-t'ien t'u*,[o] "a star chart" (or also "a celestial globe," see above, § 1),

---

[15] Sédillot, *op. cit.*, Vol. II, pp. 491ff.

with the explanatory note "a *wan-hsieh*,[p] to be made from bronze." *Wan-hsieh* means "the globe which is oblique (or sloping, slanting)," i.e. "the globe with the oblique axis." We are in the very fortunate situation of being able to tell the exact appearance of the globe that was demonstrated by Cha-ma-lu-ting. In the Mathematisch-physikalischer Salon of Dresden there is preserved a globe which is made of brass, engraved and inlaid with silver and gold, and signed "the work of Muḥammad b. Mu'ayyad al-'Urḍī," dating from 677 H./1278 A.D.[16] This Muḥammad was the son of the very same Mu'ayyad al-'Urḍī, quoted by me previously, who had described the instruments of the Observatory of Marāgha. Muḥammad's globe was probably also made for the above-named Observatory, not by order of its founder, the Īl-Khān Hūlāgū who had died in 1265, but of one of his successors, who did but little to maintain the fame of the new center of astronomical observation and research.

This celestial globe, which undoubtedly was not the first of its kind that was made at Marāgha, gives us a clear idea of Cha-ma-lu-ting's model. Composed of two hemispheres which meet at the circle of the ecliptic, it carries on its surface the forty-eight constellations as listed in the *Almagest* or in 'Abd ar-Raḥmān aṣ-Ṣūfī's "Catalogue of the fixed stars." Besides the ecliptic and the equator, which are both graduated and thus make possible the determination of the longitudes and the right ascensions of the stars, the globe presents the twelve (or we may say six) great circles of longitudes corresponding to 0°, 30°, 60°, etc. These, being perpendicular to the ecliptic, intersect one another at the poles of the ecliptic. The latter, as well as the poles of the equator, are marked by two pairs of holes. Hence it is possible to make the globe revolve either around the poles of the ecliptic or around those of the equator, in order to demonstrate the annual revolution of the sun along the zodiac, or its daily revolution around the earth. In the latter case, the whole system of longitudinal circles will be constantly inclined and, therefore, the spectator will get the impression that the whole globe "revolves aslant." It is this impression that the scribe tried to characterize by the term *wan-hsieh*, "the globe which is oblique, or sloping."

As it is not very likely that this conception of a "sloping globe" could be obtained from a picture or drawing, it may be safely inferred that the items submitted to the Emperor by Cha-ma-lu-ting, were small-scale *models* of the instruments.

6. *K'u-lai-i-a-êrh-tzŭ* = *kura-i-arḍ* (Persian pronunciation *kura-i-arz*), "a terrestrial globe." This is also corroborated by the Chinese translation *ti-li chih*,[r] which means "a geographical map." The term *chih*,[ah] it is true, does not refer to the globe proper — inevitably because terrestrial globes were unknown in China before the period in question. But in order to avoid ambiguities, the scribe added the remark: "A globe to be made from wood, upon which seven parts of water are represented in green color, three parts of land in white color, with rivers, lakes, etc."

It is up to the specialist in the history of geography to judge the importance of this reference to a 13th-century terrestrial globe manufactured in Persia. So far as I know, there is no record of a terrestrial globe, either oriental or occidental, prior to Martin Behaim's famous globe of 1492.[17] Nor does Sarton mention any such in the index to his *Introduction*, Vol. II, except for the heading of the first chapter of Sacrobosco's *Sphaera mundi*, "On the terrestrial globe," which of course deals only with the spher-

[16] Reproduced in *A Survey of Persian Art*, ed. A. U. Pope, Vol. VI, Pl. 1403 (Oxford 1939), where the artist's name is wrongly spelled. Described by A. Drechsler: *Der arabische Himmelsglobus des Mohammed ben Muyid el-'Ordhi*, Dresden 1922. See also Sarton, *Introduction* II, p. 1014. Another celestial globe dating from the save time (684 H. 1285 A.D.), which is in the Musée du Louvre, is signed Muḥ. b. Maḥmūd

b. 'Alī of Ṭabaristān. See R. Harari, "*Metalwork after the early Islamic period*," in *A Survey of Persian Art*, Vol. III, Chapter 56, p. 2518.

[17] Half a century before Behaim, Regiomontanus is said to have manufactured terrestrial globes in Nuremberg. No detailed description of these is known to me.

ical shape of the earth.[18] If I am not mistaken, then, this brief statement in the *Yüan Annals* concerning a Persian terrestrial globe is of the very greatest importance; it may throw entirely new light upon the high standard of Islamic geography in the 13th century.

The ratio "water : land = 7 : 3" comes rather close to the one accepted by modern geographers (5 : 2). The importance of this assumption, however, is of course *nil*; it was based on hardly anything but guesses.

7. It caused me some trouble to make out the meaning of the last compound of Chinese characters listed, *wu-su-tu-êrh-tzŭ* (*la*), because I followed the wrong track for quite a while, being convinced that something similar to an unusual *maṣdar* form **wuḍū'* (derived from the root *wḍ'*) must be hidden behind the two first syllables *wu-su*. The true solution, however, is as obvious, not to say trivial, as it can be.

We may safely exclude the reading *tzŭ*, because it is very unlikely that both of the two similar characters would appear in the same passage (cf. above §1, and footnote 3; in the case of the first instrument, we had to substitute *la* for *tzŭ*, as found in the text). Thus we get *wu-su-tu-êrh-la*. Being aware that the initial *w* in *wu* is so soft when pronounced by certain Chinese that *wu* sounds practically like a German or Italian *u*, one can no longer doubt that (*w*)*u-su-tu-êrh-la* is meant to render the Arabic *asṭurlāb*, which in Persian always is pronounced with an initial *u*: *usṭurlāb* (also the variant *uṣṭurlāb*, with two emphatic letters: *ṣ* and *ṭ*, is frequently encountered).[19] Although in colloquial Persian the emphatic consonants of the Arabic are not distinguished from the non-emphatic, every man who is eager to boast about his learning will try to imitate the original Arabic pronunciation. Thus our friend Cha-ma-lu-ting's mouth no doubt produced such deep, emphatic sounds when he pronounced the name of this most important instrument that it sounded almost like *uṣuṭurlāb*. No more adequate spelling in Chinese could have been found to render the word than the one chosen by the scribe. Indeed, he forgot the final *b*. The reason for it, I assume, was that his ear did not perceive it after the strongly emphasized *a* of the last syllable *-lāb*.

It would have been inconceivable for Cha's collection not to have contained an astrolabe. No "modern" observatory could be thought of in those days without at least two or three specimens of astrolabe in its possession.

No wonder the imperial scribe who had had to listen to Cha's previous lectures on the other six instruments, the equal and unequal hours, etc., had become so tired when they eventually entered upon this very complicated subject that he did not feel like cudgeling his brains for an appropriate Chinese translation of the term in question. He simply gave it up and wrote down with a final effort: "The Chinese name

---

[18] The earliest terrestrial globe mentioned in antiquity is the one devised by Crates of Mallos in Cilicia (first half of 2nd century B.C.) referred to in Strabon, *Geography*, II, 5, 10. It was based on the Pythagorean hypothesis of four continents or land masses: the οἰκουμένη plus three similar ones separated from one another by two oceans, one equatorial, and one perpendicular to it, which was thought to be situated under the first meridian, west of the Fortunate Islands. This globe had undoubtedly fallen into oblivion, although it is said that the "orb" ("Imperial Globe," "Reichsapfel") originated from a combination of it with the Cross of Christ. Moslem tradition does not seem to have been familiar with it. It is equally uncertain whether Ibn Rusta (Sarton, *Introduction* I, 635) who flourished *c.* 903, constructed a terrestrial globe. In the geographical part of his *Encyclo-*

pedia, he deals only with the theory of the celestial and terrestrial spheres.

H. Haack, *Studien am Globus. Bemerkungen z. Gesch. u. Technik der Erdgloben*, Gotha 1915, pp. 20f., states that the *Libros del saber de astronomia* of King Alfonso X, in the chapter treating of globes (which he says is a translation of a treatise by "Coszta") deals with the terrestrial globe, particularly the various materials (gold, bronze, wood, etc.) from which it can be manufactured. "Coszta" is of course Qusṭā b. Lūqā (second half of 9th century). Contrary to Haack's study, the chapter in question is devoted to the celestial globe only. It contains hardly any reference to terrestrial globes.

[19] The Turks also prefer this form starting with *u*; in modern spelling *usturlab* or even *usturlap*.

was not worked out. An instrument to be made from bronze, on which the times
(or hours) of the day and the night are engraved." By these words, two of the chief
features of the astrolabe are clearly characterized, *viz.* that it is made of bronze, and
that it carries the lines of the day and night hours on its surface. There would have
been much more to tell of this wonderful instrument. But a fairly exhaustive description of its principle and its use could certainly not be given in few words. Even the
shortest Arabic or Persian treatises on the astrolabe cover a dozen pages or more.
To make it intelligible to Chinese readers, twice as much space would have been
required because they could not be expected to possess any knowledge of stereographic
projections, spherical trigonometry, the twelve-partite zodiac, the western constellations, etc. — all that was, as Geoffrey Chaucer said, "bread and milk for children"
in the World of Islam. [20]

   The connection of Cha's instruments with those of Marāgha, as described by al-
'Urḍī, is obvious. This is by no means surprising. The first Yüan Emperor of China,
Kublai (Qubilai, according to Mongol spelling), was the brother of the Persian
Īl-Khān, Hūlāgū (Hülägü). Despite the constant internal struggle between the de-
scendants of Čingiz Khān, the two brothers lived always in close friendship with each
other. Kublai's genuine interest in the arts and sciences is beyond doubt; he had
it in common with Hūlāgū. Hence it was natural that, when Kublai heard of the new
observatory of Marāgha, he was eager to raise Chinese astronomy to an equally high
level. Accordingly, he asked his brother to supply him with all the materials suited
to his purpose. For some reason or other, the messenger, our Cha-ma-lu-ting, was
not dispatched during Hūlāgū's lifetime, but shortly after his death (8 Feb. 1265);
as evidenced by the *Yüan-shih* passage, he arrived in Khānbaliġ (Khān Balyġ,
Peking) in the course of the year 1267.[21]
   The instruments which Cha submitted to the Yüan Emperor were well selected.
Considering the circumstance that China until then had developed an astronomical
system of her own essentially different from that of the West, the astronomers of
Marāgha probably agreed that several of their new instruments, of which we are
informed by al-'Urḍī, would be much too complicated for, and devoid of any practical
value to, their eastern colleagues. For these would have to attend courses in the elements of western astronomy before they could make use of their new equipment. Thus
only such instruments were chosen as were considered indispensable: an armillary
sphere, a *triquetrum,* two different kinds of sundial, a celestial and a terrestrial globe,
and an astrolabe. Of these the armillary sphere and the *triquetrum* are both described, with several variants, by al-'Urḍī. None of his other instruments could be
considered fit for the purpose of the projected Chinese observatory because, as I
have said, there was nobody to use them properly. This observation is particularly
true of the trigonometrical instruments, such as the one mentioned by al-'Urḍī (No. 8)
for determining the sine and the azimuth, and the one for the sine and the versine
(al-'Urḍī's No. 9). At first sight it might be regarded as an omission that Cha did
not take along a quadrant (al-'Urḍī has several types); but it was probably found

[20] For the history and theory of this instrument, see W. Hartner, "The principle and use
of the astrolabe," in *A Survey of Persian Art,*
Vol. III, Chapter 57, p. 2530–54; also J. Frank
und M. Meyerhof, *Ein Astrolab aus dem in-
dischen Mogulreiche,* Heidelberg, 1925; and
more recently, O. Neugebauer, "The early history
of the astrolabe," *Isis,* Vol. 40, pp. 240–256
(1949).
   [21] A skeptic will interpret the given facts in
a different way: In spite of their friendly relations, Hūlāgū was anxious to preserve the monopoly of astronomical knowledge and therefore
did not allow his brother to compete with him.
Thus Kublai had to wait till, after his brother's
death, the astronomers of Marāgha or Hūlāgū's
disinterested successor no longer objected to
complying with Kublai's wishes. A third possibility is that Cha-ma-lu-ting simply was bribed
by one of Kublai's agents and stole the models
from the observatory of Marāgha.

dispensable because the determination of altitudes, which is its main function, could be made with the aid of either the armillary sphere or the *triquetrum*.

There still remains one problem: *Who was Cha-ma-lu-ting?* It will perhaps never be possible to solve it with any reasonable amount of probability. Of all the more or less famous contemporary bearers of the name Jamāl ad-Dīn, we know none who is reported to have traveled eastward on a special mission. Among the collaborators of the great Naṣīr ad-Dīn aṭ-Ṭūsī, who was the real founder and spirit of the observatory of Marāgha, we find, beside Mu'ayyad ad-Dīn al-'Urḍī, one Fakhr ad-Dīn al-Marāghī of Mūṣul, one Fakhr ad-Dīn al-Khalāṭī, and one 'Alī b. 'Umar Najm ad-Dīn al-Qazwīnī. All these names are composed with the word Dīn, but there is no Jamāl.

In Jourdain's translation of al-'Urḍī's treatise,[22] however, I came across a reference to a less famous and perhaps less able man by the name of Jamāl ad-Dīn; it was he whom Hūlāgū first intended to entrust with the construction of his observatory. The passage reads as follows: [23]

Il (Hūlāgū) appela près de lui Djémâl-Eddyn [*footnote*: Ses noms sont: Djémâl-Eddyn Mouhammed ben Dhaher ben Mouhammed Alzéydy Alnadjâry] pour effectuer ce projet; mais celui-ci. en trouvant l'exécution au dessus de ses forces, eut la modestie d'avouer son incapacité. L'observatoire ne fut pas commencé.

This avowal must of course have happened after Hūlāgū's accession to the throne of the Īl-Khāns in 657/1258. Although this Jamāl ad-Dīn did not have the courage to take over the responsibility for the construction of the observatory, it is certainly not unlikely that he consented to assist in some way or other, to the best of his abilities. The cognomen *an-Najjārī* [24] would indicate that he originally had been, or still was, a carpenter, or at least had something to do with carpentry. Yet the circumstance that the Īl-Khān had given him credit for carrying out such a difficult task is sufficient proof that he had at least some astronomical training. And this would certainly, after another six or seven years' additional experience, have sufficed to warrant his dispatch as a messenger to the Chinese Court. There, we learn from the Yüan-shih, he also proved his skill by devising a new calendar, which was adopted by Kublai Khān.

More information could perhaps be drawn from Khwāndamīr's *Ḥabīb as-siyar* (written in 1523), which appears to have been the source of Jourdain's statement. Unfortunately, this work is inaccessible to me.

In its further development, Chinese astronomy did not undergo any essential change through this contact with Islamic science. Some of the instruments that were constructed during the subsequent years were apparently copied after those of Cha-ma-lu-ting, whereas others definitely were not. A certain amount of Islamic influence is noticeable in the works of the distinguished Yüan astronomer Kuo Shou-ching.[z] Under the Ming[ai], it fell into complete oblivion as is evident from the deplorably low standard of science in general after the Yüan.

This decay lasted another three centuries until China eventually started assimilating Western astronomy and mathematics thanks to the efforts of Father Ricci and his successors, who commenced their activity in China in the first years of the 17th century. These European scholars, it will be well to remember, owed their knowledge

[22] See footnote 10.
[23] Jourdain, *loc. cit.* p. 49.
[24] The *ī* suffix attached to a nomen professionis (*najjār*) is quite unusual. I should not be surprised if Jourdain had misread najjārī for Bukhārī for these two words are very much alike in the Arabic script. It is probable that the scribe wrote Bukhārī and forgot the dot of the first letter or misplaced it.

to the same source which, centuries before, had been offered in vain to the Yüan astronomers. A long detour thus was required to establish a lasting scientific contact between
the Helleno-Islamic and the Far-Eastern Worlds. Through the European translations
of Oriental MSS, particularly the *Libros del saber de astronomía* of King Alfonso X
(which date from exactly the time with which we have been concerned here, and even
deal mostly with the same instruments), the new era of Western science on European
soil was inaugurated. Before two centuries had elapsed, the former disciples had
definitively become masters, and soon the rumor of their learning and skill spread all
over the world.

KEY TO CHINESE CHARACTERS

(a) 元史 (b) 西域儀象 (c) 元世祖
(d) 札馬魯丁 (e) 儀象 (f) 咱禿哈剌
(剌)吉 (g) 混天儀 (h) 咱禿朔八台 (i)
測驗周天星曜之器 (j) 魯哈麻亦
渺凹只 (k) 春秋分晷 (l) 魯哈麻亦
木思塔餘 (m) 冬夏至晷 (n) 苦來亦
撒麻 (o) 渾天圖 (p) 丸斜 (q) 苦來亦
阿兒子 (r) 地理志 (s) 兀速都兒剌
(剌)(t) 混天儀 (u) 球 (v) 圖 (w) 洛下閎
(x) 蓋天 (y) 王蕃 (z) 郭守敬 (aa) 兒
(ab) 驗 (ac) 朔 (ad) 周天星曜 (ae)
晷 (af) 春秋分 (ag) 冬夏至 (ah)
志 (ai) 明 (aj) 萬 (ak) 塔 (al) 搭

The Chinese-Uighur Calendar as Described in the Islamic Sources[*]

By E. S. Kennedy [**]

1. INTRODUCTION

MEDIEVAL astronomical handbooks (*zījes*) usually contain explanatory material and tables for the calendars and eras apt to be encountered by their users, and for converting a date given in one of these calendars into its equivalent in another. The *zījes* written in Iran or Central Asia from the thirteenth century on describe a luni-solar calendar used by the *Khaṭā* (or *Qiṭā* — both designations are used in the same texts) and the Uighur. The latter is the name of a Turkic people frequently powerful in Mongolia from the sixth century on, and whose domain was often contiguous with that of the Chinese.

The leading characteristics of this calendar have been known for a long time and are distinctively Chinese. They are

a) The division of the civil day (*nychthemeron*) into twelve *chāgh* (a Turkish word, synonymous with the German *Doppelstunde* and the Chinese *shih*) of eight *kih* (Chinese *k'o*) each, the *nychthemeron* beginning at midnight at the middle of the first *chāgh*, and the naming of the *chāgh* by a duodecimal animal cycle.[1]

b) The combination of elements of the duodecimal cycle with those of a decimal cycle to give a sexagesimal cycle used for naming years and days.[2]

c) The use of a second duodecimal cycle (*Wahlzyklus*) of days, for divination.[3]

d) The beginning of the solar year (and the solar anomaly) from, in effect, the arrival of the sun at Aquarius 15°, the four

* Study supported by the National Science Foundation.
** American University of Beirut.

[1] See Ludwig Ideler, "Über die Zeitrechnung von Chata und Igur," *Abhandlungen der Königlichen Akademie der Wissenschaften zu Berlin, Histor. philol.*, 1832. p. 275; Friedrich Karl Ginzel, *Handbuch der mathematischen und technischen Chronologie* (Leipzig: J. C. Heinrichs, 1906), Vol. 1. pp. 452, 465; Joseph Needham, *Science and Civilisation in China*. Vol. 3: *Mathematics and the Sciences of the Heavens and the Earth* (Cambridge, Eng.: University Press, 1959), p. 402.

[2] Needham, *op. cit.*, p. 396.

[3] Ginzel, *op. cit.*, p. 490.

seasons thus beginning in the middle of our seasons, and the division of the solar year into twenty-four *kancha*.[4]

In addition to these elements, already described in the literature, there are many aspects of the system which have not been studied, and the whole subject merits extensive treatment. For instance, a large number of Chinese and Uighur technical terms are transcribed into Arabic characters and scattered thickly throughout the sources. Systematic collection and collation of these should be of utility to Sinologists and Turkologists.

This paper confines itself to a single topic, one which has not as yet been commented upon, and whose interest transcends calendaric considerations — the calculation of solar and lunar equations involved in determining the instant of true conjunction, hence the beginning of the lunar month. Section 2 comments on the special variety of decimal fractions used by the Chinese in time measurement, while Section 3 cites the mean motion parameters. The following section gives the rules for calculating true conjunction, and Section 5 describes a Chinese method of generating parabolic periodic functions. In the succeeding two sections the rules are analyzed, in the course of which a very widespread ancient Babylonian period relation turns up. Section 8 comments upon the results obtained.

The number of different descriptions of this calendar is large, and these sources are listed and described in the concluding Section 9. Our main source has been the India Office copy of al-Kāshī's *Zīj-i Khāqānī*.

Although I am ignorant of Chinese, I have been extraordinarily fortunate in having been able to consult with Professor Kiyosi Yabuuti of the University of Kyoto, who controls the Chinese astronomical sources. Allusions to the Chinese material not accompanied by references to publications are due to Professor Yabuuti, who is thus largely responsible for any merit this study may have.[5]

2. DECIMAL FRACTIONS

In all the sources, time is measured in days and ten thousandths of days, the latter unit represented by a word in the Perso-Arabic characters which may be read either as FNK or FNG. No vowel marks for it have been noticed thus far. This unit, the Chinese name for which we will transcribe as *fên*, was devised in the later period of the T'ang dynasty (ninth century). In particular, it was used in two calendrical works, the *Shoushih-li* (*c.* 1280) of the Yüan (Mongolian) dynasty, and the *Tat'ung-li* (1384) of the Ming dynasty.

When a time is given in days and fên it is convenient to display it as a single decimal representation with four fractional places. The reader may

4 *Ibid.*, p. 468; Needham, *op. cit.*, pp. 396, 405.

5 Recent publications on the subject are Itaru Imai, "Ulugh Beg's Calendar" (in Japanese), *Bulletin of the Society for Western and Southern Asiatic Studies* (Kyoto University), 1962, pp. 29–37; Kiyosi Yabuuti, "Astronomical Tables in China from the Han to the T'ang Dynasties," an English section (pp. 445–538) in *Tyūgoku tyūsei kagaku-gizyutu-si no kenkyū* ("History of Chinese Science and Technology in Its Middle Ages") (Tokyo, 1963).

recall that it was Jamshīd al-Kāshī, the author of our prime source, who developed, explained, and applied decimal fractions in full generality, in so doing anticipating Simon Stevin by a century and a half.[6] One is tempted to assume that it was Kāshī's experience with the Chinese fên, a special variety of decimal fraction, which led him to invent, as he said, a variety of fractions analogous to those with base sixty. Curiously enough, in his tables and computations in the *Zīj-i Khāqānī*, the fên are invariably expressed as sexagesimals in the Arabic alphabetical numerals.

3. PARAMETERS AND MEAN CONJUNCTIONS

The length of a solar year is taken to be 365.2436 days (f.9v:7).[7] This is the length of the tropical year used in the Kantao-li, a calendar compiled *c.* 1167 during the Sung dynasty. It has been encountered solely in Chinese contexts.

The length of a mean lunation is 29.5306 days. This parameter, here rounded off to integer fên, is standard to all Chinese astronomical tables; similarly, it has been seen by us in connection with the Chinese calendar only. Since it gives the length of time required for the mean moon, having once passed the mean sun, to catch up with it and pass it again, we have for the mean rate of elongation

$$r = \dot{\lambda}_m - \dot{\lambda}_s = 360/29.5306 = 12.19074°/\text{day}, \tag{1}$$

where $\dot{\lambda}_m$ and $\dot{\lambda}_s$ stand respectively for the mean angular velocities of the moon and sun along the ecliptic.

The anomalistic month, the period of recurrence of the most marked irregularity in the moon's motion, is taken (f.11r:15) as 27.5546 days, a parameter which appears in the *Shoushih-li* and the *Tat'ung-li*, as well as in earlier Chinese works from the second century on. For short spans another value is used, 27.5556 days (f.10v:3), unknown in the Chinese material itself. It is probably a rounding off of 27.555... = 248/9, a number discussed in Section 7 below.

If, now, at any epoch, the mean positions of the sun and moon and the age of the anomalistic month are given, it is a simple matter to calculate the instant of any mean conjunction, and the age of the anomalistic month in which it occurs. The situation is displayed graphically in Figure 1. The mean longitudes of the sun and moon, $\bar{\lambda}_s$ and $\bar{\lambda}_m$ respectively, are linear functions of time. Whenever either luminary reaches a mean longitude of

[6] See Nā'ila Rajā'i, "The Invention of Decimal Fractions in the East and in the West," unpublished thesis (American University of Beirut, 1951) ; A. P. Yushkevich, *Istoriya Matematiki v Srednie Veka* (Moscow, 1961) ; Paul Luckey, *Die Rechenkunst bei Ğamšīd b. Mas'ūd al-Kāšī, mit Rückblicken auf die ältere Rechnens* (Wiesbaden: F. Steiner, 1951); B. A. Rosenfeld, V. S. Segal, and A. P. Yushkevich (translators and editors), *Dzemshid Giyaseddin al-Kashi, Klyuch Arifmetiki, Traktat ob Okruzhnosti* (Moscow: Gosudarstvennoe Izdatel'stvo Tehniko-Teoreticheskoi Literatury, 1956).

[7] This and all such designations in parentheses refer to the prime source, Jamshīd al-Kāshī, *Zīj-i Khāqānī*. The folio number is on the left-hand side of the colon, the line number on the right.

360°, having completed a trip about the ecliptic, the inclined line drops to the horizontal axis and commences rising as before. Thus the two mean-longitude functions are represented by two sequences of parallel line segments. The dotted lines indicating the mean moon are inclined more steeply than the heavy lines for the solar function, because the moon is the faster of the two. Points of intersection between the two sets of lines mark instants where the sun and the moon have the same longitude — i. e., mean conjunctions.

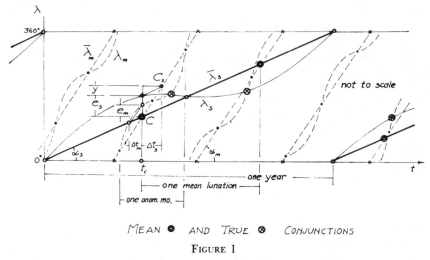

MEAN ● AND TRUE ⊗ CONJUNCTIONS

FIGURE 1

4. CALCULATING THE TRUE CONJUNCTION

Determine x_s, the *solar anomalistic argument* (*ḥiṣṣa-i āftāb*), to the nearest integer number of days by subtracting the time at which the current year began from the time of mean conjunction (f.11v:22). From x_s form x'_s as follows:

if

$$x_s \leq 182, \qquad x'_s = x_s,$$

but if

$$x_s > 182, \qquad x'_s = x_s - 182.$$

Then

$$\Delta t_s = \pm \frac{2}{9} \, x'_s (182 - x'_s) \qquad (2)$$

is the solar equation in fên, at conjunction. If $x_s > 182$, Δt_s is considered negative; otherwise it is positive.

In like manner, subtract the time at which the current anomalistic month began from the time of mean conjunction, and *multiply the result by nine* (f.11r:4). The product x_m, rounded off to integer days, is the *lunar anomalistic argument* (*ḥiṣṣa-i qamar*). Now find x'_m by the rule:

if
$$x_m \leq 124, \qquad x'_m = x_m,$$
but if
$$x_m > 124, \qquad x'_m = x_m - 124.$$
Then
$$\Delta t_m = x'_m (124 - x'_m) \tag{3}$$

is the lunar equation in fên at conjunction (f.12r:1), where, analogously to the sun, if $x_m > 124$, then $\Delta t_m < 0$, otherwise $\Delta t_m > 0$.

Now the time of true conjunction is

$$t_e + \Delta t_s + \Delta t_m, \tag{4}$$

where t_e is the time of mean conjunction.

Numerical tables of Eqs. (2) and (3) giving Δt for each integer day of the respective arguments appear on f.14r.

5. PARABOLIC PERIODIC FUNCTIONS

If one desires a periodic relation of amplitude a and period p, the function which most readily occurs to a modern is

$$S(x) = a \sin \frac{2\pi}{p} x,$$

although the basic idea is at least as old as Ptolemy. Babylonian applications of linear zigzag and periodic step functions are even earlier solutions of the same problem.

Another approach is to construct a smooth curve by piecing together on the horizontal axis successive congruent parabolic segments having zeros at $pn/2$, n integer, and opening alternately downward and upward. This is expressible as

$$P(x) = \pm a \left(\frac{4}{p} \right)^2 x' \left(\frac{p}{2} - x' \right)$$
where
$$x' = x - n \left(\frac{p}{2} \right),$$

n being the largest integer which will make $x' \geq 0$. The plus sign for $P(x)$ is taken whenever n is even; otherwise $P(x)$ is negative. Note that amplitude and period are a and p respectively, and, like the sine function, maxima and minima occur at $x = (2m + 1) p/4$, $m = 0, 1, 2, 3, \ldots$. In fact, if $S(x)$ and $P(x)$ are plotted on the same axes, they are seen to diverge from one another very little.

This technique seems to have been applied in China by the time of the T'ang dynasty (began 618 A. D.), bearing the graphic appellation *hsiang-chien hsiang-ch'êng* ("alternately-subtract alternately-multiply") .[8]

[8] See Yushkevich, *op. cit.*, pp. 100–104.

6. Examination of the Solar Equation Rule

Clearly, both equation rules are applications of the technique described in Section 5 above. It remains to identify and interpret the astronomical parameters embedded in them. For expression (2), the period $p = 2(182) = 364$ days, which is a crude approximation to the solar year, but the best which will make $p/4$ an integer. Evidently the year is so fixed that the argument of the anomaly commences with it.

The maximum of (2) occurs at $x_s = 91$ and is equal to $2(91^2)/9$ fên $= 2(91^2)/90,000$ days $=$ max Δt_s. In order to convert this into the more familiar (for us) mode of expressing a planetary equation in angular units, we refer again to Figure 1. The inclined wave form (not dotted) which has a period of a year represents the solar true longitude (λ_s), the equation $e_s(t) = \lambda_s(t) - \bar{\lambda}_s(t)$ being represented as enormously larger than it is in actual fact. If one assumes that the true moon, in the vicinity of the mean conjunction at C, is traveling along its mean path, and that the sun, although subject to its equation, has its angular velocity substantially equal to $\dot{\lambda}_s = \tan \alpha_s$, its mean angular velocity, then the true conjunction will occur at C_s and cause a delay in the conjunction of Δt_s, as shown. The approximation is good, because the maximum equation of the sun is of the order of only 2°.

Then

$$\tan \alpha_s = \frac{y}{\Delta t_s}$$

and

$$\tan \alpha_m = \frac{y + e_s}{\Delta t_s}.$$

Eliminating the y and utilizing Eq. (1) we obtain

$$\Delta t_s = \frac{e_s}{\lambda_m - \lambda_s} = \frac{e_s}{r}. \tag{5}$$

This simply asserts that at all times the speed of approach or separation of the two luminaries is approximately the mean rate of elongation, which is perhaps obvious without demonstration.

So,

$$\max e_s = r \max \Delta t_s$$
$$= (12.19°/\text{day}) \left(\frac{2(91^2)}{90,000} \text{ days} \right)$$
$$= 2.243° = 2;15°.$$

This result is quite close to 2;14°, the classical value of the maximum solar equation in Indian astronomy, and to the Ptolemaic 2;23°.[9]

[9] See, e.g., Brahmagupta, *The Khanda-khādyaka*, translated by Prabodh Chandra Sengupta (University of Calcutta, 1934), p. 156; E. S. Kennedy, *A Survey of Islamic Astronomi-cal Tables*, Philadelphia: The American Philosophical Society (Transactions, N. S. 46, pt. 2), 1956, p. 172.

7. EXAMINATION OF THE LUNAR EQUATION RULE

For expression (3) the period is 248. Recalling, however, the factor of nine which the rule for forming x_m has introduced, we have the equation

$$248 \text{ days} = 9 \text{ anomalistic months.} \tag{6}$$

This period relation has also appeared in Babylonian,[10] South Indian,[11] and Hellenistic [12] contexts.

For Eq. (3) a maximum of Δt_m is yielded at $x_m = 62$. Its value is 62^2 fên $= 62^2/10{,}000$ days.

Referring again to Figure 1 — assuming that only the lunar equation affects the difference between true and mean conjunctions — an expression analogous to (5) is obtained:

$$\Delta t_m = \frac{e_m}{r}. \tag{7}$$

Note, however, that when

$$e_m > 0, \qquad \Delta t_m < 0;$$

whereas when

$$e_s > 0, \qquad \Delta t_s > 0.$$

Substituting in (7),

$$\max e_m = r \max \Delta t_m = (12.19°/\text{day}) \left(\frac{62^2}{10{,}000} \text{ days} \right) = 4.69° = 4;41°.$$

This is not far from the Indian maximum lunar equation of 4;56°, and the Ptolemaic and Hipparchian 5;0°.[13]

The rule (4) assumes that the effect of each of the equations is independent of that of the other — another instance of approximation.

8. COMMENTS

The leading inference to be drawn from what has preceded is that, with a single exception, all the features of this calendar are distinctively indigenous Chinese. The exception is the period relation (6).

As for the period relation, one thinks immediately of the possibility of transmission. There is nothing impossible about this. An equation which traveled from Mesopotamia to South India can equally well have gone to Mongolia; and strong Nestorian, Manichean, and Buddhist influence on the Uighur is well established. The Uighur alphabet, which was passed on to the Mongols, was adapted from the Soghdian, which was in turn an Aramaic adaptation. On the other hand, we have no proof that (6) was not independently discovered in Central Asia or China.

10 Otto Neugebauer, *Astronomical Cuneiform Texts, Babylonian Ephemerides of the Seleucid Period for the Motion of the Sun, the Moon, and the Planets* (London: Published for the Institute for Advanced Study, Princeton, N. J., by Lund Humphries, 1955), Vol. I, p. 76.

11 Otto Neugebauer, "Tamil Astronomy," *Osiris*, 1952, *10*: 252–276.

12 Otto Neugebauer, "The Astronomical Treatise P. Ryl. 27," *Kongelige Danske Videnskabernes Selskabs Skrifter, Hist-filol. Medd.*, 1949, *32*, no. 2: 1–23.

13 Brahmagupta, *op. cit.*, p. 160.

What seems to be of more interest is the extent to which parabolic interpolation devices are now seen to have permeated medieval exact science. By this term we denote arithmetic schemes implicitly involving polynomials of degree two, or number sequences exhibiting runs of constant second-order differences; the geometric formulation is useful but never occurs in the texts. These are:

1) The *Khaṇḍakhādyaka* [14] interpolates for functional values between a pair of tabular entries by passing a parabola through the two functional values at the endpoints of the interval and the next preceding tabular value.

2) Al-Bīrūnī [15] employs a parabola through the same endpoints but this diverges as far on the one side of the parabola of the *Khaṇḍakhādyaka* above as does the secant for linear interpolation diverge on the other.

3) Al-Kāshī [16] explains a method for inserting a set of equally spaced functional values inside an interval by assuming that the preceding interval has already been filled in and then passing a parabola through the endpoints of the new interval and the last interpolated value in the preceding interval.

4) S. G. Ladkany, working from the *Dastūr al-Munajjimīn*,[17] has shown that the scheme of al-Kāshī above or variants of it were known to at least three scientists other than Kāshī.

5) Javad Hamadanizadeh has studied a passage in the *Zīj-i Ashrafī* [18] in which a parabola is called for which is related to the next tabular value following the interval in question rather than the one preceding it. This curve, however, passes midway between the linear interpolation secant and the parabola through the three tabular values.

Far antedating the medieval examples are instances of second-order alternating sequences occurring in the astronomical cuneiform material.[19]

Only with the investigation of further sources, Chinese, Sanscrit, and Islamic, can we hope to say whether there is any connection between these and the Chinese " alternately-subtract alternately-multiply " parabolas.

9. List of Sources

Arranged below in chronological order is a list of the writings containing descriptions of this calendar. Doubtless there exist others presently unknown to us.

a. The *Zīj-i Īlkhānī* (in Persian, *c.* 1270) by Naṣīr al-Dīn al-Ṭūsī is extant in many copies. The first chapter is a description of the Chinese-Uighur calendar; in Bodleian MS Hunt 143, it covers ff.1v–10r.

[14] *Ibid.*, p. 142.

[15] Al-Bīrūnī, *Al-Qānūn'l-Mas'ūdi* (Canon Masudicus) (Hyderabad-Deccan, 1954), Vol. 1, p. 352.

[16] E. S. Kennedy, " A Medieval Interpolation Scheme Using Second Order Differences," pp. 117–120 in *A Locust's Leg, Studies in Honour of S. H. Taqizadeh*, edited by W. B. H.

Henning and E. Yarshater (London: Lund, 1962).

[17] Paris Bibliothèque nationale MS Arabe 5968, ff.75–78.

[18] Paris Bibliothèque nationale MS Supplément Persan 1488, f.101r.

[19] See, e. g., Neugebauer, *Astronomical Cuneiform Texts, op. cit.*, Vol. 1, p. 78.

b. The *zīj* of Muḥī al-Dīn al-Maghribī (in Arabic, finished in 1276) describes this calendar in Chapter 13 of Treatise I. In the copy in the Shrine Library in Meshed, Iran, this covers ff.11v–17r. A treatise by the same author in the Bodleian Library entitled *Risālat al-Khaṭā w'al-Īghūr* may be independent of this.[20]

c. The *zīj* of Jamāl al-Dīn Abī al-Qāsim b. Maḥfūẓ al-Baghdādī (in Arabic, 1285), Paris, Bibliothèque nationale, MS Arabe 2486, has a short section on the Chinese years, f.12v.

d. The *Zīj-i Ashrafī* (in Persian, written in 1303) by Muḥammad ibn Abī 'Abdallāh Sanjar al-Kāmīlī, known as Sayf-i Munajjim. The unique extant copy is Paris, Bibliothèque nationale, Supplément Persan MS 1488. Material on the Chinese-Uighur calendar is in Treatise I, ff.15r–20v, 31v–34r.

e. The *zīj* of Muḥammad b. 'Alī, known as Shams al-Munajjim al-Wābkanwī (in Persian, finished c. 1325) is extant as Aya Sofya (Istanbul) MS 2694. The Chinese-Uighur calendar is described in Chapter 12 of Treatise II, ff.34r–45v.

f. The *Zīj-i Khāqānī* (in Persian, finished in 1413) by Jamshīd al-Kāshī, our main source, devotes Chapter 2 of Treatise I to the Chinese-Uighur calendar. In the India Office (London) copy, Persian MS 2232, this material appears on ff.8v–21r. In the Istanbul copy, Aya Sofya MS 2692, it covers ff.8r–19r.

g. The *Zīj-i Sulṭānī* (in Persian, c. 1440), by Ulugh Beg, is extant in many copies and commentaries. It is organized in four treatises, the explanatory text of which (but not the very extensive mathematical, astronomical, and astrological tables) has been published by L. P. E. A. Sedillot, the Persian text as *Prolégomènes des tables astronomiques d'Ouloug-Beg*, Paris, 1847; the translation bearing the same main title with the addition, *Traduction et commentaire*, Paris, 1853. The description of the Chinese-Uighur calendar is in Chapter 6 of Treatise I, in the text edition pp. 30–56, in the translation pp. 32–61; complete with tables in both.

---

[20] See H. Suter, *Die Mathematiker und Astronomen der Araber* (Leipzig, 1900), p. 155.

A Link in the Westward
Transmission of Chinese Anatomy
in the Later Middle Ages

By Saburō Miyasita*

I

DURING THE MIDDLE AGES teaching of anatomy in European universities was based exclusively on the text of Galen, which was canonical and indisputable. Anatomical dissections were extremely rare and never executed by physicians. However, a new trend emerged in Italy. There is a record of an autopsy at Cremona in 1286; in 1316 Mundinus of Bologna, the "Restorer of Anatomy," completed his *Anothomia*, in which he describes his systematic dissection of the human body.[1]

II

In China, physiological conceptions and anatomical knowledge of the human body were based on the *Huang ti nei ching*,[a,2] compiled in the Han[b] dynasty (probably first century A.D), but the earliest explicit documents concerning human dissection appeared in the Sung[c] dynasty.

The first anatomical dissection of the human body was performed in 1045. After a force under the command of Tu Ch'i[e] (appointed by the central administration) quelled an insurrection in Kwangsi,[d] Wu Chien,[f] judge of I-chou,[g] ordered that the abdomens of the criminals be dissected and that Sung Ching[h] sketch their viscera. The rebels dissected were Ou Hsi-fan,[i] the chief, and fifty-five members of his party.

About sixty years after this dissection further advances were made. The important work *Ts'un hsin huan chung t'u*[j] (Illustrations of Internal Organs and Circulatory Vessels) by Yang Chieh,[k] physician of Anhui,[l] appeared in 1113. The book comprises two sections; the first consists of ten anatomical drawings with ex-

* Library, Takeda Chemical Industries, Ltd., Osaka, Japan. The author wishes to thank Professor Kiyosi Yabuuti of Kyoto University and Dr. Joseph Needham of Caius College, Cambridge University, for their advice and encouragement. The Republic of Turkey gave permission to microfilm Aya Sofya MS 3596.

*[1] Charles Singer, *A Short History of**

Anatomy from the Greeks to Harvey (2nd ed., New York: Dover, 1957), pp. 74 ff.*

*[2] Appended to the end of this article is a list of the Chinese characters which represent the romanized Chinese and Japanese words appearing in the text. Superscript letters—e.g., *ching-lo*[m]—indicate the placement in the list of equivalent characters.*

planations, and the second contains illustrations and descriptions of the twelve circulatory vessels (*ching-lo*[m]). The anatomical section preserves Sung Ching's work and, in addition, records and corrects a systematic dissection performed in the Ch'ung-ning[n] era (1102–1107). This dissection, of an executed criminal, was carried out under the control of Li I-hang,[o] the governor of Ssu-chou[p] (Anhui), and probably also under the control of Chia Wei-chieh,[q] the military commander of Anhui, who wrote the preface of the book. The section on the circulatory system is entirely traditional. From the first half of the twelfth century we have information concerning other observations of human viscera, but detailed records were not published.

III

The original text of the *Ts'un hsin huan chung t'u* was lost, but a Japanese translation, with colored drawings, is found in Chapter 44 of the *Ton-isho*,[r] compiled by the monk-physician Shozen Kajiwara[s] in 1304. It also appears in Chapter 54 of the same author's *Man-anpō*[t] (1315). Three Chinese treatises—the *Nei wai erh ching t'u*[u] written by Chu Kung[v] in 1118 (the original text was lost), the Yuan[w] edition[3] of the *Hua t'o nei chao t'u*[x], and the *Kuang wei ta fa*[y] written by Wang Haoku[z] in 1294—contain only the anatomical drawings from Yang's work.[4]

The first of the ten anatomical drawings shows Ou Hsi-fan's viscera as rendered by Sung Ching (Fig. 1a). The second and third show a front and back view of the abdominal and thoracic viscera (Figs. 1b–c). The fourth and fifth show a left and right side view of the thoracic viscera (lungs and so on; Figs. 1d–e). The sixth shows the heart and its related parts (Fig. 1f). The seventh shows the diaphragm (Fig. 1g), and the eighth, the spleen and stomach (Fig. 1h). The ninth and tenth show a front and side view of the abdominal viscera (intestine, bladder, and kidney; Figs. 1i–j).

Although, as is usual in medieval works on Chinese anatomy, the pancreas and nervous system are ignored, the anatomical drawings are rather accurate. The verbal explication still relies almost entirely on ancient physiological knowledge.

Not only did Chinese physicians perform dissection on humans, but the knowledge they gained was applied practically in the earliest system of forensic medicine, as attested in the *Hsi yuan (chi) lu*[aa] written by Sung Tz'u[ab] in 1247. Chinese anatomy, however, failed to progress further until Wang Ch'ing-jen[ac] in 1797 began to observe human viscera and then published his *I lin kai ts'o*[ad] (1830).

IV

The anatomical drawings compiled by Yang Chieh in 1113 were widely circulated for a long period of time and at last were taken to West Asia. As the accompany-

---

[3] This edition contains only eight of the anatomical drawings, omitting that of Ou Hsi-fan's viscera and the left side view of the thoracic viscera. The explanations are traditional and unrelated to Yang Chieh's work.

[4] These identifications were made by the late Kozo Watanabe. See his "General Remarks on Dissection and Anatomical Figures in China" (in Japanese), *Nihon Ishigaku Zasshi*, 1956, 7:88–182.

The third book contains the left side view of the thoracic viscera, not found in the other two.

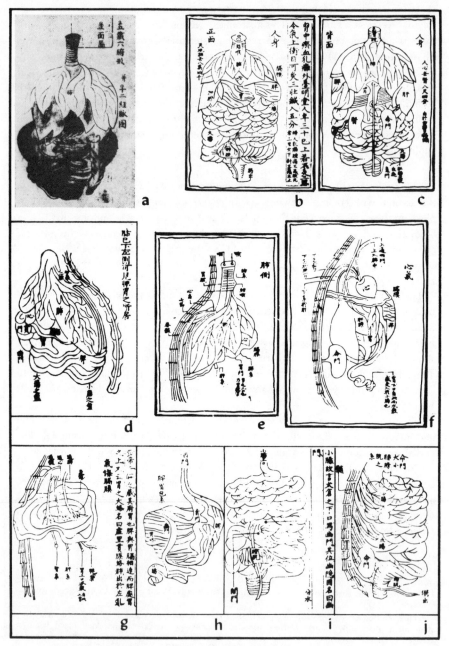

FIGURE 1. Anatomical drawings from (a) *Man-anpō*, (d) *Kuang wei ta fa*, and (the others) the Yüan edition of the *Hua t'o nei chao t'u*.

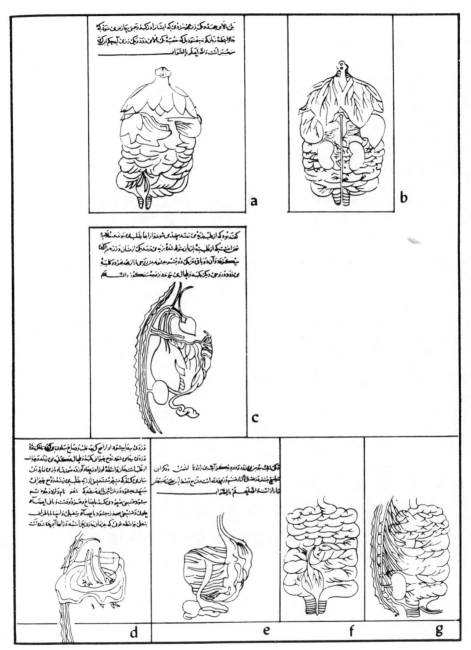

FIGURE 2. Anatomical drawings from the *Tanksuq-nāmah-i Īlkhān dar funūn-i 'ulūm-i Khiṭāi* (Aya Sofya MS 3596).

ing illustrations prove, all of the drawings of dissection in the famous Persian *Tanksuqnāmah-i Ilkhān dar funūn-i 'ulūm-i Khiṭāi* (Figs. 2a–g) are distinctly based on the drawings of the *Ts'un hsin huan chung t'u,* though probably not copied from the original book.

The manuscript of the Persian compilation (Treasure of the Ilkhan on the Sciences of Cathay), translated by order of Rashīd al-Dīn, was discovered in 1939 in the Aya Sofya Library, Istanbul.[5] It was copied at Tabriz in 1313–1314 by Muḥammad ibn Aḥmad ibn Maḥmūd Qawwām al-Kirmānī, who is known to have been one of the scribes employed by Rashīd al-Dīn al-Hamadānī (*c.* 1247–1318), a Persian physician and prime minister under the greatest of the Mongol rulers of Persia, Ghāzān Maḥmud Khān.

The Persian manuscript reproduces seven of the ten anatomical drawings compiled by Yang Chieh, in a form which closely resembles the drawings of the Yuan edition of the *Hua t'o nei chao t'u.* The drawing of Ou Hsi-fan's viscera and the left and right side views of the thoracic viscera are omitted.

It is impossible at the moment to trace the westward flow of Chinese anatomy beyond the Moslem world. We hope that it will ultimately be possible to find direct evidence from which to determine, since "Mundinus had read widely among the Arabian anatomists, and naturally borrows from them,"[6] whether Chinese medical ideas actually played a sensible role in the development of European anatomy.

[5] Abdulhak Adnan, "Sur le Tanksukname-I-Ilhani der Ulum-U-Funun-I-Khatai," *Isis,* 1940, *32*:44–47; George Sarton, *Introduction to the History of Science,* Vol. III (Baltimore: Williams and Wilkins, 1947), p. 972;

Joseph Needham, *Science and Civilisation in China,* Vol. I (Cambridge: Cambridge Univ. Press, 1954), pp. 218–219.

[6] Singer, *op. cit.,* p. 76.

APPENDIX. THE EQUIVALENT CHARACTERS FOR THE CHINESE AND JAPANESE WORDS APPEARING IN THE TEXT

a 黃帝內經　b 漢　c 宋　d 廣西　e 杜杞　f 吳簡　g 宜州推官
h 宋景　i 歐希範　j 存心環中圖　k 楊介　l 安徽　m 經絡
n 崇寧　o 李夷行　p 泗州郡守　q 賈偉節　r 頎醫抄
s 梶原性全　t 萬安方　u 內外二景圖　v 朱肱　w 元
x 華陀內照圖　y 廣爲大法　z 王好古　aa 洗寃集錄　ab 宋慈
ac 王清任　ad 醫林改錯

Hua Lo-keng (from China Reconstructs, *November 1969, p. 30).*

A Biography of Hua Lo-keng

By Stephen Salaff *

1. INTRODUCTION

THE WORK IS OF BROAD SCOPE and is more than ample to qualify him as one of the foremost mathematicians in the world."[1] So concludes a biography, published in the *Notices of the American Mathematical Society*, of the Society's only member in the People's Republic of China, Hua Lo-keng,[a] born in Kiangsu Province in 1910. Hua's research has vitally enriched the mathematical literature—there were 105 reviews of his books and papers during the years 1939–1965 in the *Mathematical Reviews*—and he has been in the midst of Chinese scientific life since his return to Peking in 1950 from the United States, where he worked for three and one-half years at the Institute for Advanced Study and the University of Illinois. For nearly forty years his singular achievements as a mathematician have made him an indispensable resource person for those concerned with China's higher educational development, and his name, although unfamiliar to nonmathematicians outside China, has long been pronounced with pride by literate Chinese in the mainland. To gain insight into Chinese contributions to contemporary mathematics, and to learn how modern science has fared in China, one must become acquainted with the life of Hua Lo-keng. This biography represents, in the first place, an appreciation of his scholarly contributions and of his literate, versatile mathematical style.

Studying his eventful career, we find that Hua's scientific and political roles cannot be uncoupled, and we will thus examine the complexities and contradictions which have governed his public life since the 1930s. His part in the drama of communism in China has been most conspicuous in the event called the Great Proletarian Cultural Revolution, 1966–1969, when he was shaken by severe political criticism at the hands of the Red Guards. His political reputation (though not his mathematical research career) remained intact, however, and was actually enhanced when he reappeared in the Peking *People's Daily* with a signed article of self-confession in June 1969. He "whole-heartedly praised Chairman Mao's line of higher education and intellectual reform," which advocates integrating the universities, suspended in 1966, with the factories. In China, where a *People's Daily* article is the result of lengthy political

* Department of Mathematics, University of Toronto, Toronto 5, Ontario, Canada. I would like to acknowledge the assistance of Professor Yau Shing-tung, of the Institute for Advanced Study, Princeton, and of Dr. Choi Man-duen, University of Toronto, in translations from the books and articles of Hua Lo-keng.

[1] Lowell Schoenfeld, "A Biographical Note on Professor Loo-keng Hua," *Notices of the American Mathematical Society*, Dec. 1959, No. 7: 729–730. For the current list of members see the *Combined Membership List, 1971–72* (Providence, R.I.:American Mathematical Society, 1971).

At the end of this article is a glossary of the Chinese characters which represent the romanized Chinese words appearing herein. Superscript letters indicate the placement in the list of equivalent characters.

preparation and where "model" figures are used to personify the goals of political campaigns, Hua Lo-keng has thus become a representative "reformed intellectual." An understanding of the effects of the Cultural Revolution on science and education thus calls for a critical analysis of Hua's recent intervention, and so I shall begin a discussion of it in Section IV below.

Since Hua's writings will be discussed here in mathematical language, mathematicians will naturally hold an advantage in reading this biography. But it has been written with other readers also in mind, and I wish to assure nonspecialists that no continuity will be lost if they simply skip over all technical notation.

II: 1910–1950

Hua Lo-keng graduated from the public lower middle school (the first two years of high school) in his native Chin-t'an[b] hsien, Kiangsu, and then attended the Shanghai Chung-hua Vocational School, where he completed one and one-half years of its two-year clerical course. He was forced by the poor circumstances of his family to leave school at the age of fifteen and return to Chin-t'an to help his father in the family's small shop. That Hua's father was unhappy with his son's intellectual preoccupations is illustrated by a winsome sketch which appeared in a popular biography of Hua.[2] The sketch shows the father chasing his son across the floor of the shop, the boy fearfully clutching to his breast some mathematics books, which his father has threatened to burn. The desk where Hua Lo-keng was made to sit while keeping the shop's accounts on an abacus goes unattended.

Hua's talent, like that of most brilliant mathematicians, was manifest early in life. He began to teach himself modern mathematics, and by the time he was nineteen his mathematical articles published in Shanghai science magazines began to attract attention. At that age, however, Hua was stricken with the often-fatal disease of typhoid, followed by arthritis, which burdened him for life with a lame left leg. But he managed to endure this affliction with the fortitude he called forth later during the years of war and hardship. Hua's writing was noticed by the chairman of the mathematics department of Tsinghua University in Peking, Hsiung Ch'ing-lai,[c] who sought to locate the author of the Shanghai articles. At first none of Tsinghua's mathematics teachers or the "Association of Students Returned from Abroad" had ever heard of anyone called Hua Lo-keng; eventually, a professor who was born in Kiangsu Province informed Hsiung that Hua was not a university graduate, not even a middle school graduate, but was merely an accountant in a small village. Determined to discount these obstacles and "discover" Hua, Hsiung visited him at home in Chin-t'an and invited him to come to Peking. After Hua's arrival there in the summer of 1931 he was able to get a job as an "assistant" in Tsinghua's mathematics department. This was a minor clerical post which he could hold without possessing any academic degrees, but he was soon promoted to the faculty as a lecturer (assistant professor).[3] His initial research interests included the theory of numbers, in which he may have

---

[2] *Shu hsüeh chia Hua Lo-keng* (The Mathematician Hua Lo-keng) (Shanghai:Jen Min Ch'u Pan She [People's Publishers], 1956), p. 5.
[3] This account of Hsiung's discovery of Hua is taken from the biography of Hua Lo-keng in Ying Tzu, *Chung kuo hsüeh shu jên wu chih* (Scholars of New China) (Hong Kong:Chih Ming Publishers, 1956), pp. 31–37.

been encouraged by Hsiung, whose Ph.D. thesis at the University of Chicago under Leonard Dickson dealt with the Waring problem in additive number theory. One of Hua's contemporaries at Tsinghua during the early 1930s was the renowned geometer Chern Shiing-shen,[d] who received an M.S. degree there in 1934 and a D.Sc. at Hamburg in 1936 and is now at the University of California, Berkeley.

These two scholars, Hua and Chern, are regarded as the leading Chinese mathematicians of today, but in order to gain this distinction both had to leave China. A brief sketch of the history of Chinese mathematics will indicate the reason why practically all capable young Chinese mathematicians at one time aspired to go abroad. Substantial work was done by Chinese mathematicians prior to the year 1300, when the following were already known: Pascal's triangle of binomial coefficients, a method for approximating the roots of polynomials (the so-called Horner's method), a technique for solving simultaneous systems of quartic equations, and the "Chinese remainder theorem" for the solution of simultaneous congruences. But progress was suspended during the Ming Dynasty (1368–1644), when creative thinking ended and when even earlier discoveries, such as the method for solving quartics, were forgotten.[4] Then, beginning in the seventeenth century, enterprising Jesuit missionaries found that they could make a favorable impression on the Manchu emperors of the Ching Dynasty court by showing them a few examples of European mathematical and astronomical knowledge. Chinese mathematicians slowly began to rediscover their long-forgotten past, and they combined it with the limited amount of European mathematics which they could glean from the texts imported by the missionaries. In the nineteenth and early twentieth centuries mathematicians in China covered the walls of their studies with elaborate, sometimes beautiful formulas—combinatorial numbers, magic squares, congruences, and other arithmetic subjects. Their results involved counting and strictly finite techniques, but not the intricate limiting processes of modern analysis and the abstract algebraic structures worked out by researchers like Lagrange and Gauss, who synthesized existing disciplines and directed and integrated mathematical knowledge to create higher-level theories. Isolated China, furthermore, differed from Japan in that it lacked the impetus to import Western scientific learning. Thus Chinese mathematics remained essentially inert until after 1900.

Starting in the current century, Chinese mathematicians began to absorb the mathematics of the West and to superimpose on this heritage their own contributions. Since there was almost no graduate work in China, advanced Chinese students, who previously had travelled only to Japan, began to enter the universities of Europe and America. They started to publish, and a stream of their papers gradually began to flow around 1928, at first in the fields of number theory and "hard" analysis (e.g., Fourier series).

The twenty-five-year-old Hua grasped a chance which was offered to make contact with the centers of mathematical thought, and he travelled to Cambridge University in the summer of 1936 on a China Cultural Foundation fellowship. This foundation was funded by a large indemnity paid out by the government of the Ching Dynasty

---

[4] "The Confucian scholars who had practiced calligraphy on all the last copies of Tsu Ch'ung-chih's *Chui shu* [a famous old mathematics book] swept back into power in the nationalist reaction of the Ming, and mathematics was again confined to the back rooms of provincial yamens." Joseph Needham, *Science and Civilisation in China*, Vol. III (London:Cambridge Univ. Press, 1959), pp. 153–154.

to eight Western powers after the Boxer Rebellion of 1900 and then returned in part. Tsinghua University, in fact, was set up by American missionaries with U.S. government remittances as a preparatory school for training students to be sent to America. It began to do this in 1911, and by 1946 it had dispatched over one thousand of them to the United States.[5] During the 1930s the Boxer Fund also supported several hundred Chinese students in Britain.

In England Hua Lo-keng joined an illustrious group of number theorists which included the Englishmen Harold Davenport, G. H. Hardy, D. E. Littlewood, and E. M. Wright, and the German émigrés T. Estermann and Hans Heilbronn. Most of Hua's papers from this period dealt with problems in additive number theory. The additive theory of numbers concerns the splitting of integers into sums of certain other integers. In the Waring problem, the most thoroughly investigated one of this subject, the specified integers are kth powers. The problem is thus to find, for a given k, the least integer s, called $g(k)$, such that the equation

$$n = x_1^k + x_2^k + \ldots + x_s^k \tag{1}$$

is solvable for every integer n.

In 1909, over a century after Edward Waring, David Hilbert proved that such a minimum $g(k)$ indeed exists for every k, but his proof was inductive rather than constructive and therefore not amenable to giving an explicit upper bound for $g(k)$. Since Hilbert many outstanding mathematicians have worked on the problem of evaluating $g(k)$. It is known, for example, that $g(2) = 4$ (Lagrange, 1770), and $g(3) = 9$ (Dickson, 1939). That is, every positive integer can be represented as the sum of four squares and of nine cubes, and the "four" and "nine" cannot be replaced by anything smaller. The attempts to find $g(k)$ explicitly for all k have not yet succeeded, although it is believed that for all but finitely many positive integers k, $g(k) = 2^k + A - 2$, where A is the largest integer not exceeding $(3/2)^k$.

Since relatively small integers may have accidentally peculiar representations which conceal to some extent the fundamental issues, $G(k)$ is defined to be the least integer for which Equation (1) is solvable with at most finitely many exceptions n. Much effort has also gone into evaluating or estimating $G(k)$, and it is known that $G(2) = 4$, $4 \leq G(3) \leq 7$, and $G(4) = 16$. Davenport proved in 1942 that $G(5) \leq 23$ and $G(6) \leq 36$, but no explicit values of $G(k)$ have been found for $k \geq 5$.

The Goldbach problem is a celebrated problem closely related to the Waring problem, in which $k = 1$ and $s = 2$ or 3, and the x's are required to be prime. The Goldbach problem reads: "Given any even integer n, can we find primes x_1 and x_2 such that $n = x_1 + x_2$?" and for $s = 3$: "Given any odd n, can we find primes x_1, x_2, x_3 such that $n = x_1 + x_2 + x_3$?"

Hua's results on the Waring and Goldbach problems meshed well with those of his European colleagues. During the 1920s Hardy and Littlewood had published a series of papers in which they used new analytic techniques to resolve the Waring problem and to show that $g(k) = O(k2^k + 1)$. They also obtained an asymptotic formula for the number $r_{k,s}(n)$ of integral solutions of Equation (1), with $x_1 \geq 0, \ldots, x_s \geq 0$.

---

[5] *A Survey of Chinese Students in American* (New York: China Institute, 1954).
Universities and Colleges in the Past 100 Years

They gave a representation for $r_{k,s}(n)$ as a function of k, s, and n, plus a term $o(n^{-1 + s/k})$ as $n \to \infty$; their result, however, was valid only for large values of s. Hua Lo-keng's best work on the Waring problem,[6] according to Heilbronn, was to demonstrate that the Hardy-Littlewood formula holds for all $s \geq 2^k + 1$. The proof of Hua's theorem depends on a hard lemma, which estimates an important trigonometric integral of the form

$$\int_0^1 |f(t)|^{2^j} \, dt, \qquad (2)$$

where $1 \leq j \leq k$ and f is a certain trigonometric sum. Hua's work is a key link in the logical chain leading to the estimates on $G(k)$ now in force; thus Davenport wrote that Hua's "very effective" bound on the trigonometric integral (2) had enabled him to derive crucial inequalities for $G(5)$ and $G(6)$.[7] Previous to Davenport, the sharpest estimate in the first case was $G(5) \leq 28$, due to Hua in 1939.[8]

Building on the results obtained by the number theorists publishing in England, but working somewhat independently, the Soviet mathematician Ivan M. Vinogradov scaled one of the peaks of number theory at this time by establishing the inequality $G(k) \leq k (6 \log k + 4 + \log 216)$. He also electrified the mathematical world by using estimates on trigonometric sums to establish that each sufficiently large odd integer can be expressed as a sum of at most three primes, thus solving in effect the Goldbach problem for odd n.

Before recounting the next stages of Hua's life, it may not be amiss to register the fact that this self-taught, independent thinker, who had already bypassed every "academic" requirement, never formally enrolled for a degree at Cambridge and hence holds no doctorate. He simply took up residence in the town and went to the university to do mathematics. He was apparently able to collect his Boxer fellowship stipend anyway. Despite this fact, which is known among mathematicians, biographers curiously bound by tradition persist in awarding him a doctorate.[9]

Hua returns twice to China and goes abroad again

The full-scale Japanese invasion of China began in the summer of 1937, and Tsinghua University, together with the National Peking and Yenching universities, fled to Kunming, capital of Yunan Province in southwest China. These universities abandoned their libraries and laboratories to the Japanese invaders, and what remained of their books and equipment dwindled to nothing when they reached the

[6] Loo-keng Hua, "On Waring's Problem," *Quarterly Journal of Mathematics*, 1938, 9:199–202.
[7] Harold Davenport, *Analytic Methods for Diophantine Equations and Diophantine Inequalities* (Ann Arbor:Ann Arbor Publishers, 1963), p. 65.
[8] "On Waring's Problem for Fifth Powers," *Proceedings of the London Mathematical Society*, 1939, Ser. 2, 45:144–160.

[9] Those who erroneously credit Hua with a doctorate in mathematics include: Chu-yuan Cheng, *Scientific and Engineering Manpower in Communist China, 1949–1963* (Washington: National Science Foundation, 1963), p. 449; Wolfgang Barthe, *Chinaköpfe* (Hannover: Verlag für Literatur und Zeitgeschehen, 1966), p. 159; William Ryan and Sam Summerlin, *The China Cloud* (Boston:Little, Brown, 1967), p. 307.

interior, most being damaged or lost on the way. The universities re-formed in Kunming as the Southwest Associated University, and Hua returned from Cambridge to become a professor there from 1938 to 1945. The wartime years were marked by material poverty and chaotic academic conditions,[10] but nevertheless there are over twenty research papers of Hua's in the literature of these years. In 1941 he completed the manuscript of his first general work, *Additive Prime Number Theory*, which deals with the Waring and Goldbach problems and some allied questions, and which unifies and improves upon the original results in his previous papers. Hua submitted this manuscript to the Institute of Mathematics, which was established in Kunming by the Academia Sinica in 1941. But the inception of this institute was attended by problems of war and disorganization, and although Hua received a Ministry of Education prize for it,[11] the manuscript was unfortunately not published at this time. (Hua offered an explanation for this controversial episode later; see Sec. III below.)

After the war there opened another chapter in Hua's international career—his close personal relationship with the Soviet Union and its mathematicians. In the second half of 1945 he journeyed to the U.S.S.R. on the invitation of the Soviet Academy of Sciences. Hua and Vinogradov had begun a correspondence in the 1930s, and their development of the method of trigonometric sums remarkably transformed the entire subject of analytic number theory. Their work provided a connected treatment of a substantial part of the results obtained in the subject since the 1920s and was to be applied in various branches of number theory. In recognition of these contributions, *Doklady*, the Soviet Academy's principal journal for research announcements, which rarely prints foreign papers, carried an article by Hua in each of the years from 1937 through 1941. Hua had already sent to the Soviet Union an English manuscript of *Additive Prime Number Theory* before his 1945 trip. He learned that Vinogradov had sharpened his methods in 1942, and Hua acknowledged these improvements in the manuscript's revision, which was translated into Russian and published in 1947 as Volume 22 of the *Proceedings of the Steklov Institute of Mathematics*.[12] Some of his results are now regarded as classic. In this Steklov memoir Hua establishes an asymptotic formula for the number of solutions of the Waring problem; his main results, however, are broader and concern the asymptotic behavior of the number of solutions of the equation

$$n = f(p_1) + f(p_2) + \ldots + f(p_s),$$

where f is a suitably restricted integer-valued polynomial of degree k. This generalizes the Waring problem, which deals with the case $f(x) = x^k$. Aside from Dickson's work, there are more papers in the literature on Waring's problem with polynomial summands by Hua than by any other author.[13]

The Soviet Academy's early and positive acknowledgment of Hua and his work

---

[10] An anecdote illustrative of the deprivation prevailing in Kunming is attributed to Hua by Ying Tzu, *op. cit.*, p. 34:

When I was in Kunming Southwest Associated University, there was a story about a beggar who met a man walking along the street. The beggar opportuned the man, who told him he had no money. The beggar did not believe him and followed along behind the man for another block. Then the man turned around and said to the beggar: "I am a professor." The beggar then walked away, because he knew that a professor would certainly have no money in his pocket.

[11] *China Handbook* (New York:Rockport Press, 1950), p. 753.

[12] *Trudy Matematiceskogo Instituta im. V. A. Steklova*, Moscow/Leningrad, 1947, Vol. 22.

[13] W. J. Ellison, "Waring's Problem," *American Mathematical Monthly*, Jan. 1971, 78:10–36.

perhaps impelled the shining preface to *Additive Prime Number Theory*, dated April 1946, in which Hua expressed:

> Profound gratitude to Academician Vinogradov . . . [and] to the Soviet Academy of Sciences for its favorable reception of my work. In these difficult days we are especially happy to see the results of our scientific study win the approval of the highest authority of a people to whom we are bound by the closest ties of friendship. . . . It is respectfully hoped that the publication of this book will strengthen the true friendship and mutual affection of our two great peoples.

In the first half of 1946 Hua returned from the Soviet Union to Kunming, where the Chinese Academy of Sciences was still located. But Hua did not remain long to work in the Academy; instead he accepted an invitation from the Institute for Advanced Study and journeyed to the United States in the autumn of the same year. Hua was one of a number of physical scientists who went from China to the United States around this time. Others included the nuclear physicist Wu Ta-you,[e] a Boxer fellow and Michigan Ph.D., 1933, and the organic chemist Tseng Chao-lun,[f] Ph.D. in chemical engineering, M.I.T.[14] Lee Tsing-tao,[g] later to share the Nobel Prize in physics for his work on nonconservation of parity, was one of the junior members of this group.

In later biographies of Tseng and Hua it is stated that in 1946 the Nationalists were interested in learning about the U.S. nuclear program through their scientists. Thus the source *Hsin chung-kuo jên wu chih* (Personalities of New China) relates that Tseng was sent by the Nationalist Government to England to study atomic energy, then went to the U.S., and returned in 1947 to Hong Kong.[15] The more recent *Chung kung jên ming lu* (Biographies of Chinese Communist Personalities) states that after the Sino-Japanese war Tseng was sent by the government to the U.S. to study atomic energy, and that he was summoned to return by the government in 1947.[16] That there was also a relationship between Hua and the Nationalist Defense Ministry is claimed in a 1956 biography of Hua by the Chinese Communists, where it is stated that "The Kuomintang said they sent Hua to the United States to investigate 'national defence mathematics' and the construction of the atomic bomb."[17] From these sources, as well as my own study of this period, I am of the opinion that Chiang Kai-shek did in fact mandate Hua and the other distinguished scientists who came to the United States to gather information on the U.S. nuclear program.

But the circumstances of 1946 were complicated, and Communist writers have advanced a different reason to explain Hua's departure for the United States— the assassination of Wen I-to.[h] A leading Chinese poet and classical scholar of the 1920s and 1930s, Wen I-to was a professor at the Southwest Associated University during the Sino-Japanese war and "a close colleague of Hua Lo-keng and one whom he deeply respected."[18] The Japanese aggression had galvanized Wen politically, and

---

[14] In the post-World War II period, Tseng Chao-lun was Professor and Dean of the Department of Chemistry at the National Peking University and was one of 81 prominent Chinese scientists and scholars elected to membership in the Academia Sinica in 1948 (*China Handbook*, p. 753).

[15] *Hsin chung-kuo jên wu chih* (Personalities of New China), Vol. II (Hong Kong: Chou Mo Pao Shê Publishers, 1950), p. 100.

[16] *Chung kung jên ming lu* (Biographies of Chinese Communist Personalities) (Taipei: Institute of International Relations, 1967), p. 496.

[17] Ying Tzu (Scholars of New China), p. 35.

[18] *People's China*, Peking, Dec. 16, 1953. (*People's China* is the predecessor of the *Peking Review*.)

he became an activist in the China Democratic League and its propaganda director for Yunnan Province.[19] On July 15, 1946, Professor Wen was killed by assassins after delivering a speech at the funeral of a murdered colleague in the Democratic League, a speech in which he accused Nationalist Government agents of committing the murder.[20] It is charged that Wen I-to himself was assassinated by special agents of the Kuomintang.[21] "It was under such conditions that Hua Lo-keng and his family went abroad in the summer of 1946," the official version concludes.[22]

The "atomic energy mission" was, to say the least, idiosyncratic. Chiang Kai-shek believed, or perhaps tried to convey the impression, that brilliant Chinese scientists could obtain access to the nuclear program. But the McMahon Act passed by the U.S. Congress in April 1946 forbade the sharing of atomic secrets, even with the closest ally, Britain. And although a number of the civilian scientists who had worked on the bomb had moved to Princeton, it is highly questionable whether they would have been willing, or able, to impart any decisive information to men they had never before met, one of whom was fresh from a winter in Moscow. To Chiang's political arrogance and/or naïveté must be added, naturally, the practical impossibility of his disorganized forces acquiring the production technology to manufacture an atomic bomb. On Hua's part the desire to maximize his research opportunities in the United States, the only big scientific nation not torn up by war, may have balanced his disquiet about the "mission." He probably also had the shrewd idea that he would never come close to nuclear data anyway. When Hua and the others arrived in the United States they dispersed into the academic world. The physicist Wu became Visiting Professor at the University of Michigan in 1947 (he afterward headed the theoretical physics section of the National Research Council in Canada and is now in the department of physics at the State University of New York, Buffalo). Tseng Chao-lun, after his return to Hong Kong in 1947, was active there as a scientist and as a Standing Committee Member of the China Democratic League.[23] (He will appear in this narrative as a national scientific administrator in Communist China.) Hua Lo-keng settled down at the Institute for Advanced Study for two years and then became a professor at the University of Illinois. During his stay in Princeton he underwent surgery at Johns Hopkins Hospital in Baltimore, which partially alleviated the lameness in his left leg. Thus the Defense Ministry "mission" evaporated.

American mathematicians who met him during those years were much impressed by Hua's clear and direct mathematical methods, the depth of his knowledge, and his ingenuity. His interests had broadened to include the theory of several complex variables, automorphic functions, and the geometry of matrices. Active mathematicians are

[19] Howard Boorman, *Biographical Dictionary of Republican China*, Vol. III (New York: Columbia Univ. Press, 1970), pp. 410, 411. Boorman adds that during this period Wen's office usually was "crowded with liberal and leftist students and professors."

[20] *Ibid.*

[21] This charge is levelled in *People's China, loc. cit.*, and also in Israel Epstein, *The Unfinished Revolution in China* (Boston: Little, Brown, 1949), p. 388:

U.S. aid to Chiang ran the whole scale from a

gift of 271 naval ships to the provision of special silent pistols.... The silent pistols developed by the Navy acquired a particularly melancholy significance. They were used by Chiang's secret police to assassinate American-trained liberal intellectuals, leaders of the Democratic League like Professor Wen I-to and Li Kung-po.

Epstein is here referring to Li Kung-p'u, whose funeral oration it was that Wen delivered.

[22] *People's China, loc. cit.*

[23] Personalities of New China, *loc. cit.*

quite familiar with the diverse and powerful contributions Hua has made to their art, since they use his results from day to day. I have mentioned the name of Hua to differential geometers and algebraists, as well as number theorists, and all of these mathematicians light up at once. When a group theorist heard me repeat this Chinese name, he remarked: "We have a famous theorem about isomorphisms called 'Hua's theorem.' That must be the same Hua!" This result is presented in *Geometric Algebra* by E. Artin:

> Hua has discovered a beautiful theorem which has a nice geometric application:
> If σ is a map of a field k into some field F which satisfies the following conditions:
> (1) σ is a homomorphism for addition,
> (2) for $a \neq 0$ we have $\sigma(a^{-1}) = [\sigma(a)]^{-1}$; i.e., we assume that σ maps the inverse of an element onto the inverse of the image,
> (3) $\sigma(1) = 1$;
> then σ is either an isomorphism or an antiisomorphism of k into F.[24]

The proof is a sequence of well-chosen field operations based upon an elementary but overlooked identity, making use of the given properties of σ.

The geometric application referred to occurs in the foundations of projective geometry, where σ is chosen as a one-to-one map of a line in a Desarguian plane into itself, which carries any four harmonic points A,B,C,D into harmonic points A',B',C',D'. It is first assumed that $\sigma(0) = 0$ and $\sigma(1) = 1$. By considering a harmonic set of points one of which is at infinity, it is shown that $\sigma(a + b) = \sigma(a) + \sigma(b)$, which means that σ is an isomorphism under addition. Then by taking $a = -1$, $b = 1$, and $c \neq 0, 1, -1$, we get $d = c^{-1}$; hence the images $-1, 1, \sigma(c), \sigma(c^{-1})$ are harmonic, showing that $\sigma(c^{-1}) = [\sigma(c)]^{-1}$. Appealing to Hua's theorem, we see that σ is either an automorphism or an antiautomorphism of the underlying field. To settle the general case where $\sigma(0) = 0$ or $\sigma(1) = 1$ may be violated, it is computed that $\sigma(x) = ax^\tau + b$, where $a \neq 0$ and where τ is either an automorphism or an antiautomorphism.

Hua's perceptive choice of direct algebraic methods enables him to give a straightforward proof of the "Cartan-Brauer-Hua" theorem: every proper normal subfield of a skew field is contained in the center.[25] Before Hua and Richard Brauer, Henri Cartan's proof used the complicated device of Galois extensions over subfields. The Cartan-Brauer-Hua theorem is used to study collineations of a field over a normal subfield and in general gives information about normal subgroups of the multiplicative group of a field.

One of the mathematicians who knew Hua, the Berkeley number theorist Derrick Lehmer, remarked to me:

> Hua had the uncanny ability of taking the best work of others and finding the exact points where their results could be sharpened. He had many tricks of his own, too. He read widely and commanded an overview of all of twentieth-century number theory. His chief interest was to improve upon the whole field; he would have, if left to himself, tried to generalize every result he came upon. His work was in some respects like that of [I.] Schur, or even [Norbert] Weiner, both of whom made deep contributions to number theory, but also branched off to other fields.[26]

---

[24] E. Artin, *Geometric Algebra* (New York: Interscience, 1957), p. 37.
[25] "Some Properties of a Sfield," *Proceedings of the National Academy of Sciences*, 1949, *35*: 533–537.
[26] Personal communication from Derrick Lehmer, Berkeley, 1970.

Lehmer points out that the new interests assumed by "generalists" like Hua and Schur were suggested to them by their former concerns. Thus, for example, estimates for the number of solutions of number-theoretic equations are provided by the coefficients of certain power series, and this fact conceivably was Hua's bridge from analytic number theory to complex variables.

One example of how well Hua scrutinized the literature is interesting, for it concerns a mistake made by two of the greatest European algebraists, O. Schreier and B. L. van der Waerden, who in 1928 determined the automorphisms of the unimodular projective group PSL_n (K) (for $n \geq 2$) over a commutative field K. Their proof contained an error in the case $n = 2$, and twenty years later Hua corrected it, delicately referring to the mistake as an "unjustified point."[27]

Hua and Tsien Hsue-sen go home to stay

During the postwar years Hua was one of many outstanding Chinese scientists working in the United States, which in addition to members of the 1946 group included the aerodynamicist Tsien Hsue-shen.[i,28] A recent book elaborates on the myth that Tsien in 1955 carried back to China with him vital secret information which accelerated China's nuclear missile program by an entire decade, and furthermore links Hua's name to Tsien's in this effort.[29] But while Tsien undoubtedly was one of the world's leading experts in the theory of supersonic flow, he was essentially an applied mathematician, whose interests were extremely theoretical. As a research superviser and project organizer in China, he has certainly given strong overall leadership to China's missile program, but like the pure mathematician Hua Lo-keng, he was not privy to American "secrets."[30]

[27] "On the Automorphisms of the Symplectic Group over Any Field," *Annals of Mathematics*, 1948, *49*:739–759. Another instance of an error pointed up by Hua is given by Jean Dieudonné, *La Géométrie des groupes classiques* (2nd ed.; Berlin:Springer, 1963), p. 96.

[28] A graduate of Chiaotung University in Shanghai (after Tsinghua and St. Johns the university which sent the most students to the U.S.), Tsien too won a Boxer fellowship. He took the Sc.D. in 1939 from the California Institute of Technology, where he studied with Theodor von Kármán. Tsien rose fast in the wartime aeronautics establishment, and by 1949 he was named Goddard Professor of Jet Propulsion at Caltech and director of its Jet Propulsion Center. In August 1950, after the outbreak of the Korean War, Professor Tsien sought to return to China, but he was forcibly detained in Honolulu with eight trunkloads of his personal books and papers which he had forwarded. FBI and Customs men seized his scientific papers, and zealous agents went so far as to confiscate his logarithm tables, which they took for code. But the U.S. government later had to admit that there had been no "secret material" in the eight trunkloads (Associated Press dispatch, *New York Times*, Sept. 13, 1955). At a Department of Immigration hearing after his arrest, two retired officers of the Los Angeles Police Department's "red squad" charged that Tsien was a member of the Communist Party of the United States in 1939 (*New York Times*, Nov. 16, 1950). Tsien was compelled to remain inside the boundaries of Los Angeles County for five years. Then, as a result of talks in Geneva between the U.S. and Chinese governments, Tsien was permitted to return home in September 1955.

[29] Ryan and Summerlin, *China Cloud*. Subtitled *America's Tragic Blunder and China's Rise to Nuclear Power*, it is riddled with factual errors and misinterpretations, for example the chronological hodgepodge on p. 78: "Hua had begun his career as a scientist at the University of Berlin in Germany as a Boxer fellow, then shifted to the University of Cambridge in England. From there he moved to the United States and spent four years at Illinois." This study of China's nuclear quest does not uncover, and in fact hopelessly obscures, the only known case in which a Chinese leader actively claimed to be seeking nuclear information in the United States— Chiang Kai-shek in the year 1946.

[30] I dispute the legend that Tsien brought American secrets to China. My views are based on several conversations in 1970 with Edmund G. Laitone, Professor of Aeronautical Sciences at the University of California, Berke-

Hua and Tsien were only two of the Chinese intellectuals in the diaspora who decided to return to the People's Republic of China. Some of them were activists, and they organized groups of returning students and teachers; others were relatively apolitical but elated that China was finally getting "a real government" which could stop the ruinous chaos and inflation and at last unify the country. Furthermore, as I understand from a number of thoughtful young Chinese mathematicians, most, if not all, Chinese scholars now living abroad suffer from solitude and have strong sentiments directed toward their country of birth and upbringing. To these general factors must be added the concern of many Chinese about the continuing menace of McCarthyism in the United States and about their ability to work happily or at all in a climate of persecution and fear. Hua Lo-keng, who in his last year or so at Illinois became active in the returning students movement, sailed for Hong Kong in February 1950, at the head of a party of Chinese students. While still on his way back he addressed an open letter to Chinese students in the United States, urging them to follow his example.

Some of the mathematicians working in the U.S. whom I interviewed in 1970 for this biography noted the serious difficulties which Hua later faced in China during successive political campaigns and responded somewhat protectively. More than one of them felt that "Hua would have done better research, published more, and in general he would have been better off if he had stayed in the United States." But in arguing this, they appear to be insensitive to the hostility directed at patriotic Chinese in the United States, where the revolution in their homeland was regarded as an incredible and traitorous disaster for the Free World. The harrassment of Tsien after Hua's departure is the best known but not the only manifestation of this hostility.

In the first instance, Hua's shift of allegiance from the Nationalists to the Communists was undoubtedly motivated by his intense desire to participate in the growth of Chinese mathematics and his recognition that the Communist Party would support this growth. Here is the gist of what he said to Professor Lehmer about Chinese mathematics: "China is a big country and a great country. So why then should we remain so far behind in mathematics? We must catch up, and I think that we can catch up." Much later he wrote, not at variance, "I wanted to devote the rest of my life to contributing my technical knowledge to the revolution and the people."[31] If we grant the patriotic sentiment embodied in these expressions, while taking into account Hua's poor family origins (a propitious factor as far as the Communists were concerned) and

---

ley. Laitone knew Tsien while the latter was Goddard Professor, and he is familiar with Tsien's technical work. He remarks: "Tsien was not at all inclined toward hardware. There were many engineering details, concerning, for example, low- and high-temperature gases and low-temperature liquids, which he didn't know. I would say that he probably didn't even know the difference between a welded joint and a riveted joint." This quip was not meant to slight Tsien's mathematical achievements but only to point up his shortcomings as an engineer. In Boorman's biography of Tsien (*Biographical Dictionary*, Vol. III, pp. 312–316), which stresses his role in

initiating "long-range U.S. aeronautical programming," including guided missile weapons and the "theoretical possibilities of nuclear fission and fusion for interplanetary and interstellar flight," Tsien's concern is nevertheless described as "primaily conceptual," for "he had little interest in (or patience with) routine engineering problems" (p. 315).

[31] Hua Lo-keng, "Chairman Mao Points Out the Road of Advance for Me," *China Reconstructs*, Nov. 1969, pp. 30–31, 41. (*China Reconstructs* is an English-language monthly published in Peking which reports on the economic and social life in China.)

218   STEPHEN SALAFF

the vexatious political atmosphere in the United States, we perhaps have the explanation for his turning-point decision.

There exists, however, a much less sympathetic interpretation of Hua's course of
action, which stresses the "half-promising, half-threatening" letters he allegedly received from the mainland. According to this interpretation, which appeared in a 1952
New York Times article, precisely none of the six thousand Chinese studying in the
United States would have gone home, except for the "threats, intimidations, and subtle
attempts at persuasion" in letters from China.[32] The *Times* reported that when Hua
landed at Hong Kong, "The Chinese Communists handed him a statement that he was
forced to broadcast over the radio, saying that he had been mistreated in America, that
our country was filled with race hatred." It then made the startling but unsubstantiated
assertion that "Not long ago, he jumped out of the window in a suicide attempt." I
shall not comment at this point on the question of suicide, but suffice it to say that this
brassy newspaper story in itself lays open all the hostility which the principal U.S.
media bore toward the People's Republic of China. The fact that the *New York Times*
could ignorantly or maliciously slight the patriotic sentiments of the returning Chinese
testifies to the correctness of Hua's "forced" accusation of race hatred.

III: 1950–1966

Prompt and full recognition was accorded Hua Lo-keng after he returned to China.
Tsinghua University, which had re-established itself in Peking after 1946 and which
was soon to be transformed into an advanced polytechnic institute, welcomed him
back: he became chairman of the mathematics department. The Academy of Sciences
was reorganized one month after the birth of the People's Republic, in Peking in
November 1949 under the Marxist historian and writer Kuo Mo-jo (the former
Academy had been disbanded by Chiang Kai-shek and moved to Taiwan), and in 1950
Hua took part in the preparatory work for the establishment of its Institute of Mathematics. When the Institute was opened in July 1952 he was appointed director. He was
invited to join a delegation organized by the Academy of Sciences to visit the countries
of Eastern Europe, and on this tour he attended the first postwar congress of Hungarian
mathematicians.

Under the new government, university professors participated in political discussions to guide them toward the "red and expert" goal.[33] During the thought reform
campaign of 1951–1952 students and faculty members were organized into small
groups and assigned to study Marxism-Leninism, the writings of Mao Tse-tung, and
the doctrines of the "New Democracy" and to apply these writings in analyzing their
own behavior and attitudes. An important theme of the campaign was the denunciation
of the previous glorification of things Western, a theme given special emphasis by the
concurrence of the Korean War.

From this campaign emerged Hua's article "We should have only one tradition, the
tradition of serving the people."[34] It dealt mainly with the situation at Tsinghua, whose

[32] *New York Times*, May 11, 1952.

[33] "Red and expert" is a standard Communist
precept: leadership based on "expertise" means
that leadership exercised through some organizational office or role; by contrast, "red" leader-

ship is based upon personal solidarity relationships.

[34] *Kuangming Daily*, Peking, Nov. 15, 1951.
Reprinted in *Ssu hsiang kai tsao wen hsüan* (Col-
lected Articles in Thought Reform), Vol. I
(Peking: Kuangming Daily, 1951), pp. 22–28.

narrow class of graduates did not recognize its obligations toward the masses. The mathematics department, which had been in existence for about twenty years, and which Hua could truthfully say he "knew very well," produced only sixty-one graduates and seven graduate students during these two decades. It was wrong for Tsinghua, which had sufficient facilities to train a greater number of students, to single out only a few "geniuses" (*t'ien ts'ai*[i]) for special attention. Hua further cited the "painful" example of Tsinghua's department of aeronautical engineering, in which ten students were enrolled before the Liberation. One of them died, three went to Taiwan, and four had left for the United States. Hua urged the students to break with these selfish traditions and turn to the service of China, which required many more trained people. He criticized the old "semi-colonial" research system, in which a few mathematicians borrowed problems from foreign researchers and sent their results, which were patterned after the foreigners, to overseas journals for publication. If their papers were published, they were overwhelmed with joy. The kind of research that qualified was like a "flower on a foreign head": it was not only rootless, but a mere decoration for others. Because of U.S. hostility and attempts to isolate China by a naval blockade and a trade embargo, some people in the universities feared that Chinese research would be hampered by the shortage of American journals. Hua claimed that such fears were groundless. It transpired that China's alternative channel for the importation of modern science and technology was the Soviet Union.

From a stance partially that of a one-time outsider, Hua could speak with much bitterness against the old regime and its elitist traditions of higher education, and in this central article, at any rate, he was not required to criticize his own former connections with the Kuomintang. The self-examination which Hua does offer indicated that he was at least conscious of many of the fundamental issues of ideological rectification. He acknowledges the leadership of the Communist Party in the struggle for personal transformation through the Bolshevik weapons of criticism and self-criticism, and he apologizes for his inevitable mistakes. "This is the first time that I am writing with these weapons," he explains. Hua had taken the initial steps toward acquiring a socialist outlook, and his constructive, officially supported programs for the overall development of Chinese mathematics[35] imparted great momentum to his work at the start of his life in the People's Republic. China had embarked upon profound social changes, involving vast educational experiments designed to teach the essential principles of modern science and practical technology to the masses, and it needed Hua's help. Never in history has a scientist of his caliber been entrusted with as much responsibility for the mathematical education of an entire people.

The Academy of Sciences, from the date of its reconstitution, was built with Soviet assistance, and for the leading Chinese scientists a central task of "political activism" was the forging of strong ties with their Soviet colleagues. In 1953 a twenty-six-member delegation left Peking for the U.S.S.R. and in three months toured scientific research and educational institutions in Moscow, Leningrad, Kiev, Tashkent, and Novosibirsk. Hua was part of this delegation, and late in the year his report on the current position of mathematics in China was published by the Soviet Academy.[36] The report begins with a survey of Chinese mathematics through the early twentieth cen-

[35] E.g., see his plan for training mathematical "cadres" in the field of classical groups (discussed below).

[36] "The Present Position of Mathematics in China," *Vestnik Akademii Nauk SSSR*, 1953, 6: 14–20 (Russian).

tury. Under the Nationalists Chinese mathematics was well developed in several fields, such as analytic number theory, Fourier series, and topology. But Chinese research interests varied with the fashion in the West and developed slowly, spontaneously, and without a plan. As Hua puts it, developing the theme he enunciated in his Tsinghua article:

> Our colleagues did not have a definite direction of work, and they changed directions depending only on their personal interests, which varied with the "mode" in bourgeois countries. Soviet work was neglected. Many mathematical works were without practical purpose (with "game" mentality), and the fact that mathematics arises from the practical requirements of experience was not acknowledged.

Under the impact of Liberation, many of China's scientific problems were on their way to solution. There was a national plan for the expansion of scientific education and the all-round development of mathematics. Hua stated that the gaps in Chinese mathematics were being closed by speeding up, for example, the development of functional analysis, vital to both pure and applied mathematics, and also by accelerating the development of probability and statistics. Soviet literature is particularly rich in these areas, and Hua cited a number of first-class Soviet textbooks which were being translated into Chinese.

There was, however, the internal Chinese problem of the departure of a number of leading mathematicians in 1949. Hua mentioned "with regret" that some of the topologists in the Nationalist Academy of Sciences, the foremost among them Chern Shiing-shen, had left for the United States when Chiang Kai-shek moved the Academy to Taiwan, and he expressed hope for their return to the mainland—a hope that was not realized.

In 1955 the Academy of Sciences in Peking was empowered by the State Council to encourage scientists through a system of national science awards, which included medals and cash prizes (10,000 yuan, or $4,000 for the "first-clsss awards") for "practical advances in science and technology as applicable to building socialism." Three first-class-award winners were honored in January 1957: Hua Lo-keng, the topologist Wu Wen-tsun,[k] and Tsien Hsue-shen, for his work on engineering cybernetics. Hua's prizewinning monograph was entitled *Harmonic Analysis of Functions of Several Complex Variables in Classical Domains.*[37] In this work, which was a collection and systematization of results he published in numerous articles over the preceding years, Hua studied the four classical types of irreducible bounded symmetric domains, which are the higher-dimensional analogues of the unit disk and other domains in the complex (Argand) plane. The four classical domains R_I, R_{II}, R_{III}, and R_{IV} are defined as follows:

$$R_I = \left\{ m \times n \text{ matrices } Z \text{ satisfying } I_m - ZZ^* > 0 \right\},$$

$$R_{II} = \left\{ \text{symmetric matrices } Z \text{ of order } n \text{ satisfying } I_n - ZZ^* > 0 \right\},$$

$$R_{III} = \left\{ \text{skew-symmetric matrices } Z \text{ of order } n \text{ satisfying } I_n - ZZ^* > 0 \right\},$$

$$R_{IV} = \left\{ z = (z_1, ..., z_n) \epsilon C^n; |zz'|^2 + 1 - 2\bar{z}\bar{z}' > 0, |zz'| < 1 \right\}.$$

[37] Hua Lo-keng, *To-fu-pien han-shu-lun chung ti tien-hsing-yü ti tiao-ho fan-hsi,* (Harmonic Analysis of Functions of Several Complex Variables in Classical Domains) (Peking: Science Publishing Co., 1957; 2nd ed., 1965).

where I_m denotes the identity matrix of order m, Z^* is the complex conjugate of the transpose Z' of Z, and z' is the transpose of the vector z.

Hua calculates various natural geometric quantities, such as metrics, volumes, and curvatures in each of these domains. For example, the analogue of the integral

$$\int_D (1 - |z|^2)^\lambda \, dx \, dy,$$

where D is the unit disk, is explicitly computed for each domain, giving its volume. Having established a geometric structure, he considers the Laplacian, which leads to harmonic functions and general theorems about the Bergman, Cauchy, and Poisson kernels of a homogeneous circular domain with a sufficiently regular boundary. These kernels are explicitly determined for the classical domains. Hua describes orthonormal systems in the space of holomorphic L^2 functions of each domain and orthonormal systems in the L^2 space of the distinguished boundary. The boundary behavior of the Poisson kernel is studied, and the Dirichlet problem is solved. (It must be remarked that the classical domains are today considered as particular cases of bounded symmetric domains and are thus studied using general theorems about Lie groups and Lie algebras. Therefore Hua's original results can now be obtained using fewer direct computations.)

In addition to this monograph, two of Hua's books on number theory were published in China during the 1950s. *Additive Prime Number Theory* appeared in its first Chinese edition in 1953. The preface explained that the Nationalist Academy had lost the manuscript Hua submitted back in Kunming. Because of this, and apparently because he did not keep a Chinese copy, Hua claimed that the Chinese edition had to be retranslated from the Russian edition of 1947. He used the incident didactically, to show "how little the Nationalists really cared about science, and how much more attention the People's Government paid to the establishment of science."[38]

There followed in 1957 the 652-page *Introduction to Number Theory*, written "to fill the long-standing need for a basic Chinese textbook in number theory."[39] It embodies numerous unpublished results, as well as basic material on trigonometric sums, Diophantine equations, modular transformations, and the Waring and Tarry problems. The Preface contains Hua's statement on the teaching of number theory, where he decries the lack of reference texts in number theory suitable for Chinese students. In mathematical history the ideas and methods of number theory influenced the development of other fields, and conversely, number-theoretic problems can be solved by the use of methods from various disciplines: "Other introductory textbooks do not exactly show these relationships, and some 'self-contained' introductory textbooks in number theory give the reader an inexact picture, that is, they imply that number theory is an isolated subject." Hua cites three concrete examples of the interplay between number theory and other fields: (1) the prime number theorem and the Fourier integral; (2) the problem of splitting natural numbers and the problem of representing natural numbers as sums of four squares, and modular function theory; and (3) quadratic forms, modular transformations, and Lobachevskian geometry.

A second pedagogical theme in the Preface is the progression from concrete to

[38] *Tui lei su shu lun* (Additive Prime Number Theory) (Peking:Chinese Academy of Sciences, 1953; rev. ed., 1957), Preface.

[39] *Shu lun tao yin* (Introduction to Number Theory) (Peking:Science Publishing Co., 1957; 2nd ed., 1965), Preface.

abstract results. This progression is fundamental in the teaching of mathematics, and Hua explains how he will use it in approaching the prime number theorem—the estimation of the number $\pi(n)$ of primes less than n. Coarse estimates for an asymptotic value of log $\pi(n)$ can be obtained algebraically; Tauberian methods for an asymptotic value of $\pi(n)$ are somewhat deeper; finally, modular function theory and analytic methods are much deeper still and yield the proof that $\pi(n)$ is asymptotic to $n/\log n$.

Hua also traces the genesis of the book:

> I remember clearly teaching number theory in Kunming in 1940. It was then that I began the present book. I wrote 80,000–90,000 characters, based on my original notes, as well as some newer ones, and I thought that 20,000–30,000 more characters would suffice for publication. But where could I publish it? I was discouraged. When I taught in the United States, I added to and revised the manuscript, for teaching purposes only, but I did not consider the matter of publication. I really only began to work on preparing this textbook after Liberation. I had much assistance and I worked faster than before [in Kunming]. As a consequence, the contents have been expanded and I have been able to add new results.[40]

Mathematical Reviews is unusually lavish in its praise of the book: "This is a valuable and important textbook, somewhat along the lines of Hardy and Wright's *Introduction to the Theory of Numbers*, but going far beyond it in scope." Also lauded was the "very clear and easy" literary style of the text, which recommends it as an excellent introduction for those wishing to learn mathematical Chinese.[41]

During the mid-1950s phase of encouraging scientific talent, the Academy of Sciences initiated a competitive examination for school students, modeled upon the Mathematical Olympiads in Hungary and similar competitions in the Soviet Union. The Olympiad problems were thought-provoking and difficult, although elementary. They could be solved by students with only a modest amount of formal training, provided they could think rigorously and imaginatively. Hua certainly expected the examination to help Chinese mathematics, for according to the respected mathematician and pedagogue George Polya, the Hungarian Olympiad had "essentially contributed" to the development of mathematics in Hungary.[42] Hua tried hard to encourage the study of mathematics through such competitions and in writings and talks offered to the youth. In July 1956 he wrote an article for *Chinese Youth*, the official semi-monthly organ of the Communist Youth League, entitled "Wisdom Comes from Study, Genius is Accumulated."[1] In this article Hua claims: "No trace of genius can be discovered in me," and he affirms that "so-called genius" really depends on study. He goes on to explain the methodological problems of studying science, which must include close attention to the scientific method. Our knowledge of reality is organized logically, and conquering science is thus like climbing stairs. "If we try to climb up four or five steps at a time, 'from the earth to the sky,' then we shall fall down and hurt our head."[43] Although the personal disclaimer is not entirely convincing, Hua did perhaps instill confidence in his mathematically minded readers that they could indeed learn higher mathematics—whose difficulties are sometimes overrated—by steady, well-directed

---

[40] *Ibid.*

[41] *Mathematical Reviews*, 1959, *20*, #829. The reviewer was K. Mahler, a prominent number theorist fluent in Chinese.

[42] George Polya, *Mathematical Discovery*,

Vol. II (New York: McNally, 1964), p. 186.

[43] "Ts'ung ming tsai yü hsüeh hsi, t'ien ts'ai yu yü lei chi¹" (Wisdom Comes from Study, Genius Is Accumulated), *Chinese Youth*, July 1956, pp. 15–17.

work. Mathematics is not the province of a few "geniuses," and the fears of mathematical abstraction which afflict noninitiates can be nullified by systematic exposure to the definitions and axiomatic methods of modern physical science.

But learning gradually is not learning by repetition, and Hua inveighs against rote-learning methods, a traditional impediment to creative thinking in China as well as elsewhere. "For example, a student was learning mathematics. Many calculus books on the same level were brought before him. He read every one, and worked every exercise many times. This is a book-worm reading method." Instead, the student should choose a good book, finish it carefully under the supervision of a capable teacher, and then should read more advanced books. A student with concrete supervision can acquire fundamental knowledge in mathematics and simultaneously begin research. To do good research, the student must think independently (*tu li ssu kao*).[m] Since the objective world is always changing, scientific work is constantly developing, and this calls for ever-new and constructive methods, and the courage to be creative. Hua goes on to censure the rote-learning patterns prevalent among even advanced university students:

> When I visted Democratic Germany, our students there told me that because our universities do not train students well enough in the skills of independent thinking, they find they have serious difficulties. . . . They were no worse in reading and aural ability than the German students, but when they participated in seminars they were at a loss. They did not even know how to find reference materials. Even when they could find references, they had no new ideas. . . . In our country, some universities have used old teaching methods wherein students who do not understand something ask the teacher; if they do not understand a second time, they ask again; if they do not understand a third time, they ask again, until they understand it thoroughly. Although this is the method which saves strength, you unfortunately cannot teach everything this way—if the teacher could include all research know-how in this way of teaching, then no student would need to do research. The purpose of the supervisor is to indicate the general direction so that the student does not have to fumble around. But the student must personally experience the obstacles along the way.[44]

Hua thus advocates a close, collegial student-teacher relationship, which helps to guide the young mathematician through the literature, to make him aware of what is already known, and then to illuminate the broad avenues of research. His advice is sound, and although it was intended for Chinese students, it would be meaningful and well received in any country. In this remarkable article, furthermore, we gain insight into his own mathematical personality—steady work, creative growth, and innovation.

In addition to his mathematical offices, Hua was active in national and international affairs. He was a deputy from Peking to the National People's Congress, the formal governing body of the Chinese People's Republic, and he addressed its first Congress in 1954, where he was elected to the Presidium. In deputy Hua's short address, two points are notable.[45] First, he acknowledges that "In the past five years . . . the scientists have not been able to catch up with the rapid progress of our country . . . we still cannot accommodate the needs of production and construction." He thus anticipates on the Congress rostrum his criticism of erratic and one-sided science which he elaborated at length in his 1956 *Chinese Youth* article. It is not the province of scientists,

---

[44] *Ibid.*
[45] *Jên min shou ts'e* (People's Handbook) (Tienstin:Ta Kung Pao, 1955), pp. 165–166.

he says, to avoid hard work by creating "startling accomplishments by chance through innate cleverness (ts'ung ming)[n] and gimmicks." Instead, the scientists must work industriously and in accord with a directive of the Culture and Education Committee of the State Council: "reform and stabilize, develop [the most] important aspects, improve quality and increase quantity, and progress with steady paces."

A second topic of the 1954 speech is the relationship of science and culture to the socialist state. Hua quotes Lenin's dictum to the effect that "Only when we enrich our minds with the whole treasure of human knowledge can we consider ourselves communists." Lenin's precept of universality is cited as a guide not only to the practice of balanced scientific research, but by implication also to the cultural policy of the State. Central to Lenin's attitude was an appreciation of "expertness"—the accumulation of scientific and cultural knowledge in all previous society. During the Congress discussion many representatives approvingly cited article 95 of the new constitution to illustrate the consideration of the people's democratic government for science, literature, and the arts. "As a scientist, I am certainly overwhelmed with the same feeling. Especially when I recall the time when I was unable to publish my work during the reactionary regime, this feeling becomes more intimate," adds Hua. The overcoming of "poverty in science, literature, and the arts," he states, "is the zealous hope of the masses." He continues,

> Naturally, it is the responsibility of every one of us who are working in scientific, literary, and artistic fields to rectify these [our] faults. But on the other hand, we also hope that the government will properly safeguard (chao ku)[o] the time used for their specialized work by those specialists with more social activities (she hui huo-tung)[p] so that we can have time to study, research, and write.

The comparative "more" (chiao)[q] of the last sentence is indefinite. Hua is saying gently but unmistakably that the old experts inherited from the pre-socialist society should be employed to the maximum, by relieving them of excessive "social activities," a polite term for political meetings and administrative functions. He thereby presages the conflicts engendered in the Hundred Flowers campaign.

Soon after his return to China, Hua became an officer of the China Peace Committee and then of the Sino-Soviet Friendship Association. When the Peace Conference of the Asian and Pacific Regions convened in Peking in October 1952, Hua whom the Chinese press referred to as a peace partisan, was present. He represented China at conferences of the World Council of Peace in Stockholm and East Berlin in 1954, and he was named a member of the Chinese Preparatory Committee for the Conference of Asian countries, convened in the spirit of Bandung. In April 1955 he was a member of the delegation to New Delhi, headed by Kuo Mo-jo, to attend the conference which called for Asian nations to "coordinate their efforts in tackling their scientific, technical and engineering problems."[46]

In February 1956 Hua was elected a member of both the Second Central Committee and the Standing Committee of the China Democratic League. The CDL was an organization of intellectuals opposed to the Kuomintang, and the first issue of its organ Kuangming Pao (Light) appeared in Hong Kong in 1941. Its successor, the Peking Kuangming Jih-pao (Kuangming Daily) is still one of China's daily newspapers.

[46] People's China, May 1, 1955, Supplement II, p. 6.

After the Liberation, the CDL helped to unite the intellectuals and link them to the government, and its leaders, many of them returned professionals, represented various important occupational groups.

The One Hundred Flowers Campaign and the Great Leap Forward

The China Democratic League's leadership became deeply involved in the One Hundred Flowers "blooming and contending" movement of 1956 and the first half of 1957.[47] This new policy toward intellectuals was announced in Chou En-lai's *Report on the Question of Intellectuals*, delivered on January 14, 1956, at a meeting held under Communist Party auspices. Premier Chou admitted the existence of "certain irrational features in our present employment and treatment of intellectuals, and in particular, certain sectarian attitudes on the part of some of our comrades toward intellectuals outside the party." One of the irrational features was assigning research scientists to administrative posts in offices and schools, and another was withholding the trust the intellectuals deserved by refusing to let them visit factories and "to look at some material they are entitled to see."[48] The issue of scientific research was urgently put in the *People's Daily*:

> Our scientific techniques are seriously backward. . . . To raise the level of our science and technology, we must, in matters of scholarship, thoroughly carry out the policy of free discussion and letting one hundred schools contend. We must not only learn the advanced scientific techniques of the Soviet Union and the People's Democracies, but we must also learn the advanced scientific techniques of the capitalist countries, especially those of the United States, Britain, and France. If there is any scientific or technical knowledge that can be applied to our socialist construction, we must make a business of studying it.[49]

As an integral part of the new flexibility, Chou En-lai promised the intellectuals improvements in working and living conditions (better housing, salaries, and promotions) as well as the "five-sixths" formula, wherein they would have at least five-sixths of the working day available for their own work, with the rest of the time to be spent on political study, meetings, and so forth. Library and research facilities were expanded, and the library of the Academy of Sciences initiated a number of foreign exchanges of books and periodicals, as China began to import large quantities of scientific literature and equipment. A drive to admit intellectuals to the Communist Party gained momentum in the spring of 1956, and even non-Marxists were encouraged to express criticisms of the Party and their reactions to the new course of "liberalization." The period 1956 to June 1957—by which time it was claimed that the basic goals of the First Five Year Plan, 1953–1957, had been achieved—was the most optimistic and confident one since the Liberation. Many intellectuals responded to the invitation by taking up the suggested role of social critics.

As we have seen, Hua Lo-keng had already petitioned the government, on behalf of "scientific, literary, and cultural workers,"to reduce administrative claims on their time and to broaden their opportunities to serve socialist construction. During the One Hundred Flowers he reinforced his plea and identified himself with an "alliance" initiated by two of the senior leaders of the China Democratic League, Chang Po-

[47] As a metaphor "one hundred flowers" refers to very many tendencies of thought.

[48] Chou En-lai, *Report on the Question of*

Intellectuals (Peking: Foreign Languages Press, 1956) (English), p. 18.

[49] *People's Daily*, Peking, May 9, 1956.

chun[r] and Lo Lung-chi.[s,50] Lo Lung-chi summed up the case of the "displaced intellectuals":

> There are students of philosophy who work on the compilation of catalogues in libraries, students of law who take up bookkeeping work in offices, students of dye chemistry who teach languages in middle schools, students of mechanical engineering who teach history. Among the higher intellectuals there are also returned students from Britain who earn their living as cart-pullers and returned students from the United States who run cigarette stalls.[51]

At the alliance's six-professor conference of June 6, 1957, Tseng Chao-lun, the Vice-Minister of Higher Education is alleged to have said:

> One must not think that literary men cannot effect a rebellion. It is in the tradition of Chinese intellectuals to stir up disturbances. The students of the Imperial College in the Han Dynasty, and the students who started the May the Fourth Movement in 1919 both made big trouble in China.[52]

Tseng was the leading figure in the CDL's Scientific Research Committee, which on June 9 published under five signatures a set of relatively audacious suggestions to the State Council (Tseng's name was listed first, followed by Ch'ien Chia-chü, Hua Lo-keng, T'ung Ti-chou, and Ch'ien Wei-ch'ang):

> We propose (1) that with few exceptions, scientists with the ability to lead scientific research should as far as possible not perform any administrative work, especially those older scientists over sixty who are needed to pass on their knowledge to the next generation. (2) We should reserve to each scientist a definite period in every year when he is completely free to do research work. We should like the Government to consider regulating the system of holidays and advanced studies for professors and research workers. (3) With a few exceptions, those scientists who are at the same time People's Representatives or Members of the Political Consultative Conference, etc., should not normally occupy more than one outside position. Those holding regional positions should not also hold national positions, and vice versa. (4) Due to the need for developing scientific research work, scientists should be granted a long-term exemption from social activities (*she hui huo-tung*) and administrative work. (5) If it is not absolutely necessary, the entertainment of foreign guests should not be the duty of the scientists.[53]

There followed, among other suggestions, appeals for the freedom of scientists to choose their own assistants, for an end to secrecy surrounding professors' expertise, and for helping scientists who had been misdirected to the wrong profession to "return to the colors."

[50] Chang Po-chün studied philosophy in Berlin from 1922 to 1925. He participated in the Chinese Communist insurrection at Nanchang in August 1927 and thereafter emerged as a leader of the anti-Kuomintang forces which were to found the CDL. He became the managing director of the *Kuangming Daily* in June 1949 and held his government post as minister of communications without interruption until early 1958 (Boorman, *Biographical Dictionary*, Vol. II, pp. 98–100).

Lo Lung-chi, after graduation from Tsinghua in 1921, went to the U.S. on a Boxer scholarship to study political science and received a Ph.D. degree from Columbia University in 1928. After his return to China and a period of imprisonment for strong criticism of the government, he became a crusading journalist and, like Chang, a leader of minority political parties. He taught at the Southwest Associated University until he was dismissed, and he then became head of the CDL's Kunming branch. He was active in foreign affairs under the People's Government, and in May 1956 he was named minister of timber industry (*ibid.*, pp. 435–438).

[51] *People's Daily*, Mar. 23, 1957. Quoted in Edgar Snow, *The Other Side of the River* (New York: Random House, 1962), pp. 393–394.

[52] Hu Yu-chih, "Rightwing Coterie Exposed—The Story of the Chang-Lo Alliance," *People's China*, Aug. 16, 1957.

[53] *Kuangming Daily*, June 9, 1956, p. 1.

During the second week of June, just as Tseng, Hua, and others were advancing the CDL's program, the political tide turned, and this program was labeled anti-socialist. Beginning on June 8, the *People's Daily* featured news of workers' meetings to "denounce the rightists,"[54] and a rectification campaign was in full swing by the middle of June. To many intellectuals it came as a shock when the CDL's science program was called an attack on the Party and Government and when the CDL leaders were accused of trying to provoke students, seize the leadership of all professionals, and set up an Anglo-Saxon type of parliamentary system. At the Second National People's Congress, which began in late June, Chou En-lai replied to the CDL science program in a speech that was defensive in tone compared to his 1956 report. He denied that communists were ignorant of science, or that they could be called "laymen" relative to specialized professionals:

> If the idea that the "laymen" are not qualified to lead the "experts" implies that only specialists are qualified in their own field, this not only negates leadership over science but is tantamount to ruling out the possibility of any unified leadership in scientific research, because there could not be a leader of scientific work who was himself a master of all branches of science.[55]

One representative to the People's Congress upheld the Communist Party's leadership of science, in line with Chou:

> The idea that so-called "laymen" (meaning the party) are incapable of leading the "professionals" (words of Ch'ien Wei-ch'ang) is ridiculous and laughable. . . . It is now universally recognized that the divisions of scientific work are so minute that not only is the omniscient scientist nonexistent, but even one who is familiar with all the branches of one discipline is very rare.

> Our Communist Party is a Marxist party, and since Marxism is general scientific truth, the party itself is the product of science. Only because of its prominent scientific character can it lead [the struggle on all fronts] and be victorious every time.

> Scientific advance without organization is unthinkable. [It] needs a general counseling office and a general commanding headquarters. These are indeed the responsibilities our Communist Party can and ought to bear.

These words are attributed to Hua Lo-keng,[56] who had in the June 23 issue of the *Kuangming Daily* pledged to study more about Marxism-Leninism, to gradually establish a Communist *Weltanschauung*, and to determinedly rebuff the schemes of the "rightists."

In both of his articles of retribution, he explained that his political sense was not keen enough to detect the intrigues of rightist, ambitionist politicians within the CDL who were making use of scientists. He accused Tseng Chao-lun of publishing the June 9 program without his full assent. According to the June 23 article, a few days before the program was dispatched to the Science Planning Committee of the State Council there was one last meeting which Hua and T'ung could not attend "because of other business." They were therefore absent when the crucial document was completed.

---

[54] As reported by Hu Yu-chih, "Rightwing Coterie Exposed," p. 17.

[55] New China News Agency (NCNA), June 26, 1957 (English).

[56] "The Party is Capable of Leading Science and Education," summary of the speech by representative Hua Lo-keng, *Kuangming Daily*, July 14, 1957. It should be noted that Hua's exact words are not given in this edited summary.

This disavowal of responsibility and charge that Tseng engaged in foul play meant that along with Ch'ien Wei-ch'ang, Tseng was to become the scapegoat.

These backtracking maneuvers, accompanied by contentious and entangling struggle meetings, exacted their price on the CDL leadership. The attacks on Chang and Lo were characterized as "merciless" by Edgar Snow, who adds that hundreds of intellectuals with views which could be linked with theirs were called to account and disgraced.[57] Hua did not have to undergo all of these attacks, but he still must have been depressed by the whole affair. We referred above to a 1952 *New York Times* dispatch about Hua's attempted suicide. It seems possible that this article was inspired by alarmist accounts of Hua's intervention in the 1951–1952 ideological reform campaign and can be safely discounted on the basis that his involvement in it was modest. But Chinese intellectuals now in the West report that he tried suicide in 1957, and this does admit concern, for in the June factional infighting Hua was required to criticize himself abruptly and sternly for political omissions, and he could have suffered disorientation and considerable humiliation during those fitful weeks.

A succession of increasingly sharp ideological rectification campaigns, dating back to 1942, characterizes the Chinese Communist Party's policy toward intellectuals. The anti-rightist campaign of 1957, though not particularly lengthy, was the most ironic one of these, because it condemned intellectuals for taking a course of action which they believed was constructive and consistent with Communist Party policy. The CDL's proposals, the science program included, although they were interpreted as the basis of an incipient opposition to the Communist Party, were intellectually coherent and defensible in terms of scientific and technological advance, and it is difficult to detect the spirit of Westminster in them.

In his essay "Three Currents in Communism" Isaac Deutscher criticizes the view that the One Hundred Flowers was a trick designed to deceive the elements of opposition and provoke them to expose themselves.[58] The available evidence supports Deutscher's contention that the One Hundred Flowers campaign was "too weighty and too self-consistent" to be dismissed as mere fraud. Deutscher theorizes that in every revolution there occur "critical and tragic moments" when revolutionary governments become terrified of their real or apparent isolation, and he believes that such a moment was June 1957, when Mao Tse-tung "took fright" at the spate of hostile criticism coming from the intellectuals. Deutscher's interpretation of the Chinese events has a ring of truth, but precisely as a result of the turn "inward" in Party policy which was effected in the summer of 1957 and in a sense has not been reversed, we lack the factual and explanatory material with which to improve his or other external analyses. "The Chinese themselves," underlined Deutscher, "have failed to give a frank and convincing explanation of their behaviour" during the anti-rightist campaign.

But in the course of the relentless struggles against his past, which commenced in 1957, Hua Lo-keng learned how to regain his poise, and he has shown great capacity for personal recovery and reconciliation. It is apparent that his statesmanship as a

---

[57] Snow, *The Other Side*, p. 401.

[58] *Ironies of History, Essays in Contemporary Communism* (London: Oxford Univ. Press, 1966), pp. 68–87. Although not a sinologist, Deutscher is regarded by some within the Chinese studies field as having reached the right conclusions, long before many others, on the political movements of Chinese communism and their relation to the international communist movement.

renowned self-made scholar and the leader of China's mathematical community won him a way out when his ideological faults were judged. His self-criticism has in the end always been accepted, and in 1957 he got the green light to continue on the Central Committee of a reorganized China Democratic League and in the Sino-Soviet Friendship Association. (Not even Tseng Chao-lun was completely purged, and he continued to function as an academic administrator into the 1960s.) Hua participated in the sessions of the Second and Third National People's congresses held in 1959 and December 1964–January 1965, and he was on both occasions re-elected to the Congress Standing Committee.

Although the anti-rightist movement was at first targeted on the critical intellectuals, it turned into a campaign against the entire professional intelligentsia. Many professional economists and other incipient critics of Mao Tse-tung's economic policies were thereby eliminated from positions of influence. It was in this atmosphere that the Great Leap Forward was launched,[59] calling for swiftly storming the heights of science. The University of Science and Technology was founded by the Academy of Sciences in September 1958 with Kuo Mo-jo as president. The teaching staff consisted of the "top research workers" in the Academy's research institutes, including Hua Lo-keng, chairman of the applied mathematics and electronic computer department, and Tsien Hsue-shen, chairman of the mechanics department. The 1,600 students, "about 70 percent of whom come from families of workers, peasants and revolutionaries," were enrolled after "strict nationwide examinations."[60] Vice Premier Nieh Jung-chen, with overall responsibility for scientific planning, described this school as

A new type of university located near scientific research institutes with the most advanced departments of science and technology as its main academic curriculum. Such a university is to pick up a group of the best senior middle school graduates . . . the training of new scientific workers will be accelerated so that, within a short period, China will catch up with the advanced nations.[61]

From its inception the University of Science and Technology became an intensive training base for and a ladder of promotion to the Academy of Sciences, which began to make it a national center of scientific learning. In 1961 Hua became a vice president (dean) of this university, and it is believed that he has concentrated his remaining mathematical activities there.

During the Great Leap Forward scientific communication between China and the Western world slowed down, and Hua, who from November 1947 to December 1959 was a reviewer for *Mathematical Reviews*, resigned without explanation and returned

[59] Franz Schurman, *Ideology and Organization in Communist China* (Berkeley: Univ. of California, 1968), p. 91. Schurman points out that the anti-professionalism was worked into the overall program of the Great Leap Forward, and he draws the connection between the One Hundred Flowers aftermath and the Great Leap Forward: "There can be no doubt that some crucial policies leading to the Great Leap Forward were decided during this meeting" (the 3rd Communist Party plenum, Sept.–Oct., 1957). Edgar Snow, too, concludes his chapter on the Communist Party's counterattack after the Hundred Flowers with a paragraph (*The Other Side*, p. 406) on how Mao Tse-tung and Liu Shiao-chi went on a fact-finding tour to a province where the rectification had removed many cadres and officials.

Of all the consequences of bloom-and-contend and "search for unity" none was more momentous than this. Here in a province of party dissension there now began an experiment—the people's rural communes—which would in less than a year shake Chinese society more fundamentally than anything that had happened since the revolution.

[60] NCNA dispatch, Sept. 20, 1958.

[61] *Ibid*

his not-yet-reviewed papers to the journal's editors in Providence, Rhode Island.[62] The theme of self-reliance enunciated during the Great Leap Forward undoubtedly influenced the Academy's policy on scientific communication, but there was a partial revival of contacts during the retrenchment period of 1960–1962, when communication with the scientific world outside China again increased. This was limited and selective, however, geared to finding substitute sources for scientific and technical assistance from the socialist countries. The volume of Chinese literature imports increased somewhat, and exchange of publication arrangements were made between Chinese libraries and Western university and institute libraries. In 1960 Sydney Gould, translations editor of *Mathematical Reviews*, went around the world to compare notes in Japan, India, the U.S.S.R., and Poland with others concerned with scientific exchanges with China. As a result, *Chinese Mathematics*, a cover-to-cover translation of the *Acta Mathematica Sinica*, the leading journal of the Mathematics Institute of the Chinese Academy of Sciences, was begun in 1962 under Gould's editorship. Also in the early 1960s an arrangement was worked out by the British Royal Society and the Chinese Academy of Sciences which provided for exchanges of students and scholars.

The year 1963 saw the publication of Hua's latest book, the twelve-chapter *Classical Groups*, co-authored with Wan Chieh-hsien.[t, 63] We can gauge the scope of Hua's contributions to research in this field by the extensive bibliography of his publications in Dieudonné's *La Géométrie des groupes classiques*. In Chapter 4, Dieudonné attributes to Hua many of the important results on the structure and automorphisms of the classical groups. The preface to *Classical Groups*, dated August 1962, once again informatively establishes the context for Hua's enlivening pedagogy and shows that the book had been on his mind for a long time. He writes that in 1949 he conceived the project of a seminar for graduates and senior-year students which would conduct research in classical groups, projective geometry, matrix theory, and the theory of group representations. Beginning early in 1950, when he arrived at Tsinghua University, through the summer of 1951, Hua organized the seminar and drafted Chapters 1–6 of the present work. His teaching considerations are brought to light:

> One of the reasons why the subject of this book was selected [by Hua] was because it is an easy one in which to train cadres (*kanpu*).[u, 64] This is because the prerequisites are few and instruction can begin on the simple, concrete level. The perspectives of this development are in no way limited. By means of this planned research, we can familiarize ourselves with many branches of algebra and geometry and then devote ourselves to the broad and the abstract.[65]

Hua twice reported on the main contents of the first six chapters in an algebra seminar of the Academy of Sciences in the second half of 1951 and again in the first half of

---

[62] Personal communication, Chandler Davis, 1970. Professor Davis, of the Department of Mathematics, University of Toronto, was formerly an editor of the *Mathematical Reviews*.

[63] This substantive and thorough opus has regrettably been slighted by *Mathematical Reviews*. Not until four years after its publication did a notice of *Classical Groups* appear in the reviewing journal, and then (1967, ⧣2715) only title, co-authors, and publisher were given, minus a review. Furthermore, the title was incorrectly

translated as *Group Representation Theory*.

[64] Strictly speaking, a cadre is someone who holds a formal leadership position in an organization; however, the concept has been extended to Party members and other activists who exercise leadership roles.

[65] Hua Lo-keng and Wan Chieh-hsien, *Tien hsing chün* (Classical Groups) (Shanghai: Shanghai Science and Technology Press, 1963), Preface.

1957. After Hua, Wan also talked about part of the contents of the first six chapters in the seminar, and then completed the remaining chapters, following both the spirit and the methods of the preceding material. Still, "after ten years we have completed only the very first part [the first part of the first part] of our project. More important work lies ahead."

The group of invertible $n \times n$ matrices over a field K is called the general linear group over K, and is written $GL_n(K)$. Matrix groups are a subclass of locally compact groups. Since the study of general topological groups is very recent compared with that of matrices, the customary word *classical* is used for groups of matrices. In this spirit a classical group is defined to be either a subgroup H of $GL_n(K)$ or else a quotient group H_1/H_2 of two such groups. The projective group $PGL_{n-1}(K)$ is the quotient of $GL_n(K)$ by the group of $n \times n$ nonsingular scalar matrices (those having a nonzero element a along the diagonal and zeros elsewhere). The special linear group $SL_n(K)$ is the subgroup of $GL_n(K)$ consisting of all unimodular matrices—those whose determinant is equal to 1. Another example of a classical group is the unitary group. Dieudonné defines $U_n(K,f)$ to be the group of nonsingular $n \times n$ matrices over K which preserve a given form f. This becomes the group of all unitary matrices ($UU^* = I$) when K is the field C of complex numbers, and $f(x,y) = \sum_{j=1}^{n} x_j \bar{y}_j$. The orthogonal group is a special case of the unitary group. A final example is the symplectic group, which for commutative fields K is the subgroup of $U_n(K,f)$ consisting of all matrices which leave an alternating form f invariant.

A geometer tells me that another reason why the adjective *classical* appears is that when $n = 2$ or 3 the groups considered are groups of transformations associated with the classical geometries such as Euclidean, projective, hyperbolic, and elliptic. Hua is thus on the mark in claiming that his material intersects many topics in geometry and also in algebra. He points out that the interesting parts of the theory of associative rings, Lie rings, and Jordan rings have matrix forms. And the "classical domains" in the theory of several complex variables, as was mentioned in Hua's monograph on harmonic analysis, have representations in matrix form.

In *Classical Groups* he begins by introducing the real and complex fields, finite fields, and the quaternion ring; progression is then to groups of matrices over skew fields and to group representations. The first five chapters treat projective geometry and the geometry of rectangular matrices, while Chapter 6 is devoted to the structure of $GL_n(K)$ and its automorphisms. The second half of the book deals with the structure of the unitary, orthogonal, and symplectic groups and their automorphisms, making the necessary distinction for fields with characteristic 2.

The linear programming campaign

Industrial policy was modified in China after December 1958. The high and unbalanced targets for economic growth set during the Great Leap Forward in the various nonagricultural sectors were not achieved, although exaggerated industrial and economic claims were made, partly on the basis of "inflated and highly inaccurate" statistics.[66] The "crash industrialization" program for building communism in a few

---

[66] Barry Richman, *Industrial Society in Communist China* (New York: Random House, 1969), p. 14. Apart from Richman, the fact of statistical nonreporting and over-reporting along the line of the political and production movements of the Great Leap Forward is well known and not seriously open to question. It is a sign of possible confusion among the officials responsible

years (of which the 1958 rural backyard steel-making drive was an early-abandoned part) is considered by most scholars of China writing in the West to have been a relative and possibly absolute failure, although some argue favorably for the resulting economic decentralization, indigenous technology, and the development of local small industry. After an excellent harvest in 1958 China was overtaken by a three-year agricultural crisis which severely affected the harvests of 1959–1961. The country's economic priorities were altered in favor of boosting agricultural production, and the mathematicians were called from other tasks to participate in this work.

Hua Lo-keng was commissioned to participate in the wheat harvest in the Peking suburbs, and in 1960 he contributed a long article to the *Kuangming Daily* entitled "Yün Ch'ou Hsüeh,"[v] A New Branch of Science Applied to National Economic Construction. He led off on a highly tentative note:

> First of all, I have to make it clear that I am a raw hand at linear programming, but . . . now that all trades are giving big aid to agriculture and mathematical workers are constantly considering how to aid agriculture, I will mainly introduce, in this article, the use of linear programming in agriculture. It is, I am sure, not quite right in many respects, and comrades are requested to make corrections.[67]

Yün ch'ou hsüeh literally translated means "science of operation and programming," and it entailed the application of linear programming and other optimization techniques to economic decision-making. According to Hua, more than four hundred thousand persons took part in popularizing applied linear programming in a short period in Shantung Province. They included college, middle school, and primary school teachers, workers, peasants, and managerial personnel. "Through the diversified media of broadcasting, ballads, peep-shows, indigenous 'movies,' and songs, they made known linear programming to more than eight million people in the province." Certain efficiencies were achieved, and the Academy of Sciences called an on-the-spot meeting in Tsinan, the provincial capital, to introduce Shantung's experience "to all mathematical workers in China."

In his exposition Hua introduces transportation and distribution problems, where the main application of linear programming is in eliminating the waste involved in cross-hauling and detouring. These irrational phenomena are not easy to detect in complicated commodity transportation. His emphasis is on the devising of what he calls "indigenous" graphical methods for transportation scheduling which are "simple and easy to follow" by the masses, yet which are as general as the "modern" algebraic method he named the "tableau" method (which is undoubtedly the well-known Dantzig simplex tableau).

The representative problem he chooses for an application of programming methods in agriculture is the wheat-threshing problem. Given several wheat fields, which are "not big in area" and which are joined by various roads "without loops," where should the threshing ground be located in order to minimize the weight of wheat transported? Hua answers with a formula whose discovery he attributes to the Ch'ü-fou[w] Normal College of Shantung Province: "The best place for the threshing ground is where the quantity of wheat brought in along each road is less than half the total

---

for statistical disciplines. Richman reports, however, that significantly exaggerated statistical claims have not been issued since the period of the Great Leap Forward.

[67] Hua Lo-keng, "Yün ch'ou hsüeh," A New Branch of Science Applied to National Economic Construction, *Kuangming Daily*, Nov. 1, 1960.

amount of wheat." For a mathematician not familiar with this type of problem it would require several readings and some imagination to understand how to apply this rule of thumb, which looks right, since like the rest of the article it is presented in everyday nontechnical language. Knowing Hua, it is safe to assume that the principle is valid under the appropriate conditions, but the Yün ch'ou hsüeh workers could not have discovered what these were from this article alone.

Hua is clearly concerned here with popularizing Yün ch'ou hsüeh ("propagandizing for this new branch of science," he calls it) and not rigor. In addition to the threshing problem, he talks about other interesting problems where mathematical methods can serve production practice. His list includes the operation of reservoirs, where the parameters to be regulated are the rate of flow for navigation, the volume of water released for power generation, and a seasonal "water budget" for husbanding the heavy autumn rains. "A set of methods and theories concerning reservoir operation can be evolved by summing up and generalizing indigenous methods of operating small mass-built reservoirs . . . it will then be easy to operate the medium reservoirs which undertake irrigation and power generation. With [this] experience, it will be easy to operate big reservoirs." Other problems mentioned are those of minimizing crop damage when weather data are available in advance, determining the best sowing schemes for various crops under different soil and irrigation conditions, and estimating commune output in advance by selecting square meters of land and taking averages of the number of grains of wheat growing in these squares.

In sum, the emphasis of the article is on the possibilities for rational decision-making guided not so much by what we know as linear programming, but rather by elementary maximum-minimum methods:

> It is not difficult for us to find out that a cylinder-shaped can of given volume whose height equals the diameter of the base minimizes the area of tinplate required. . . . Manufacturers in capitalist countries make slender and long cans, and squat and short cans in order to give customers the illusion of substantiality . . . [but] we have found that in the case of some cans, as much as 10% or more of tinplate can be saved.

Although Hua's examples are admittedly heuristics to motivate public interest in mathematical methods and do not necessarily reflect the mathematical procedures used during the campaign, we still await details about how agricultural workers and mathematicians interrelated in the use of more complicated programming methods. This question remains outstanding even after study of a professional paper produced during the campaign.

This paper followed the participation of the Institute of Mathematics in the suburban wheat harvest, along with the University of Science and Technology and other academic organizations in Peking. The object of this work was stated to be the experimental use of mathematical models in the selection of the threshing site most economical for transportation. Material on the experiment was assembled by Hua Lo-keng and others and was read at the National Operations Research On-the-Spot Conference held in Tsinan in July 1960 and published in the *Acta Mathematica Sinica*.[68] The first part is headed "Selection of threshing site for string-of-grapes type wheat fields." To guide the peasants in choosing the best sites for setting up their threshing floors in

---

[68] Hua Lo-keng, *et al.*, "Application of Mathematical Methods to Wheat Harvesting," *Acta* *Mathematica Sinica*, 1961, *1*, 2:77–91 (translated in *Chinese Mathematics*, 1961, *1*:77–91).

the fields, mnemonic rhymes, considered indispensable in popularization activity, were devised. The simplest solution is where there are two tracts of wheat, in which case the best procedure is obviously to transport the smaller tract of wheat in toto to the larger tract and then to thresh both tracts at the site of the larger. No point along the route is as good for establishing the threshing floor. A mnemonic rhyme generalizes this:

> When all the routes have no loops,
> Take all the ends into consideration,
> The smallest advances one station.

As shown in working diagrams, an "end" is a tract of wheat lying on only one road; a "loop," on the other hand, is shown as a closed road containing several tracts.

> When the routes do have loops,
> A branch is dropped from each one,
> Until there are no loops,
> Then calculation as before is done.

That is, the road is opened so that the tracts on it become "ends." There are many ways of dropping branches, and the calculation for each must be performed before deciding which loops are the best to break.

This analysis goes beyond the simplified formula given in the *Kuangming Daily*, which is not mentioned here, although reference to a proof of the principle translated as "The larger half is for setting up the threshing floor, the smaller half leans toward the inside" appears in a footnote. Nevertheless, one cannot read this article as an ordinary mathematical paper, because "the emphasis is on methods and conclusions" and not on proofs. However, the language is just as terse as the typical mathematics journal article. Moreover, the terms of the original problem are not stated in full, and reliance is laid on skeletal diagrams to carry the argument. The diagrams illustrate certain principles in well-chosen cases, but no literature references are given other than the one cited, and perplexed Yün ch'ou hsüeh workers in the field would most likely have required some form of contact with the authors of this article in order to apply the techniques successfully.

In Part 2, "Selection of threshing site for large tracts of wheat," it is assumed that the wheat grows evenly in large rectangular tracts. The conclusion is not surprising: when setting up the threshing floor in the interior of the tract, the center is best, and for a threshing floor on the edge of the tract, the midpoint of the long edge is best. But of course, "the problem of selection of a site in irregular tracts of wheat cannot be solved so simply." It is possible that this paper was presented to the *Acta* in a hurry, for the authors remark: "Due to the short time allowed for the work and to the low level of knowledge of those comrades undertaking to put these materials in order, the contents are extremely unpolished, and we hope that all comrades will provide criticism."[69]

Additional mnemonic rhymes and suggestions of elementary means for solving real-life problems were presented in an article carried by the English-language monthly *China Reconstructs* in its August 1961 issue. But here too, as in each of the articles I have seen on the programming mathematics campaign, the mathematical content

_____

[69] *Ibid.*, p. 78.

itself is slight—maximum-minimum problems in plane geometry and easy algebra— and the pedagogic flair characteristic of Hua's popularizations does not come across. There were *pro forma* aspects to this work, and Hua, try as he might, could not generate much enthusiasm for it.

It was never claimed in the linear programming campaign that any new formal discoveries in mathematics had taken place. Rather, the goal was to help the working masses to apply quantitative techniques by developing imaginative teaching methods based on indigenous formulas originating at the point of production and then elaborated for other productive units with the help of mathematicians. The campaign must be evaluated, therefore, on the merits of these new teaching methods, the extent to which they were applied, and on the quality of the cooperative relationships entered into by the scientists and the people. We have hardly any such information, but on Hua's side there is self-criticism for his reluctance to get involved with the task of programming.

What are the barriers to participation in practical work by mathematicians in a developing society? For some relevant insights into this question we are obligated to another campaign for integrating the Chinese mathematicians into socialist construction. This campaign took place in 1964–1965, when Hua spent about six months conducting experiments in popularizing the method of overall planning (*t'ung ch'ou*)^x in civil engineering and other industrial and manufacturing projects. "Overall planning" is a nontechnical term, meaning planning from start to finish, and on paper it involves finding the shortest path through a system of industrial tasks with a great number of links between them. Like the "string-of-grapes" type of wheat fields, this problem calls for treatment by methods akin to graph theory. The overall planning efforts resulted in a pair of articles by Hua, in *Red Flag*, the Communist Party's theoretical organ, and the *People's Daily*.[70] "Taking the road of revolution" is a course beset with obstacles for Hua, because he is continually bedeviled by old ideas. The question of method is particularly important, for example. When in the past someone came to him with a problem, he would deal with the problem strictly as such and give the solution without going into the whys and wherefores of the method used—perhaps with the excuse that it involved abstruse mathematics. It is better, he now realizes, to actively seek out problems, and make the steps to the solution clear in a thorough and simple manner. As more people acquire mathematical knowledge, the mathematicians no longer have a monopoly over their speciality, but it can be more readily made to serve production reality and "can also be raised more easily to a higher plane." Writing in the *People's Daily* when the movement to emulate dedicated, self-sacrificing heroes was underway, Hua sizes up the obstacles that "a man from the old society, one who was late in awakening and low in consciousness" faces when he is asked to use his abilities in production practice. Having acquired a method is not enough; when it comes to actual work, frustrations occur at every step.

> When I first came in touch with the method of programming[71] I had a mind to turn it down. Was I not a quite good mathematician even if I never bothered myself with the method of programming? Since the method originated in capitalist countries, was it not

---

[70] "Notes on Experimental Work in the Method of Overall Planning," *Red Flag*, Oct. 1, 1965. "Resolutely Take the Road of Revolution," *People's Daily*, Dec. 4, 1965.

[71] I do not know whether this means the "Yün ch'ou hsüeh" of 1960 or "overall planning."

tainted with the odor of the bourgeoisie? If I engaged myself in working on this method I might easily get myself contaminated. Why should I do that? Let me turn it away. Let me pass it on to other academic departments and other people.

The second barrier was "fear" $(p'a)$.[y] The first time I went to a selected spot to make a trial, I was daring and full of confidence . . . arriving at the spot, I saw what an enormous and complex project it was. There were many things that I did not understand. All of a sudden I was struck by fear . . . . Fortunately for us, the local Party organization did some timely ideological work on us. It arranged for us to go deep into the construction site to find out facts for ourselves and listen to stories about exemplary persons and exemplary deeds. It gave us encouragement.

I had another "fear." I was afraid of what other mathematicians might say about the popularization of the science. They might say that this work involved no advanced mathematics, was "uninteresting," was simple mathematics, or even that it was no mathematics at all. Such imagined criticism was in fact an illusion. I was judging a new generation of people by standards of my old self.[72]

"Selflessness" is advanced as the way to integrate with the worker and peasant masses, and Hua vows that his "egotistic" considerations will gradually vanish, that he will fight together with his comrades, become used to their ways, and work hard for the revolution with "no care for my personal well-being."

His well-being, as he once might have defined it, was indeed to suffer for a time. His job in the Institute of Mathematics at the Academy of Science goes unmentioned in both the *Red Flag* and the *People's Daily*, and he is identified in the latter only as Vice President, China University of Technology. This suggests that his position as the head of the Institute may have been in question.

The final *People's Daily* paragraph deserves quotation in full, since it foretells much of what is ahead for him:

"I will do whatever Chairman Mao tells me to do." This famous saying of comrade Wang Chieh[z,73] rings in my ears. Although my youth is gone and there is little time left, Chairman Mao's thinking has "stayed the sun" for me. Under the brilliance of this sun that never sets, I will resolutely and ceaselessly walk on the road of revolution!

IV: 1966–1970

The events of the Cultural Revolution now force us to alter the character of this narrative, for there has been no mathematical work published after 1965 which can be consulted as a primary source on Hua. Thus in order to confront at least some of the issues involved in the recent reconstruction of his career, we shall examine the Cultural Revolution's impact on education and science. Its inroads into these fields can be demarcated by the assault on educational structures by the Red Guards and "revolutionary rebels," starting in mid-1966, the first attempts to restore unity in the Academy

---

[72] "Resolutely Take the Road of Revolution."

[73] Wang Chieh and Lei Feng were the national exemplary heroes of the pre-Cultural Revolution year. From the *Diary of Wang Chieh* (Peking: Foreign Languages Publishing House, 1967), pp. 1 and 9:
Wang Chieh, a twenty-three-year-old squad leader in an engineering company of the Chinese People's Liberation Army died a glorious death on July 14, 1965, in saving the lives of eleven militiamen and a cadre during a blasting operation.
In his illuminating diary he expressed again and again the desire to be a servant of the people, a "willing ox" for the revolution.

of Sciences in mid-1967, and the entry of "worker-army" propaganda teams into the universities in July 1968. Hua's contribution to the *People's Daily* in 1969 was a direct outcome of the regular "heart-to-heart" talks these propaganda teams held with him, and it reveals the stand he took as a result.

The Red Guard attacks, 1966

The Cultural Revolution was in full swing in May 1966 when Peking Red Guards placed big-character posters on the walls of their middle schools and universities and organized mass meetings to besiege the "capitalist renegades and revisionists" holding power in the Communist Party and in the school administrations. The anti-democratic aspects of the examination system were criticized for having created an elite of culti-vated men divorced from the Chinese masses, and the influence of cultural traditions from China's past and from other countries was decried. The "four olds" condemned by the Red Guards were the old ideology, culture, customs, and habits of the bour-geoisie and all other exploiting classes. The Red Guards seem to have opposed the establishment of all educational standards, believing that as soon as standards were set, individuals would begin selfishly to plot their careers.

Enrollment in university and senior middle school classes was postponed for six months in a notice issued on June 13 by the Central Committee of the Communist Party and the State Council, to "completely reform the system of examination" and to "work out new methods of enrollment."[74] Instruction remained suspended, how-ever, and school facilities were used instead to demonstrate against the deposed school authorities. The Red Guards launched attacks on both the "open" and the "hidden" representatives of the bourgeoisie." Many of the authorities resisted them, including Hua Lo-keng, who wrote: "At the beginning of the cultural revolution, I felt that the masses' criticism of me was uncalled for and could not accept it."[75]

Those who were the objects of denunciation attempted, through various forms of cooptation and other means, to circumvent the rising Red Guard movement, and some old revolutionaries who did not cave in even questioned the students' right to demon-strate. "It is right to rebel against reactionaries!" they retorted, and with his strong personal encouragement, Mao Tse-tung's imprecation[76] helped them to defy and "drag out" their teachers.

Beginning with the first massive Red Guard rally in the capital on August 18, 1966, Mao received eleven million "revolutionary students and teachers and Red Guards" who had come to "establish revolutionary ties" and "exchange revolutionary ex-periences."[77] Toppling first the Peking Municipal Party Committee and the adminis-tration of Peking University, the Red Guards were able to upset the power holders in the educational system, but in the process they created a vast labyrinth of factional organizations, all of which claimed to be faithful to Chairman Mao. The internecine struggles they waged through wall posters and ideological debates often dealt with

---

[74] *Peking Review*, June 24, 1966, p. 3.

[75] *China Reconstructs*, Nov. 1969. See below for the relation between Hua's *People's Daily* and *China Reconstructs* articles.

[76] In Yenan to celebrate Stalin's sixtieth birth-day, Mao said: "In the last analysis, all the truths of Marxism can be summed up in one sentence:

'To rebel is justified.' ... According to this principle, stand up and resist, struggle, build socialism." Quoted in *The Great Proletarian Cultural Revolution in China* (No. 10) (Peking: Foreign Languages Press, 1967) (English), Front-ispiece.

[77] *Peking Review*, Dec. 2, 1966, p. 6.

complex issues relating not only to foreign policy but also to the conduct of officials during campaigns of previous years, and regulations were enforced prohibiting any foreign reports relaying the contents of wall posters and newsletters because of their sensitive nature. Red Guards also broke into armories to seize weapons, and as a result running factional battles were fought. The struggle seems to have polarized by 1967 into a see-saw between the Red Flag and East Wind Red Guard Groups, as they have been called, each of whom sought adherents nationally, competing for influence by lobbying in Peking and by setting up liaison stations in other provinces and circulating newsletters to them. In January 1967 the front was widened, as the "revolutionary rebels" were formed from among youth of post-middle school age to carry the Cultural Revolution out in factories, offices, and other productive units.

The official sponsorship of the Red Guards resided in the Cultural Revolution Group of the Communist Party Central Committee, which after a certain date was composed of leaders close to Mao. Two of their names will appear in this study: one is Mao's wife, Chiang Ch'ing; the other is Yao Wen-yüan, who held a relatively minor position in the Shanghai Party Committee prior to the summer of 1966, and who rode his sharp pen to power as a strict Cultural Revolutionary ideologue. The Cultural Revolution Group was largely independent of the Central Committee, and under it a national network sprang up to render aid to the various Red Guard factions, in which the People's Liberation Army played a major part. Guidelines for the Cultural Revolution were issued only ex-post facto, at the Eleventh Plenum of the Central Committee, which met August 1–12, 1966. The Plenum adopted a sixteen-point decision, the first official Party document of the Cultural Revolution. Article 12 read:

> As regards scientists, technicians and ordinary members of working staffs, as long as they are patriotic, work energetically, are not against the Party and socialism, and maintain no illicit relations with any foreign country, we should in the present moment continue to apply the policy of "unity, criticism, unity." Special care should be taken of those scientists and scientific and technical personnel who have made contributions. Efforts should be made to help them gradually transform their world outlook and style of work.[78]

Article 12 clearly applied to individuals and not to institutions, and September 1966 was the last month in which the Academy of Sciences is known to have worked as in the past. Its journals appeared regularly until the summer of 1966 but were then uniformly terminated. The Academy's book publishing was interrupted earlier and has likewise not yet been resumed (as of the time of writing, April 1972). No other press in China has published professional books or journals in the physical sciences (mathematics, physics, chemistry, and closely related disciplines) since the Cultural Revolution. The *Acta Mathematica Sinica*, a bimonthly which contained an English translation of the table of contents on its inside back cover, printed its first two issues of 1966 as usual, up to page 282. Then at the head of the third issue appeared two editorials from the *Liberation Army Daily*: "Hold High the Red Banner of Mao Tse-tung's Thought and Actively Join the Great Proletarian Cultural Revolution" (April 18, 1966) and "Never Forget the Class Struggle" (May 4, 1966), as well as a key polemic of the Cultural Revolution, "The Reactionary Nature of 'Evening Chats at Yenshan' and' Notes from the Three Family Village'" by Yao Wen-yüan (origi-

---

[78] *Peking Review*, Aug. 12, 1966, p. 10.

nally published in the *Shanghai Wen Hui Pao*, May 10, 1966).[79] The pages on which these three articles appeared were numbered beginning with page 1, whereas the page numbers of the mathematical articles which followed them were 283–424; that is, the editorials and Yao's essay had been clearly demarcated from the mathematical contents. The English translation of the table of contents was omitted from the inside back cover, which remained blank. The fourth and final issue of the *Acta* began with two more editorials from the *People's Daily*, those of June 1 and 2.[80] These editorials, in contrast, were paginated in sequence, beginning on page 425, and the translation of the table of contents was restored to the inside back cover. Thus the polemic had been fully incorporated into the pages of the *Acta Mathematica Sinica*.

All foreign students in China were ordered to leave the country within fifteen days by a Ministry of Higher Education directive of September 20, 1966, and student exchange programs stopped on this date if not before. The institutions arranging foreign scientific relationships were in disarray, and only a few scientists have visited the country since 1966. One of them was the topologist Ioan James, of Oxford, who traveled to China in August–September 1966, representing the Royal Society. Professor James did not meet Hua and did not learn any news about his position, but he did meet all the mathematicians at the Academy, including Wu Wen-tsun, who at that time worked in his field. He has informed me that he knows of no other mathematician who has visited China in 1966–1969. But in the summer of 1971 Chandler Davis, a prominent functional analyst of the University of Toronto, was warmly received by a group of well-known Chinese mathematicians, including Hua, at the Academy of Sciences.[81]

The quadrennial International Congress of Mathematicians took place in Moscow, August 6–16, 1966, and was attended by more than four thousand mathematicians, comprising the most representative gathering of mathematicians ever held. A strong delegation from North Vietnam, led by the geometer Nguyen Canh Toan, took an active part in the congress, and there were also mathematicians representing North Korea. No mathematicians from China were present, however, and the Chinese did not participate at Nice, the venue of the next International Congress, in August 1970.

In China, as would be expected in any scientifically active nation, there were many researchers and teachers who were in the midst of scholarly activity at the universities and in the institutes of the Academy of Sciences when they were drawn into the Cultural Revolution. If their scientific behavior resembles that of their professional colleagues in other nations—and there is considerable past evidence that it does—then they would certainly have desired to complete at least their year's work. But if, as was seemingly the case, their seminars, the publication of scholarly papers, and informal scientific contacts with students and colleagues were put to an end, it is hard to see how they could have continued in their former occupations.

[79] The two *Liberation Army Daily* editorials and Yao's polemic were published as *The Great Socialist Cultural Revolution in China* (No. 1) (Peking: Foreign Languages Press, 1966).

[80] The June 2 editorial, "A Great Revolution that Touches People to Their Very Souls," signalled the testing of the leadership in the Academy with the lines: "Whether or not you are genuinely in favor of the socialist revolution or whether you are even against the socialist revolution is bound to manifest itself in your attitude toward the proletarian cultural revolution. This is a question that touches people to their souls. . . ."

[81] Davis published an account of his visit in *Notes of the Canadian Mathematical Congress*, Jan. 1972, 4:4.

The efforts at unification, from 1967

By mid-1967 efforts had begun to unify the country and to reconstruct the educational system in the midst of the Red Guard turbulence, and new organizational forms were called into being to replace state and Party organs. Thus the "three-in-one" alliance was created, consisting of (1) the People's Liberation Army, (2) the "revolutionary cadres"—former Communist Party and "working class" cadres, and (3) representatives of the "revolutionary masses"—the Red Guards, "revolutionary rebels," and other non-Party groupings. An important goal was to promote three-in-one alliances in every province to supersede the provincial governments and to assume the administration of all schools, factories, and firms. The first provincial three-in-one alliance, called a Revolutionary Committee and described as a "provisional organ of power," was set up in January 1967, and the last two were inaugurated in September 1968.

The Academy of Sciences and its research institutes had been the scene of Red Guard strife, which led the *People's Daily* to comment on July 14, 1967, that a vigorous "civil war" (*nei chan*)[aa] among the "revolutionary mass organizations" in the Academy, which lasted over twenty days, had caused a "deviation from the main orientation of struggle." Further details were not provided. The formation of a Revolutionary Committee in the Academy was an event of considerable importance, and it merited the presence of Premier Chou En-lai. According to Peking Radio, August 2, 1967, those attending the July 30 celebration for the new Committee also included Kuo Mo-jo and Deputy Premier Nieh Jung-chen, who has remained the country's leading scientific administrator. Also present were Hsiao Hua, Director of the General Political Department of the People's Liberation Army, and Su Yu, a Deputy Minister of National Defense, a reminder of the military significance of the Academy. In his speech Chou En-lai said:

> The handful of Party persons in authority taking the capitalist road now have been brought down . . . and the Revolutionary Committee is now inaugurated. This signifies that the Chinese Academy of Sciences will be turned into a red, red, great school of Mao Tse-tung's thought and will follow a course pointed out by Chairman Mao to catch up with and surpass world levels in science and technology and come forward with the most advanced science and technology in the world.

The nature of the three forces which united in the Academy's Revolutionary Committee is not known, and the names of former Institute directors such as Hua Lo-keng are absent from reports of the ceremony. Hua was apparently not mentioned in any of the official media during this period, and it is impossible to tell what role he played during this period of the Academy's reorganization. I learned from Chandler Davis that Hua eventually regained his broad leadership functions in the Academy and that in 1971, as one of the few mathematicians travelling from the capital to other academic institutions, he was again coordinating Chinese mathematical activity, albeit of a vastly different nature.

There are, of course, Red Guard papers concerning personalities like Hua, and a few of them are extant. One of these was referred to by *Facts and Features*, a publication of the Taiwan Institute of International Affairs, as follows: "According to recent Red Guard newspapers, the noted Chinese mathematician Hua Lo-keng has been subject to fierce struggle by 'Peihang Red Flag Fighting Corps,' a Red Guard organization directed by Chiang Ch'ing."[82]

Peihang[bb] is the popular name for the Peking Aeronautical Institute, whose Red Flag Fighting Corps was part of the Red Flag faction and is known to have published the Red Guard newspaper *Red Flag* (*Hung ch'i*) and circulated it to a liaison station in Shanghai and beyond. Japanese correspondents reported from Peking at the height of the Cultural Revolution that the Red Flag Fighting Corps was one of the "ultra-left" Red Guard groups which seemed to be allied with Chiang Ch'ing. A list of charges, focusing on Hua's activities as far back as the 1930s, is attributed to the Peihang group by *Facts and Features*, but their wording, although suggestive, is vague. I tried unsuccessfully to obtain a xerox copy of the Peihang *Red Flag* newspaper, and I surmise that *Facts and Features* itself does not have a copy of it. The phrasing of the *Facts and Features* article ("according to recent Red Guard newspapers . . .") seems to suggest that it has taken the Hua criticism from some other Red Guard newspapers.[83] In the absence of other evidence, this intelligence and one or two dispatches filed by European news agency correspondents in September 1966, which quoted wall posters put up by Red Guards from the Institute of Mathematics, are not sufficient material for a solid commentary on Hua's experiences in 1966–1967. It was admitted by Hua himself that he received "sharp criticism" and a "good shakeup" from the Red Guards,[84] but he had weighty allies and the overall picture is far from clear.

The Worker-Army teams and Hua Lo-keng

To fill the vacuum in the educational system left by the early Red Guard purges, and to effect the three-in-one takeover in the universities, where it met intense opposition, Mao Tse-tung issued a directive on July 21, 1968, which praised the Shanghai Machine Tools Plant, a large factory producing precision grinding machines, for on-the-job training of its own technicians from among the workers:

> It is still necessary to have universities; here I refer mainly to colleges of science and engineering. However, it is essential to shorten the length of schooling, revolutionize education, put proletarian politics in command and take the road of the Shanghai Machine Tools Plant in training technicians from among the workers. Students should be selected from among workers and peasants with practical experience, and they should return to production after a few years' study.[85]

The Shanghai plant urged the transformation of colleges of science and technology into "not only schools but also factories and scientific research units" whose teaching ranks should be a three-in-one combination comprising "workers with a high proletarian political consciousness and practical experience, worker and peasant students with practical experience, and revolutionary intellectuals." The main role of the full-time teachers was to arrange "an organic link among the colleges, factories, and scientific research units, which will help the students raise their practical knowledge to the theoretical level and then apply it in practice.[86]

This program, the model for Mao's instruction to tighten the relationship between research and production, was predictably difficult to realize, and to implement it still

---

[82] *Facts and Features*, Apr. 17, 1968, p. 15 (English).

[83] I am indebted to Richard Sorich, China Bibliographer, East Asian Institute, Columbia University, for this pertinent observation.

[84] *China Reconstructs*, Nov. 1969, p. 31.

[85] *People's Daily*, July 22, 1968.

[86] *Peking Review*, Aug. 2, 1968.

another unit of the Cultural Revolution was called into being. One week after the key directive, on July 27, 1968, a propaganda team made up of "industrial workers and army men" entered Tsinghua University. There the team "thoroughly smashed the control of bourgeois intellectuals," in a deed which was called "mounting the political stage of struggle-criticism-transformation in the realm of the superstructure."[87] Although Hua makes no mention of Red Guard violence at Tsinghua, the worker-army intervention was directed at least in part at the students. Correspondent Norman Webster visited there two years later and reported that after the Red Guards had ousted the Tsinghua professors and administrators, they split into warring factions for big "gang rumbles." The size of the occupation force of soldiers and workers from Peking factories which was needed to stop the internal bloodletting is given by Webster as thirty thousand men.[88] Chou En-lai told visiting U.S. scholars that the number of wounded in the Tsinghua battle was seven to eight hundred.[89] Tsinghua's Revolutionary Committee was not set up until January 25, 1969.[90] Later in that year the Tsinghua worker-army team was called to Peking University, the scene of "extremely complicated and sharp class struggle,"[91] and Peking University's Revolutionary Committee was finally established in September 1969.

Hua wrote that when the worker-army teams entered the University of Science and Technology in August 1968:

> They organized the teachers and students in a living study and application of Chairman Mao's works and helped them carry out revolutionary mass criticism with Chairman Mao's thinking in educational revolution as the weapon.

> The members . . . carried out the Party's policy toward intellectuals—unite with, educate, and remold the great majority of them. Their attitude toward revolutionary and patriotic intellectuals has not been one of discrimination or humiliation or just "wait and see." They took the initiative to be friendly, helped us warmly and patiently, and united with us . . . . They saw both our shortcomings and the fact that we had made some ideological progress. They made a concrete analysis of the historical and social backgrounds in which our ideology was formed.[92]

In the course of frequent comradely "heart-to-heart" talks with Hua, the team was apparently able to explain to him why the Red Guards had rebelled and why their rebellion was good for the intellectuals. Referring to the Red Guards:

> With their sharp criticism they pulled up, and thus saved, many intellectuals like me who had made serious mistakes. . . . It was out of real concern and love for me that the Red Guards had given me a good shakeup. If they had let my wrong ideas develop, I cannot imagine what would have become of me. Finally they brought me around to the right way of thinking and I felt much happier.

The document resulting from Hua's confrontation with the propaganda team is unique. It is one of the few, if not the only one, broadcast internationally by a Chinese intellectual of world standing which endorses Mao's "July 21" line of "educational revolution."

[87] "Struggle-criticism-transformation" refers to "the struggle to seize power from the Party capitalist-roaders, the criticism of bourgeois and revisionist ideology, and the socialist transformation of all institutions of the superstructure" (China Reconstructs, Nov. 1969, p. 31).
[88] Toronto Globe and Mail, June 16, 1971.

[89] Committee of Concerned Asian Scholars Interview with Chou En-lai (Pacific News Service, San Francisco, July 19, 1971).
[90] NCNA, Jan. 29, 1969.
[91] NCNA, Oct. 6, 1969.
[92] China Reconstructs, Nov. 1969, p. 31.

Under the headline "Learn Again To Dedicate Strength to the Reform in Education (Ch'ung hsin hsüeh hsi wei chiao yü ko ming hsien li lang),^{cc} Hua's self-criticism was first published in the Peking *People's Daily* of June 8, 1969. Its context was the debate then taking place about restructuring the universities "under working class leadership." In the course of the debate drastic revisions in the university mathematics curriculum were decided upon which have eliminated most subjects hitherto considered indispensable, such as $\epsilon - \delta$ analysis in freshman calculus, in favor of relatively brief training in methods judged most useful for current production practice. Hua's piece was disseminated widely by the media as a condemnation of the old state of affairs which necessitated the sweeping overhaul. *China Reconstructs* featured a version of it, "Chairman Mao Points Out the Road of Advance for Me," from which we have been quoting, and two patriotic Chinese dailies in Hong Kong, the *Wen Wei Pao* and the *Ta Kung Pao*, carried respectively the June 8 article and a Chinese version of the *China Reconstructs* article.[93]

Let us first consider the way in which Hua is identified for the readers of these publications. Contrary to former practice, none of the four cited in the previous paragraph mentioned the Academy's Institute of Mathematics. In its August 1961 issue, when introducing Hua's discussion of operations research in agriculture, *China Reconstructs* referred to him as head of this Institute; in 1969, however, the same magazine identified him only as a "well-known mathematician and a professor in the University of Science and Technology of China." Since the traditional system of academic ranks founded on "professor" may well have been discontinued in China even before 1966, the title is perhaps deferential. The *People's Daily* and the Hong Kong *Wen Wei Pao* in fact handled the problem by simply printing in Chinese characters "Hua Lo-keng, University of Science and Technology of China." The *Ta Kung Pao*, which followed the English of the *China Reconstricts* article, and hence did not adopt this notation, dropped all attempts to identify Hua. Its omission was not necessarily awkward journalistically, for as we have seen, Hua was already a legend in China; but it does provide further evidence that at the time of writing Hua Lo-keng was not officially of professorial rank.

We now turn to Hua's ideological self-portrait. He says in *China Reconstructs*[94] that he was an intellectual who "grew up in the old society, deeply influenced by its ideology," but that he has now learned that "intellectuals who were trained to serve the bourgeoisie can never be of one heart and mind with the proletariat . . . unless they completely remould their bourgeois world outlook." He thereby defines a basic problem of the reform in education: How can the older intellectuals be made of "one heart and one mind with the proletariat"? This is no mean problem, since as late as 1963 the overwhelming majority of the Board Members (similar to Soviet Academicians) in the Academy of Sciences Department of Physics, Mathematics, and Chemistry were educated abroad. All of them were thus trained to serve the bourgeoisie, and so twenty years after Liberation, the reform, to be successful in Hua's lights, would need to totally remold China's scientific leadership.

But what of these twenty years? By suggesting that the patterns of the old bourgeois society persisted immutably until the start of the Cultural Revolution, Hua's text

---

[93] *Wen Wei Pao*, Hong Kong, June 11, 1969; *Ta Kung Pao*, Hong Kong Dec. 10, 1969.

[94] All subsequent quotations will be from this magazine article, pp. 30–31.

blurs many of the accomplishments of the People's Government and seems to renounce his own work in charting the course of Chinese mathematics after the Liberation. Nowhere is his attempt to group together all pre-1966 developments more evident than in the following paragraph (italics added):

> Under this revisionist line, the school [University of Science and Technology] was actually training successors to the bourgeois scholars and authorities, driving the young people away from proletarian politics, reality and the masses. *And I, who for 40 years had buried myself in research and teaching behind closed doors*, became a victim of this line, a living model of the kind who struggles all alone to win personal fame as a specialist, a living tool being used to poison the rising generation. The very thought of this filled me with deep shame.

Western mathematicians, too, are intimate with the "closed-door" or ivory tower syndrome, and in their heart of hearts many of them are embarrassed or even guilty about their professional isolation from productive and social life. But a closer look shows that it is not just Hua's scientific life which is said to have been conducted behind closed doors. Before 1949, as we know, Hua lived and worked in England, the Soviet Union, and the United States. In the first years after he returned to China he continued to travel widely, and he represented China at important gatherings in Scandinavia, the Eastern European countries, the U.S.S.R., and India. He was in short one of the most international Chinese during those years. Now his participation in the world community of scholars and in antiwar meetings is either left out of account or conceivably even evaluated negatively. What little I have been able to learn about Hua in human terms bespeaks a concern not only for his professional associates and students but for all of humanity. What factors, it must be asked, now impel a socialist-minded man like Hua Lo-keng to make an equation of twenty years behind closed doors working for the Kuomintang and twenty years behind closed doors in cooperation with the Communist Party?

There were on any account basic differences between research and teaching at Tsinghua University when he was there in 1931–1936 and in 1950–1957. At least three national political campaigns swept the Chinese intelligentsia into acute criticism and self-criticism struggles. These occurred, as we have outlined, in 1951–1952, 1957, and 1964–1965. The long-term *People's Daily* reader remembers Hua's involvement in these campaigns and knows about his efforts to apply mathematics in the service of socialist construction during the wheat harvests and in various projects initiated during the Great Leap Forward, and how Hua went deep into construction sites during six-months work with industry. During the Cultural Revolution it was conceivably emphasized that intellectuals such as Hua did not do much toward remolding their bourgeois world outlook and were allowed to escape with only limited or token participation in the thought reform and rectification campaigns, that they had always avoided making a really clean break with their past, and that they had to be prodded into all their applied work. It is still difficult to fathom how anyone could accept the assertion that the experiences of two eventful decades transpired "behind closed doors."

After the Liberation, Hua produced superlative, wide-ranging works of mathematics and pedagogy and taught many outstanding students. For this writing and teaching he was highly rewarded, materially and morally, by China's workers, peasants, and soldiers, and in labelling the "revisionist line" as "counter-revolutionary" Hua thus singles out positions taken by Liu Shiao-chi that were accepted and implemented by all Communists. Serious differences, to be sure, emerged in the Communist Party leadership in 1957, but was the struggle between the forces of proletarian revolution and capitalist restoration or were the divergences about alternative policies for socialist construction? The regrettable device of branding differences as counter-revolutionary was central to the oppressive features of the Stalin era in the Soviet Union, but it was repeated throughout the polemics of the Cultural Revolution.

Although the worker-army propaganda team stationed in the University of Science and Technology was said by Hua to have exposed "countless shocking facts" condemning Liu Shiao-chi and his agents in education, only two of these "facts" actually find their places in the *People's Daily* and *China Reconstructs*. One concerns the University itself:

> They [Liu's agents] declared that the aim of the University of Science and Technology was to train cadres to work with the newest and most advanced science and technology. By advocating a "white and expert" road, they were leading students to concentrate on technical education and pay no attention to politics.

Under attack, in other words, are the guidelines for the University framed by Nieh Jung-chen in September 1958. But it is not unreasonable to infer from Nieh's slogan of raising Chinese technology up to world levels "in a very few years" that a new form of elitism, based on the encouragement of exceedingly expert and applications-minded "candidate scientists," was inevitable there from the start and hence cannot be blamed entirely on the Liu Shiao-chi educational line.

The other "shocking fact" concerns the examinations in mathematics:

> I organized competitions in mathematics and promoted them throughout the country as an "advanced experience" from abroad. Such a practice was really an open call for young people to strive for personal achievement through concentration on technical knowledge only.

We have already seen, however, that in 1956 in *Chinese Youth* Hua went to great pains to stress the necessity for independent, creative thinking on the part of science students. The learning-by-repetition patterns which he criticized in 1956 included extremely standard, stereotyped questions on school examinations; and while the competitions may well have encouraged personal achievement, their documented, constructive learning function makes it improbable that they were administered solely for achievement's sake.

While it is amply clear that Hua Lo-keng, far from being politically ruined by the Cultural Revolution, was in the end elevated in stature as an exponent of the changes in Chinese educational policy developed since 1969, it is equally apparent that this policy does not fully accommodate the mathematical creativity of his colleagues and students. The consequences within China are yet to be determined, but there has been a loss to world science and culture as a whole. That this loss to all of us is great, I believe even Hua Lo-keng will agree.

GLOSSARY

a 華羅庚

b 金壇

c 熊慶來

d 陳省身

e 吳大猷

f 曾昭掄

g 李政道

h 聞一多

i 錢學森

j 天才

k 吳文俊

l 聰明在于學習
天才由于積累

m 獨立思考

n 聰明

o 照顧

p 社會活動

q 較多

r 章伯鈞

s 羅隆基

t 萬哲先

u 幹部

v 運籌學

w 曲阜

x 統籌

y 怕

z 王杰

aa 內戰

bb 北航

cc 重新學習爲教
育革命獻力量

BIBLIOGRAPHY

THE LONGER MATHEMATICAL WORKS OF HUA LO-KENG

Additive Prime Number Theory, Volume 22 of *Trudy Matematiceskogo Instituta im.* *V. A. Steklova*, Moscow/Leningrad, 1947. (Russian; English summary.)

Tui lei su shu lun (Additive Prime Number Theory), Peking:Chinese Academy of Sciences, 1953; rev. ed., 1957.

Shu lun tao yin (Introduction to Number Theory), Peking:Science Publishing Co., 1957; 2nd ed., 1965.

To-fu-pien han-shu-lun chung ti tien-hsing-yü ti tiao-ho fan-hsi (Harmonic Analysis of Functions of Several Complex Variables in Classical Domains), Peking: Science Publishing Co., 1957; 2nd ed., 1965.

Garmonicheskii analiz funktsii mnogikh kompleksnykh peremennykh v klassicheskikh oblastiakh (Harmonic Analysis of Functions of Several Complex Variables in Classical Domains), Moscow: Izdatel'stvo Inostrannoi Literatury, 1959.

Additive Primzahltheorie, Leipzig:Teubner, 1959.

"Die Abschätzung von Exponentialsummen und ihre Anwendung in der Zahlentheorie," *Enzyklopäde der mathematischen Wissenschaften*, Band 1, 2, Heft 13, Teil I, Leipzig: Teubner, 1959.

(With Wan Chieh-hsien), *Tien hsing chün* (Classical Groups), Shanghai: Shanghai Science and Technology Press, 1963.

Harmonic Analysis of Functions of Several Complex Variables in Classical Domains, translated from the Russian by Leo Ebner and Adam Korányi, Providence, R.I.: American Mathematical Society, 1963.

Additive Prime Number Theory, translated from the Chinese by N. H. Ng, Providence, R.I.:American Mathematical Society, 1965.

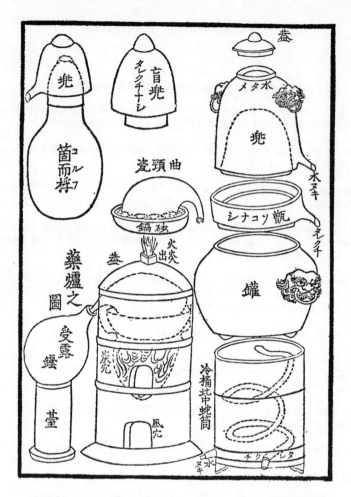

Chemical apparatus in Ranka naigai sanbō hōten *(Handbook of the Three Aspects of Dutch Internal and External Medicine), a pioneer translation by Hashimoto Sōkichi (1805) which helped to introduce chemistry to Japan.*

The Reception of Lavoisier's Chemistry in Japan

By Eikoh Shimao*

WHEN THE EUROPEAN SCIENCES INFILTRATED into Japan in the Edo period, mainly through Dutch books, the order of their acceptance seems to epitomize the evolution of the sciences in Europe. Astronomy and anatomy were introduced at the earliest stage. Physics began with mechanics, and medicine with anatomy. Chemistry was noticed as iatrochemistry first, before it enjoyed recognition as a science independent from medicine. In *Ranka naigai sanbō hōten*[a] (1805), or "Handbook of the Three Aspects of Dutch Internal and External Medicine," by Hashimoto Sōkichi[b,1] (1763–1836), a translation of a pharmacopoeia written in Dutch and Latin, we see the first introduction of European iatrochemistry significantly different from the traditional pharmaceutical natural histories in Japan.[2] Besides iatrochemical discussions, the book includes diagrams of chemical instruments, ninety alchemical symbols with Japanese transliteration, and Dutch and Latin terms expressed phonetically by Sino-Japanese characters. That Hashimoto translated *Scheikonst* (*chimia*) as *seiyaku*,[c] or pharmacy, indicates that chemistry as a science independent from medicine was not yet known. Hashimoto's book indeed reflects early-eighteenth-century knowledge. Its original text has been recently identified as Wouter van Lis' *Pharmacopoeia Galeno-Chemico-Medica* . . . (Rotterdam, 1747).[3]

Hashimoto's pioneer effort aided the development of pharmacology and eventually led to the discovery of chemistry as a new science. Crafts such as copper-plate etching and glass production had been introduced, arousing an interest in nonmedical chemistry. The systematic study of chemistry began in the early nineteenth century, with the introduction of the work of Antoine Lavoisier.

Received June 1971: revised/accepted Oct. 1971.

* Kansai University, Suita, Osaka, Japan. This work was supported by the Ministry of Education Grant (General Research C) for 1971. I am very grateful to Drs. Kiyoshi Yabuuchi, O. Theodor Benfey, Saburo Miyashita, Nathan Sivin, and Joseph Needham for valuable discussions.

[1] Hashimoto Sōkichi, *Ranka naigai sanbō hōten* (Handbook of the Three Aspects of Dutch Internal and External Medicine) (Naniwa [Osaka], 1805).

With the exception of the author's name, Japanese names in this article are written surname first. Persons are referred to by surname except where the courtesy names of members of the Udagawa family (Kaien, Shinsai, and Yōan) are occasionally used to avoid confusion. At the end of the article is a list of the Sino-Japanese characters representing the romanized words appearing herein. Superscript letters indicate the placement in the list.

[2] S. Miyashita, "Hashimoto Sōkichi, Pioneer Pharmacologist," *Kagakushi Kenkyū* (Journal of the History of Science, Japan), 1971, *100*:208.

[3] I am grateful to Dr. Derek J. de Solla Price, and to Mr. F. Gyorgyey, Yale Medical Library, for their help in the identification.

It was almost concurrently that Lavoisier was introduced into America, and analogously, American physicians took part in establishing Lavoisier's new nomenclature.[4] The difference was that in America there was a period of phlogiston chemistry before the appearance of Lavoisier, while in Japan the phlogiston theory was never established. In America manuals based upon *Méthode de nomenclature chimique* (1787), by Guyton de Morveau, Lavoisier, Berthollet, and Fourcroy, were circulated to spread Lavoisier's new nomenclature, and Robert Kerr's *Elements of Chemistry* (1790), an English version of Lavoisier's *Traité élémentaire de chimie* (1789), was adopted without attempting a new American translation. In Japan, in contrast, only a Dutch version of the *Traité* was available; no Dutch version of the *Méthode* existed.[5] Translating the *Traité* into Japanese required tremendous effort to create a whole system of chemical terminology, not only the new nomenclature of elements and compounds by Lavoisier, but also basic chemical terminology.

It is probable that one of Lavoisier's articles was read in Japan in his lifetime. The Dutch journal *Nieuwe Genees- Natuur en Huishoud-Kundige Jaar Boeken*, Volume 4 (1784), includes an important article by Lavoisier, in which he stated that water is not an element and is subject to decomposition and synthesis. This journal is supposed to have been presented by C. P. Thunberg, the Swedish botanist who introduced Linnaeus' taxonomy into Japan, to Katsuragawa Hoshu,[d] one of the physicians-in-ordinary to the shogun, when Thunberg visited Edo in 1787.[6] But the reaction of Katsuragawa to the article is not known.

Kikai kanran[e] (1827), or "Contemplation of Waves in the Sea of Air," by Aochi Rinsō[f] (1772–1833) is supposed to have been the first publication on Western physics in Japan.[7] We should note that it includes also the first published presentation of Lavoisier's chemistry. It is not clear how Aochi chose the title. It reminds us of Torricelli's words, "we live submerged at the bottom of an ocean of the element air." Since Aochi remarks, "the earth is a great body in the sea of air," and also "the sun is the greatest illuminating body in the sea of air," his "sea of air" seems to confuse space and the atmosphere. The book as a whole, however, deals with the physics and chemistry of the atmosphere.

Let us examine three characteristic aspects of Lavoisier's chemistry in early Japanese writings: caloric theory, oxygen theory of acids and combustion, and chemical nomenclature.

In *Kikai kanran*, caloric was translated by Aochi as *onshitsu*,[g] or "matter of warmth." The concept of caloric in *Kikai kanran* may be summarized as follows: (1) it is the matter of fire; (2) it is a repulsive force; (3) it is a subtle fluid; (4) it tends to expand uniformly; (5) materials exhibit different conductivity to caloric. Aochi could have obtained these concepts of caloric not only from Lavoisier but from books on physics. However, when he describes four species of gas—*chikki*,[h] or "suffocating air" (nitrogen

[4] D. I. Duveen and H. S. Klickstein, "The Introduction of Lavoisier's Chemical Nomenclature into America," *Isis*, 1955, *46*:278–292, 368–382.

[5] D. I. Duveen and H. S. Klickstein, *A Bibliography of the Works of Antoine Laurent Lavoisier* (London:Dawson & Weill, 1954).

[6] G. Imaizumi, *Katsuragawa no hitobito* (The Katsuragawas), (Tokyo: Shinozaki Shorin, 1964), p. 279.

[7] Aochi Rinsō, *Kikai kanran* (Contemplation of Waves in the Sea of Air) (Edo [Tokyo]: Meizankaku, 1827). A modern printing is included in *Bunmei genryū sōsho* (Tokyo:Kokusho Kankōkai, 1914), Vol. II, pp. 256–283.

gas); *seiki*,[i] or "pure air" (oxygen gas); *nenki*,[j] or "inflammable air" (hydrogen gas); and *kōki*,[k] or "hard air" (carbon dioxide), explaining that *seiki* is produced from *sanshitsu*,[l] or "acid matter" (oxygen) and *onshitsu*, and that *nenki* is produced from *suishitsu*,[m] or "water matter" (hydrogen) and *onshitsu*—he is undoubtedly relying on Lavoisier's chemical theory. His further statements that *seiki* revives life, makes fire burn, combines with substances to cause acids, and makes copper and iron rust, and that water is composed of *suishitsu* and *sanshitsu* are also Lavoisier's chemistry. As for *nenki*, however, Aochi attributes a bad smell to it, confusing it with methane gas.

It is quite probable that Aochi saw *Grondbeginselen der Scheikunde* (1800), a Dutch version of Lavoisier's *Traité*. But it is doubtful whether he studied it carefully, for notwithstanding the reception of the core of Lavoisier's chemical ideas, the new method of chemical nomenclature, which is inseparable from Lavoisier's chemistry and had been pronouncedly emphasized in the *Traité*, was not adopted in *Kikai kanran*. Aochi remarks that *seiki*[i] may be expressed *seiki*[n] (lit., "vital air"), using different Sino-Japanese characters, and also *sanki*,[o] or "acid air." If he had followed Lavoisier's nomenclature, he would have adopted *sanki* for oxygen gas, instead of *seiki*,[i] an obsolete term. Likewise he adopted the pre-Lavoisierian term *nenki*. In short, Lavoisier's chemistry is indeed found in *Kikai kanran*, but its presentation is quite fragmentary and incomplete. It is doubtful whether the knowledge came directly from Lavoisiers' book. However, since *Kikai kanran* was published, unlike most of Aochi's and other's work in those days, it must have inspired scholars to become interested in chemistry, a science new to the Japanese.

To find the chemistry in *Kikai kanran*, we have to search for it. As for *Ensei suishitsu ron*,[p] by Takano Chōei[q] (1804–1850), a short work of unknown date but probably a little later than *Kikai kanran*, the whole content is chemistry; in fact it consists of an introduction to Lavoisier's chemistry.[8] *Ensei suishitsu ron*, or "Far Western Theory on the Component Matter of Water," treats the universal elements *kagen*[r] (lit., "fire principle," caloric), *sangen*[s] (lit., "acid principle," oxygen), and *suigen*[t] (lit., "water principle," hydrogen) which compose water. While *Kikai kanran* contains no discussion of elements, *Ensei suishitsu ron* begins with a definition of "element," repudiating the four Aristotelian elements. It is followed by a statement that experiments on the synthesis of water were performed by Lavoisier, Monge, and Cavendish independently, and the same result was obtained by all of them. The article further treats the formation of acids by *seiki*[i] (oxygen gas); oxidation of metals and augmentation of weight; decomposition of the atmosphere and of water; properties of oxygen, hydrogen, and nitrogen; gunpowder and fermentation. It ends with some biochemical discussions on animals and plants. In the article, Takano also displays some knowledge of the historical background of chemistry, referring to fourteen chemists including Jean Rey, Boerhaave, Stahl, van Helmont, Scheele, and Berthollet besides the three mentioned above.

For names of gases, Takano followed Aochi's terminology (for instance *seiki*[i] and *nenki*[j]) while for names of elements Takano coined new words, such as *kagen*, *sangen*, and *tangen*[u] (lit., "coal principle," carbon). For oxide of metal he adopted the term *sō*[v]

---

[8] Takano Chōei, *Ensei suishitsu ron* (Far Western Theory on the Component Matter of Water), in *Takano Chōei zenshū* (Complete Works of Takano Chōei) (Tokyo: Takano Chōei Den Kyōdō Kankōkai, 1931), Vol. IV, pp. 61–73.

(lit., "frost") from Chinese pharmaceutical natural histories, where it was an equivalent of metallic oxide (in ancient China, however, this was a technical term for a sublimate, not oxidation products). It is only with *yuōsan*,[w] or "acid of sulfur," and *phosphorus-san*,[x] or "acid of phosphorus," that an attempt at using Lavoisier's nomenclature is recognized in naming compounds. Nor was a term for chemistry yet determined. Besides *seiyaku*,[c] Hashimoto Sōkichi had referred to chemistry as *yōshaku henka no hō*,[y] or art of melting and transforming. Aochi referred to chemistry as *bunsekijutsu*,[z] or the art of analysis, which must have come from the Dutch *Scheikunde*. Takano referred to it as *bungō jutsu*,[aa] or the art of analysis and synthesis. It is to be regretted that Takano's voluminous manuscripts entitled *Scheikunde* have been lost.

In *Igen sūyō*[ab] (1832), or "Essence of Medical Principles," the first printed physiological treatise in Japan, the author Takano discussed the physiology of respiration in terms of Lavoisier's chemistry. At the same period Ozeki San-ei[ac] (1787–1839) also discussed the physiology of respiration using Aochi's terminology for pneumatic chemistry.

Udagawa Yōan[ad] (1798–1846) is the first person who fully came to grips with Lavoisier. It is not well known even in Japan that Udagawa, author of the celebrated *Semi kaiso*[ae] (1837), the first printed textbook on chemistry in Japan,[9] devoted himself to the study of Lavoisier. In the rich Udagawa Yōan manuscript collection possessed by Kenkensai Library,[af, 10] we find Udagawa's translation drafts of Lavoisier's *Traité*, based on a Dutch version.

The manuscripts are scattered among several itemized volumes, each being juxtaposed with translations from other authors, William Henry, Johann Bartholomä Trommsdorf, Adolphus Ipey, and so on. When we search for Lavoisier in these manuscripts, which are no longer complete, we find that the *Traité* is translated in an abridged form, but its major parts are covered. The date of translation is problematic. The Udagawa manuscripts do not bear a date except for one entitled *Lavoisier, dōsan semika*,[ag] or "Lavoisier's Animal Acid Chemistry," which has the following date at the end: "The 13th year of the Bunsei reign period, the year of the tiger, 5th day of the 8th month," corresponding to 1830. This was three years after the publication of Aochi's *Kikai kanran*, and preceded Udagawa's own *Semi kaiso* by seven years.

Kikai kanran was written in classical Chinese, but the Udagawa manuscripts and *Semi kaiso* were written in Japanese. We shall compare Udagawa's treatment of caloric with that in *Kikai kanran*. Caloric was translated by Aochi as *onshitsu*,[g] and was treated in four pages in *Kikai kanran*. Udagawa translated it *danso*,[ak] or "warmth element," and treated it in forty-two pages of a booklet entitled *Danso*. In general, Udagawa did not follow the terminology of *Kikai kanran*.

Information about caloric not found in *Kikai kanran* but developed in *Danso* and other Udagawa manuscripts can be summarized as follows: (1) universality of caloric;

[9] Udagawa Yōan, *Semi kaiso* (Foundations of Chemistry) (Edo [Tokyo]: Seireikaku, 1837).

[10] The following are included in the Udagawa Yōan Manuscripts, Kenkensai Library: *Dōsan semika*[ag] (Animal Acid Chemistry), 1830; *Kōso semika*[ah] (Chemistry of the Matter of Light);

Sansan semika[ai] (Mineral Acid Chemistry); *Gas semika*[aj] (Gas Chemistry); *Danso*[ak] (Caloric); *Chūseien semika*[al] (Chemistry of Neutral Salt); *Seikō rui*[am] (Spirits and Fermentation); *Semi kikai zui*[an] (Illustrations of Chemical Instruments).

(2) two types of existence, manifest and latent; (3) the most subtle of all elements, and imponderable; (4) the degree of expansion caused by caloric depends upon the particular substance heated; (5) the concept of specific caloric; (6) the relationship of boiling point, atmospheric pressure, and caloric; (7) substances combine with each other by means of caloric; (8) change of state and caloric; (9) calorimeter; (10) the combination of light and caloric. This last subject is included in the Udagawa manuscript *Kōso semika*,[ah] or "Chemistry of the Matter of Light," which is a translation of Section III, Part II of the *Traité*.

It is regrettable that the preface of the *Traité* is missing in the Udagawa manuscripts, for we are eager to see whether Udagawa read Lavoisier's views on elements and his philosophy of nomenclature. Fortunately we find the well-known list of elements in Part II of the *Traité* translated in *Sansan semika*,[ai] or "Mineral Acid Chemistry." As is well known, Lavoisier's list enumerates thirty-three elements,[11] while Udagawa lists forty elements, following strictly the form of Lavoisier's list.[12] This is because *Grond-beginselen der Scheikunde*, the Dutch version upon which Udagawa's translation was based, enumerated the forty elements.[13] The order of elements in the list of the *Grondbeginselen* is somewhat different from that of the *Traité*. Udagawa followed the latter faithfully.

Not only is the philosophy of nomenclature in the preface of the *Traité* missing in the Udagawa manuscripts, but discussions on nomenclature in other parts of the *Traité* were strangely enough not translated either. For instance, the explanations of the nomenclature of oxygen and nitrogen are omitted, although we find the title of Chapter 4, Part I, in the Udagawa manuscript *Gas semika*,[aj] or "Gas Chemistry." Neither is the nomenclature of acids in Chapter 6 to be found in the Udagawa collection.

Although the philosophy of nomenclature was not translated, Udagawa's effort to render the chemical terminology in the *Traité* into Japanese was indeed an introduction of the Guyton-Lavoisier chemical nomenclature into Japan. Udagawa coined *danso*[ak] (caloric), *sanso*[ao] ("acid element," oxygen), *sasso*[ap] ("lethal element, "nitrogen), *suiso*[aq] ("water element," hydrogen), and *tanso*[ar] ("coal element," carbon). These terms, except for *danso* and *sasso*, have been used up to the present day. Since the translation was from the Dutch, *so*[as] (lit., "element") bears a trace of Germano-Dutch *Stoff* or *Stof*.

When *oxygène* was rendered into *Sauerstoff* in German and then into *Zuurstof* in Dutch, the implication of *gène*, the latter half of *oxygène*, was lost. The same is true of *sanso* in Japanese, which was influenced by the Germano-Dutch nomenclature. In *Hua hsüeh chien yuan*[at] (1871), "Introduction to Chemistry," the first comprehensive monograph on modern chemistry in China, we find the Chinese term *suanmu*[au] (lit., "acid mother") for oxygen.[14] This is definitely more faithful to the Lavoisierian nomen-

[11] Antoine Lavoisier, *Traité élémentaire de chimie* (Paris, 1789), p. 192.

[12] The additional "elements" are chromium, tellurium, titanium, uranium, strontia, zirconia, and glucina.

[13] Antoine Lavoisier, *Grondbeginselen der Scheikunde* (Utrecht, 1800), translation into

Dutch of *Traité élémentaire de chimie*, p. 190.

[14] *Hua-hsüeh chien-yüan*, translation into Chinese by John Fryer and Hsü Shou[av] of David A. Wells, *Principles and Applications of Chemistry* (New York, 1858); Vol. I, Section 32, "Origin of *yang ch'i*[aw] (oxygen)."

clature than *sanso*. *Suanmu*, however, was never adopted in China; instead, *yang ch'i*[aw] (lit., "nourishing air") and then the newly coined symbol *yang*[ax] were used. The latter is current. In Japanese the same Sino-Japanese character *san*[ay] or "acid" is used both in *sanso* (oxygen) and in *san* (acid). Accordingly, Lavoisier's error in nomenclature has been immortalized in Japanese chemical nomenclature as well.

Udagawa Yōan's neologisms launched a new nomenclature after Lavoisier in Japanese. *Sanso* gas, *suiso* gas, and *sasso* gas are produced when *danso* combines with *sanso*, *suiso*, and *sasso* respectively. The term "gas," which Lavoisier established as a generic name, was expressed phonetically in two Chinese characters[az] by Udagawa and was used in the same way as Lavoisier had done. *Rinsan*[ba] (phosphoric acid), *kōsan*[bb] (sulfuric acid), and *tansan*[bc] (carbonic acid) are produced when *sanso* (oxygen) combines with *rin*[bd] (phosphorus), *kō*[be] (sulfur), and *tanso* (carbon) respectively. This is nothing but the adoption of Lavoisier's chemical theory, which in turn was combined with his nomenclature.

Lavoisier enumerated forty-eight acids in Chapter 17, Part I of the *Traité*.[15] Udagawa enumerated fifty-four acids in the same place, the list being supplemented as in the case of the list of elements.[16] Here again Udagawa followed the *Grondbeginselen*.[17] With inorganic acids, terms to express degrees of oxidation were created according to Lavoisier's system, for instance, *misei shōsan*[bf] (lit., "immature acid of niter," nitrous acid), *zensei shōsan*[bg] (lit., "mature acid of niter," nitric acid), and *kasan shōsan*[bh] (lit., "overacidified acid of niter," oxygenated nitric acid). These expressions were improved by Udagawa himself into *ashōsan*[bi] (lit., "secondary acid of niter," nitrous acid) and *shōsan*[bj] (nitric acid), which appeared in his later publication *Semi kaiso* and have been used ever since. Organic acids had not been named according to the new nomenclature in the *Traité*, as Lavoisier had admitted, but Udagawa created Japanese terms for the twenty organic acids enumerated in the *Traité*.

In Udagawa's Lavoisier studies, we should not overlook his *Semi kikai zui*,[an] or "Illustrations of Chemical Instruments." Although the manuscript bears no account of the sources of the plates, it is clear that the major ten plates are from Lavoisier's *Traité*. The thirteen plates in the *Traité* were reduced to ten in the Dutch version *Grondbeginselen der Scheikunde*. Udagawa's are beautiful copies of these ten plates, with the shading rendered in Chinese ink. The size of each figure is precisely the same as the original in the *Traité*. Plates I, II, and a part of Plate III of the *Traité* are omitted in *Grondbeginselen*. The remaining plates are in general associated with Lavoisierian pneumatic chemistry. A chemistry book with such exhaustive illustrations of chemical instruments must have been unique even in Europe in those days; we can imagine how deeply Udagawa was impressed by them. Part I of the *Traité* would be unreadable without these illustrations, and in fact the Udagawa manuscript includes descriptions of most of the plates. We find that Udagawa picked up accounts from Part III of the *Traité* on major instruments such as the gasometer, calorimeter, and apparatus for fermentation, although most of Part III was not translated. Figures 1 and 2 are from the Udagawa illustrations.

---

[15] Lavoisier, *Traité*, pp. 180, 181.
[16] The additional acids are suberic, zoonic, chromic, telluric, titanic, and uranic acids.
[17] Lavoisier, *Grondbeginselen*, pp. 179, 180.

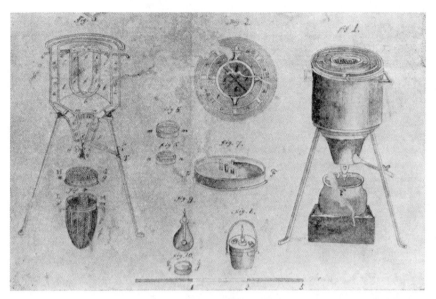

Figure 1. The calorimeter in Udagawa Yōan's Semi kikai zui.

The Udagawas were physicians-in-ordinary to the Lord of Tsuyama for generations, Yōan being the fifth generation in the profession. Since the time of Udagawa Genzui (Kaien)[bk] (1775–1797), their specialities—medicine and "Dutch learning"—were inherited and enhanced by adoption. Yōan, a son of a physician, was adopted in 1811 by Udagawa Genshin (Shinsai)[bl] (1769–1834), who in turn had been adopted by Kaien. Thus Yōan, when young, absorbed the scholarship of his adoptive father who enjoyed a reputation as an authority on medicine and Dutch learning. Owing to his competence in Dutch, Yōan like his father was appointed in 1826 as official translator at the government astronomical observatory. He participated in the biggest translation enterprise of the whole Edo period—the officially sponsored *Kōsei shinpen*[bm] (1811–1839), or "A New Book of Welfare," from *Algemeen Huishaudelyk, Natuur-, Zedekundig, en Konst-Woordenboek* (Amsterdam, 1743), a Dutch version of Chomel's encyclopedia.[18] Udagawa Shinsai and Udagawa Yōan made a major contribution to *Kōsei shinpen*. Since Yōan took charge mainly of natural history and chemistry, the work undoubtedly drove him to intensive studies on these subjects.

Shinsai published two books on pharmacology, *Oranda Yakkyō*[bn] (1820), or "Dutch Pharmacology," and *Ensei ihō meibutsukō*[bo] (1822–1825), or "Treatise on Far Western Medicaments."[19] Western pharmacology, which had been initiated by Hashimoto

[18] *Kōsei shinpen* (A New Book of Welfare), translated in 1811–1839 (Shizuoka: Aoi Bunko, 1937).
[19] Udagawa Shinsai, *Oranda yakkyō* (Dutch Pharmacology) (Edo [Tokyo], 1820); Udagawa Shinsai, *Ensei ihō meibutsukō* (Treatise on the Far Western Medicaments) (Edo [Tokyo]: Seireikaku, 1822–1825).

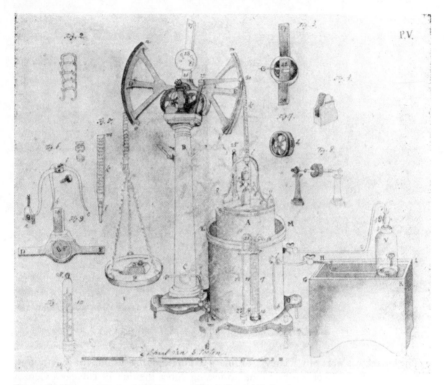

Figure 2. The gasometer in Udagawa Yōan's Semi kikai zui.

Sōkichi, culminated in these two monumental works of the Edo period. Both were revised and enlarged by Yōan, in 1828–1835 and in 1833–1834 respectively. It was in his *Ensei ihō meibutsukō hoi*[bp] (1833–1834), or "Supplement to the Treatise on Far Western Medicaments" (hereafter referred to as the *Supplement*) that Yōan's Lavoisier studies were published for the first time.[20]

The "Treatise on Elements," which constitutes Chapters 7, 8, and 9 of the *Supplement* by Yōan, is drastically different from the rest. In contrast to the *olla podrida* of pharmaceutical knowledge in the remainder of the book, the "Treatise on Elements" is a systematic presentation of pneumatic chemistry. Its rubrics were element, caloric, matter of light, gas, oxygen, nitrogen, hydrogen, carbon, carbonic acid, atmosphere, hygrometer, and earths. A mere glimpse of these headings reminds us of Lavoisier's chemistry. This is, indeed, the first publication of modern chemical theory and of Lavoisier's chemistry in Japan, although Lavoisier is not referred to at all.

[20] Udagawa Shinsai and Yōan, *Ensei ihō meibutsukō hoi* (Supplement to the Treatise on Far Western Medicaments) (Edo [Tokyo]: Seireikaku, 1833–1834).

Aochi and Takano referred to "element" as *genshitsu*.[bq] Shizuki Tadao[br] (1760–1806) translated it as *jisso*.[bs] Udagawa proposed in the beginning of Chapter 7 of the *Supplement* his newly coined term *genso*[bt] for "element," which had already appeared in the Udagawa manuscripts and has been used ever since in Japan, and in China too since its appearance later on. Here we can find Lavoisier's concept of an element, which is missing in the Udagawa manuscripts. The "Treatise on Elements" reads "a simple substance remains simple and pure, and is mixed with no different substance, no matter how many times a chemist may analyze it. Such a substance is defined as an element." This is an echo of "element" as viewed by Lavoisier, who regarded an element provisionally as the last point which analysis was capable of reaching. As to the number of elements, Udagawa enumerated forty in the *Sansan semika*, but here in the *Supplement* he mentioned "more than fifty."

At one time Lavoisier took electricity as an element, besides caloric and light. But it was not included in the list of elements in the *Traité*, and Udagawa did not mention it either in his *Sansan semika*. Strangely enough, it reappeared as a term *esso*,[bu] an abbreviated form of *erekiteruso*,[bv] or "electricity element," in the *Supplement* and then in *Semi kaiso*, his final work on chemistry.

Nomenclature was indeed improved in succeeding works. For instance, the term "caloric" changed from *danso* in the Udagawa manuscripts into *onso*,[bw] or literally "warmth element," in the *Supplement* and in *Semi kaiso*. The term for nitrogen changed from *sasso*[ap] in the Udagawa manuscripts, which means "lethal element," into *chisso*,[bx] or "suffocating element," in the *Supplement* and in *Semi kaiso*. The latter term is still used.

As to oxygen, the *Supplement* reads, "it produces acids and the acidic taste by combining with other substances, therefore it is named *sanso*[ao] or acid element." This is an explanation of the nomenclature of oyxgen which we could not find in the Udagawa manuscripts. As to hydrochloric acid, however, it reads, "it produces the acidic taste without oxygen." In this statement the *Supplement* explicitly went beyond Lavoisier.

That Udagawa's *Shokugaku keigen*,[by] or "Introduction to Botany," was published in 1835, the year after the publication of the *Supplement*, may indicate that he had been studying chemistry and botany concurrently.[21] In a preface to *Shokugaku keigen* Udagawa allocated to chemistry a place in a system of *ban-yū gaku*,[bz] or "universal science," which consists of three branches: *benbutsu gaku*,[ca] or "taxonomic natural history," *hisika*,[cb] or physics, and *semika*,[cc] or chemistry. He compared natural history to the gate and chemistry to the main hall of universal science. Udagawa further classified natural history into botany, zoology, and mineralogy. In *Shokugaku keigen* he treats Linnaeus' taxonomy of plants, and in *Dōgaku keigenkō*[cd] (1835), or "Draft of the Introduction to Zoology," which he intended as a sister volume to the "Introduction to Botany" but which remained unpublished, he treats of Linnaeus' taxonomy of insects and animals. Part II of Lavoisier's *Traité* consists of tables of taxonomy in chemistry. We should note that Lavoisier must have been inspired by Linnaeus in classifying chemical compounds through binomial nomenclature. We

---

[21] Udagawa Yōan, *Shokugaku keigen* (Introduction to Botany) (Edo [Tokyo], 1835). A modern printing is included in *Bunmei genryū sōsho*, Vol. II, pp. 284–323.

have found in Udagawa's memorandum biographical sketches in Dutch of Linnaeus and Lavoisier.

In a preface to *Semi kaiso* we find a brief historical survey of chemistry. Since Udagawa had established for chemistry a place in the system of sciences and had elucidated its taxonomic and systematic foundations, it is natural that he now should turn to an historical survey. The survey merely treats of a division into four periods: the first (300–1650) is the period of chemical chaos, the second (1650–1782) the period of phlogiston, the third (1783–1807) the anti-phlogistic period, and the fourth (1808—) the period of electrochemistry. Stahl, Lavoisier, and Davy are referred to as founders of the second, third, and fourth periods respectively. Concerning the origins of the four elements, Thales, Anaximenes, and Empedocles are mentioned.

Semi kaiso, the first published monograph on chemistry in Japan, was closer to a textbook in style. In the preface it enumerates authors, titles in Japanese, and dates of publication of twenty-four Dutch books as references, including Lavoisier and Guyton de Morveau.[22] The text includes many annotations by Udagawa and citations of various authors, without, however, specifying bibliographic details. Although Lavoisier is often cited, *Semi kaiso* is not based upon the *Traité*, but upon a Dutch version of William Henry's *An Epitome of Chemistry* (1801).[23] In *Semi kaiso* Lavoisier's chemistry is taken for granted. Berzelius and Davy's electro-dualism is introduced, but Dalton's atomic theory is not yet to be found. Nomenclature has been definitely improved, and the basic principles of Japanese chemical nomenclature were established. Thus *Semi kaiso*, or "Foundations of Chemistry," literally founded a new science of chemistry in Japan. The term *semi*,[ce] which was a phonetical expression in two Chinese characters for *Chemie*, would be replaced by *kagaku*,[cf] or "the study of change," a term introduced from China about twenty years after *Semi kaiso* and still current.

Why did Udagawa choose Henry instead of Lavoisier? The reason could be that Henry's *Epitome* seemed to Udagawa more appropriate for beginners than Lavoisier's *Traité*. Had Udagawa decided differently, a Japanese version of the *Traité* would have appeared in the early 1830s. Lavoisier was advocating his new chemical theory too vigorously to give a full account of elements and compounds; moreover, he was preoccupied with pneumatic chemistry. In the Udagawa manuscripts we have found complementary translations from various authors.

Generally, sciences were introduced into Japan not through the original classics but through popularizations and primers. Thus Newton was introduced not through his *Principia* but through John Keill's *Inleiding tot de waare Natuur- en Sterrenkunde* (1741). Vesalius' anatomy was received not through *De humanis corporis fabrica* but through Johan Adam Kulmus' little book on anatomy, *Ontleedkundige Tafelen* (1734). The classics of science, after all, seemed too unapproachable to the novice.

---

[22] Guyton's work referred to here is a treatise on atmosphere, since there was no Dutch version of *Méthode de nomenclature chimique* (Paris, 1787).

[23] A second edition of *Epitome* was translated into a German version by J. B. Trommsdorf, which in turn was translated and enlarged by A. Ipey into a Dutch version, *Chemie voor Beginnende Liefhebbers* (Amsterdam, 1803), which was further translated and enlarged by Udagawa into *Semi kaiso*. See M. Sakaguchi, "On the European Translations of Henry's *Epitome of Chemistry* and *Semi Kaiso*," *Kagakushi Kenkyu*, 1966, *80*: 171.

a	蘭科内外三法方典	w	硫黄酸
b	橋本宗吉	x	忽私忽略私酸
c	製藥	y	鎔鑠変化之法
d	桂川甫周	z	分析術
e	気海観瀾	aa	分合術
f	青地林宗	ab	医原枢要
g	温質	ac	小関三英
h	窒気	ad	宇田川榕庵
i	清気	ae	舍密開宗
j	燃気	af	乾々斉文庫
k	硬気	ag	動酸舍密加
l	酸質	ah	光素舍密加
m	水質	ai	山酸舍密加
n	生気	aj	瓦斯舍密加
o	酸気	ak	煖素
p	遠西水質論	al	中性塩舍密加
q	高野長英	am	精酵類
r	火原	an	舍密器械図彙
s	酸原	ao	酸素
t	水原	ap	殺素
u	炭原	aq	水素
v	霜	ar	炭素

as 素

at 化学鑑原

au 酸母

av 徐壽

aw 養気

ax 氧

ay 酸

az 瓦斯

ba 燐酸

bb 礦酸

bc 炭酸

bd 燐

be 礦

bf 未成消酸

bg 全成消酸

bh 過酸消酸

bi 亜消酸

bj 消酸

bk 宇田川玄隨(槐園)

bl 宇田川玄眞(榛齋)

bm 厚生新編

bn 和蘭藥鏡

bo 遠西医方名物考

bp 遠西医方名物考補遺

bq 原質

br 志筑忠雄

bs 實素

bt 元素

bu 越素

bv 越列吉的爾素

bw 温素

bx 窒素

by 植学啓原

bz 萬有学

ca 觕物学

cb 費西加

cc 舎密加

cd 動学啓原稿

ce 舎密

cf 化学